LONDON MATHEMATICAL SOCIETY LECTURE NOTE SERIES

Managing Editor: Professor J.W.S. Cassels, Department of Pure Mathematics and Mathematical Statistics, University of Cambridge, 16 Mill Lane, Cambridge CB2 1SB, England

The titles below are available from booksellers, or, in case of difficulty, from Cambridge University Press.

London Mathematical Society Lecture Note Series. 237

Sieve Methods, Exponential Sums, and their Applications in Number Theory

Proceedings of a symposium held in Cardiff, July 1995

Edited by

G. R. H. Greaves
University of Wales, College of Cardiff

G. Harman
University of Wales, College of Cardiff

M. N. Huxley
University of Wales, College of Cardiff

CAMBRIDGE
UNIVERSITY PRESS

PUBLISHED BY THE PRESS SYNDICATE OF THE UNIVERSITY OF CAMBRIDGE
The Pitt Building, Trumpington Street, Cambridge CB2 1RP, United Kingdom

CAMBRIDGE UNIVERSITY PRESS
The Edinburgh Building, Cambridge, CB2 2RU, United Kingdom
40 West 20th Street, New York, NY 10011-4211, USA
10 Stamford Road, Oakleigh, Melbourne 3166, Australia

First published 1997

A catalogue record for this book is available from the British Library

Library of Congress Cataloguing in Publication data available

ISBN   0 521 58957 6 paperback

Transferred to digital printing 2004

Contents

Foreword

Between the 17th and the 21st of July, 1995 a group of about fifty mathematicians gathered in Cardiff for the Symposium on Sieve Methods, Exponential Sums and their Applications in Number Theory. They exchanged lectures and informal discussions and posed problems in the broad subject area defined by the title of the Symposium. This volume of papers gives an account of work described at the Symposium. Certain articles include a description of work done after the Symposium closed, some of this being prompted by questions posed during the Symposium.

Many of the contributions involve connections between the additive and multiplicative sides of analytic number theory, an interplay which the Symposium had been planned both to reflect and to promote. The reader will find represented here most of the branches of this subject which, as this volume demonstrates, has its roots in antiquity with the Sieve of Eratosthenes, yet is currently vibrant and receiving fresh stimulus from such diverse topics as trace formulae and elliptic curves.

Many of the participants at the Symposium were supported by the Engineering and Physical Sciences Research Council of the U.K. Others were able to bring their own support with them. The administrative costs were underwritten by the London Mathematical Society. The organisers, who are also the editors of this volume, are grateful to all the sources of support which made the Symposium possible.

The papers in this volume have been refereed to the standards required by leading journals; we take this opportunity to thank the referees for their work. The book was typeset by the editors using EMTEX software, for the most part on the basis of source code supplied by the authors.

The Symposium marked the twenty-eighth year during which Professor C. Hooley, F.R.S. had been head of a university mathematics department in Cardiff, during which time he accomplished much distinguished work in the subject area of the Symposium. In particular his pioneering work on the circle method and on applications of sieves, involving research on additive divisor problems and exponential sums (including Kloosterman sums), has been a constant source of inspiration to all the participants. Authors of papers printed in this volume dedicated their articles to him, in various styles. We have not reproduced these dedications at the head of each article; we record them collectively here, and take the opportunity to dedicate the entire book to Christopher Hooley.

Cardiff, Wales   G.R.H.G.
September, 1996   G.H.
                                                                      M.N.H.

Index of Authors

R. R. Hall 143
Department of Mathematics, University of York, Heslington,
York YO1 5DD, U.K.

G. Harman 1, 161
School of Mathematics, University of Wales Cardiff, 23 Senghennydd
Road, P.O. Box 926, Cardiff CF2 4YH, Wales, U.K.

C. Hooley 175
School of Mathematics, University of Wales Cardiff, 23 Senghennydd
Road, P.O. Box 926, Cardiff CF2 4YH, Wales, U.K.

M. N. Huxley 187
School of Mathematics, University of Wales Cardiff, 23 Senghennydd
Road, P.O. Box 926, Cardiff CF2 4YH, Wales, U.K.

A. Ivić 205
Katedra Matematike RGF-A, Universitet u Beogradu, Dousina 7,
11000 Beograd, Serbia (Yugoslavia)

H. Iwaniec 109
Mathematics Department, Rutgers University, New Brunswick,
New Jersey 08903, U.S.A.

M. Jutila 245
Department of Mathematics, University of Turku, SF-20500 Turku,
Finland.

I. Kiuchi 255
Department of Mathematics, Faculty of Science, Yamaguchi University,
Yoshida, Yamaguchi 753, Japan

J. McKee 275
Department of Pure Mathematics and Mathematical Statistics, 16 Mill
Lane, Cambridge CB2 1SB, U.K.

K. Matsumoto 255
Graduate School of Polymathematics, Nagoya University, Chikusa-ku,
Nagoya 464-01, Japan

B. Z. Moroz 283
Max-Planck-Institut für Mathematik, Gottfried-Claren-Strasse 26,
D-53225 Bonn, Germany

Y. Motohashi 293, 309
Department of Mathematics, Nihon University, Surugadai, Tokyo 101,
Japan

M. R. Murty 325
Department of Mathematics, McGill University, Montreal,
Quebec H3A 2K6, Canada

x

Participants in the Symposium

J. Andersson	H. Halberstam	R. Murty
R. Astley	R. R. Hall	G. Myerson
R. C. Baker	G. Hanrot	A. Plagne
A. Balog	G. Harman	A. Pollington
J. Brüdern	A. Hildebrand	R. A. Rankin
D. A. Burgess	C. Hooley	H. E. Rose
J. W. S. Cassels	M. N. Huxley	A. Schinzel
J. H. Coates	A. Ivić	W. Schmidt
M. D. Coleman	H. Iwaniec	E. J. Scourfield
R. J. Cook	M. Jutila	M. Sheingorn
G. Coppola	J. McKee	S. Sperber
J.-M. Deshouillers	G. Martin	S. Stepanov
M. Filaseta	K. Matsumoto	O. Trifonov
E. Fouvry	J. R. Merriman	R. C. Vaughan
J. Friedlander	P. Morée	N. Watt
G. R. H. Greaves	B. Moroz	S. M. J. Wilson
J. A. Haight	Y. Motohashi	T .D. Wooley

1. The Exceptional Set for Goldbach's Problem in Short Intervals

R. C. Baker, G. Harman and J. Pintz

1. Introduction

Define a *Goldbach number* to be an even number which can be written as the sum of two primes. Ramachandra [26] proved that almost all even numbers in an interval of the form $[\,x,\, x + x^\theta\,]$ are Goldbach numbers, provided that

$$\theta > \tfrac{3}{5}\,. \tag{1.1}$$

Here and below 'almost all' means 'with less than $x^\theta / \log^A x$ exceptions, provided that $A > 0$ and $N > C_1(A, \theta)$.'

Recently there has been a flurry of papers on this problem, by Perelli and Pintz [23], Mikawa [20], Jia [16], [17] and Li [18]. In successive steps the condition (1.1) has been weakened to $\theta > \frac{7}{81}$ (Li, [18]). In the present paper we shall push the method close to what is possible with our existing knowledge of mean and large values of Dirichlet polynomials.

Theorem 1. *For* $\theta \geq \frac{11}{160}$, *almost all even numbers in* $[\,x,\, x + x^\theta\,]$ *are Goldbach numbers.*

Remark. We note that $\frac{7}{81} = 0.08641\ldots$ whereas $\frac{11}{160} = 0.06875$.

The idea in all these papers is to show that, for almost all even integers $2n$ in $K = [\,x,\, x + x^{\theta_1 \theta_2}\,]$,

$$S(n) := \sum_{\substack{k+m=2n \\ k \in I,\ m \in J}} \rho(k)\rho(m) \ > 0 \tag{1.2}$$

where ρ is the indicator function of the prime numbers,

$$I = (\,x - 2Y,\, x\,] \quad \text{with} \quad Y = x^{\theta_1}, \qquad J = (\,Y,\, 2Y\,].$$

Thus, for example, Perelli and Pintz [23] get an asymptotic formula for $S(n)$ with $\theta_1 \geq \frac{7}{12} + \epsilon$, $\theta_2 \geq \frac{1}{3} + \epsilon$. We shall take

$$\theta_1 = \tfrac{11}{20} + \epsilon, \qquad \theta_2 = 2\big(\tfrac{1}{16} - 10^{-5}\big).$$

R.C. Baker is partly supported by a National Science Foundation grant.

Sieve Methods, Exponential Sums, and their Applications in Number Theory
Greaves, G.R.H., Harman, G., Huxley, M.N., Eds. ©Cambridge University Press, 1996

Here and below, ϵ is a sufficiently small positive absolute constant. Obviously Theorem 1 follows from (1.2) with this θ_1 and θ_2. In our proof of (1.2) we assume (as we may) that $x - \frac{1}{2}$ is an integer.

To prove (1.2) we borrow a simple but effective inequality from Brüdern and Fouvry [4] (a similar inequality had been given earlier by Iwaniec [15]). Suppose that

$$a_0(k) \leq \rho(k) \leq a_1(k) \quad \text{if} \quad k \in I; \tag{1.3}$$
$$b_0(m) \leq \rho(m) \leq b_1(m) \quad \text{if} \quad m \in J; \tag{1.4}$$

then

$$\rho(k)\rho(m) \geq a_0(k)b_1(m) + a_1(k)b_0(m) - a_1(k)b_1(m)$$

for $k \in I$, $m \in J$, and accordingly

$$S \geq S_{0,1} + S_{1,0} - S_{1,1}$$

where

$$S_{i,j} = \sum_{m+k=2n} a_i(k)b_j(m).$$

Here and below, *summations over k and m will always run over I and J respectively.* It now suffices to prove that

$$S_{i,j} = u_i v_j \frac{Y}{\mathcal{L}\mathcal{L}'} \mathfrak{S}(2n)\big(1 + O(\mathcal{L}^{-1})\big)$$

for almost all even $2n$ in $[\, x, \, x + x^{\theta_1 \theta_2} \,]$, with $\mathcal{L} = \log x$, $\mathcal{L}' = \log Y$,

$$u_0 > 0.99, \quad u_1 < 1.01, \quad v_0 > 0.05, \quad v_1 < 2.2. \tag{1.5}$$

(The definition of the singular series $\mathfrak{S}(2n)$, which is always positive, is given in §2.) Obviously it is crucial that the constants satisfy

$$u_0 v_1 + u_1 v_0 - u_1 v_1 > 0$$

and this is an easy consequence of (1.5).

The authors of [23], [20], [16], [17], [18] 'sieved J but not I,' in other words, used the simpler inequality

$$\rho(k)\rho(m) \geq \rho(k)b_0(m).$$

Our choice of functions a_0, a_1, b_0, b_1 is based on the sieve method of Harman [7]. In order to make this choice we need to establish classes of sequences $a(k)$ $(k \in I)$, $b(m)$ $(m \in J)$ for which an asymptotic formula

$$\sum_{k+m=2n} a(k)b(m) = uv \frac{Y}{\mathcal{L}\mathcal{L}'} \mathfrak{S}(2n)\big(1 + O(\mathcal{L}^{-1})\big) \tag{1.6}$$

holds for almost all $2n$ in $[x, x + x^{\theta_1 \theta_2}]$. We apply the Hardy-Littlewood method, and follow [23] quite closely, to obtain the following result. We write

$$H = Y^{\theta_2}, \quad Q = [H^{\frac{1}{2}}/2], \quad P(z) = \prod_{p < z} p,$$

$$\delta_\chi = \begin{cases} 1 & \text{if } \chi \text{ is the principal character } \chi_0 \pmod q \\ 0 & \text{if } \chi \text{ is non principal } \pmod q. \end{cases}$$

(The letter p is reserved for prime numbers.) By B we denote an absolute constant (not always the same one); ϵ is chosen so that $B\epsilon$ is sufficiently small (whenever necessary).

Theorem 2. *Suppose that the sequences $a(k)$ $(k \in I)$ and $b(m)$ $(m \in J)$ satisfy the following, for every $A > 0$ and $N > C_2(A)$:*
  (i) *we have*

$$\sum_{k \in I, \, k \leq t} \left(a(k)\chi(k) - \frac{\delta_\chi u}{\mathcal{L}} \right) \ll Y\mathcal{L}^{-A} \qquad (1.7)$$

for $t \leq x$ and any character $\chi \pmod q$ when $q \leq \mathcal{L}^A$;
  (ii) *we have*

$$a(k) = O\big(\tau(k)^B\big), \quad b(m) = O\big(\tau(m)^B\big)$$

where τ is the divisor function, and $a(k) = 0$ unless $\big(k, P(\mathcal{L}^A)\big) = 1$;
  (iii) *we have*

$$\sum_{m \in J, \, m \leq t} \left(b(m)\chi(m) - \frac{\delta_\chi v}{\mathcal{L}'} \right) \ll Y\mathcal{L}^{-A} \qquad (1.8)$$

for $t \leq 2Y$ and any character $\chi \pmod q$ when $q \leq \mathcal{L}^A$;
  (iv) *for any $q \leq Q$, and any $z \in \left[\dfrac{qQ}{6Y}, \dfrac{6qQ}{Y} \right]$, we have*

$$\sum_{\substack{\chi \,(\text{mod } q) \\ \chi \notin E_q}} \int_{Y/2}^{3Y} \left| \sum_{n \in J_y} \beta(m)\chi(m) - \frac{\delta_\chi v}{\mathcal{L}'} \right|^2 dy \ll (qQ)^2 Y\mathcal{L}^{-A}. \qquad (1.9)$$

Here $J_y = [y, y + yz]$ and E_q is a set of $O\big(q^{1/2}\mathcal{L}^{-2A}\big)$ characters $\pmod q$. Then (1.6) holds for almost all even integers $2n$ in $[N, N + N^{\theta_1 \theta_2}]$.

Remark. Here and later, implied constants depend at most on A.

We shall prove Theorem 2 in §2. In §4, we shall find two sequences $a_0(k)$, $a_1(k)$ which satisfy (i) and (ii), with associated constants u_0, u_1, in the role of u, such that (1.3) and the first two inequalities in (1.5) hold. In §5, we shall find sequences $b_0(m)$, $b_1(m)$ which satisfy (ii), (iii), (iv), with associated constants v_0, v_1 in the role of v, such that (1.4) and the last two inequalities in (1.5) hold. This will establish Theorem 1.

It is worth noting the following consequences of our construction.

Theorem 3. (I) *Let* $x^{0.55+\epsilon} \le M \le x\mathcal{L}^{-1}$. *For all* $q \le \mathcal{L}^A$, $N > C_3(A)$,

$$\frac{0.99M}{\mathcal{L}} < \pi(x;q,a) - \pi(x-M;q,a) < \frac{1.01M}{\mathcal{L}}$$

whenever $(a,q)=1$.

(II) *Let* $\lambda \ge \frac{1}{16} - 10^{-5}$. *For all integers* $h \le Y$ *with* $O(Y\mathcal{L}^{-A})$ *exceptions,*

$$\pi(h) - \pi(h - h^\lambda) > \frac{0.05h^\lambda}{\log h}.$$

(III) *For* $x > C_4$, *the interval* $[x, x+x^\mu]$ *contains Goldbach numbers. Here* $\mu = 0.0335$.

A result of the form (I) with $q = 1$ is claimed by Lou and Yao [19]. However, there are serious errors in this paper, including the use of the bound

$$\pi(x) - \pi(x-y) \ll \frac{y}{\log x}$$

with y tending to zero. Nevertheless, (I) is not new for $q = 1$, having been found earlier by D.R. Heath-Brown by a method similar to that in his paper [11]. We thank Roger Heath-Brown for making available to us his unpublished notes; our approach is somewhat different.

For results of the type (II), see [21], [7], [8], [33], [16], [17], [18]. The condition on λ in [33] is $\lambda \ge \frac{1}{14} + \epsilon$. Li (work in preparation) replaces $\frac{1}{14}$ by $\frac{1}{15}$. K.C. Wong [34], using a theorem of Watt [32] which would not be helpful in the context of §5, is able to get $\lambda = \frac{1}{18} + \epsilon$. Jia is able to improve this to $\lambda = \frac{1}{20} + \epsilon$ but the details are quite formidable.

To prove Theorem 3 (I), we simply use the fact (already explained above) that (i) of Theorem 2 holds for sequences $a_0(k)$, $a_1(k)$, lying below and above $\rho(k)$, with associated constants in $(0.99, 1.01)$. We then pick out integers congruent to $a \pmod q$ using characters in standard fashion.

To prove Theorem 3 (II), we use the sequence $b_0(k)$ constructed in §6 having the properties

$$\rho(m) \ge b_0(m) \quad \text{when} \quad m \sim Y,$$

$$\int_Y^{(1+\epsilon)Y} \left(\sum_{m \in J_y} \left(b_0(m) - \frac{v}{\mathcal{L}'} \right) \right)^2 dy \ll Q^2 Y \mathcal{L}^{-A}$$

with $v > 0.05$. Here we use property (iv) with $q = 1$ (the set E_1 is obviously empty), and we employ slight variants

$$Q = Y^\lambda, \quad J_y = [y, y + yQ/Y]$$

of earlier definitions. Thus

$$\sum_{y \leq k < y + y^\lambda} \rho(k) \geq \sum_{y \leq k < y + yQ/Y} b_0(k) > \frac{(v - 2\epsilon)y^\lambda}{L}$$

for all $y \in [Y, (1 + \epsilon)Y]$ except for a set of measure $O(Y\mathcal{L}^{-A})$. It is now easy to complete the proof of Theorem 3 (II).

To prove Theorem 3 (III) we argue as in Montgomery and Vaughan [22], employing Theorem 3 (II) in conjunction with the lower bound

$$\pi(x) - \pi(x - y) \gg \frac{y}{\log x}$$

for $x^{0.535} \leq y \leq x$ (Baker and Harman [2]). The constant 0.0335 is larger than $0.535/16$.

2. Proof of Theorem 2

We write $e(\theta) = e^{2\pi i\theta}$,

$$S_1(\alpha) = \sum_k a(k)e(k\alpha), \quad S_2(\alpha) = \sum_m b(m)e(m\alpha).$$

Thus

$$\sum_{k+m=2n} a(k)b(m) = \int_{1/Q}^{1+1/Q} S_1(\alpha)S_2(\alpha)e(-2n\alpha)\, d\alpha.$$

We divide up the interval $[1/Q, 1 + 1/Q]$ into Farey arcs of order Q, writing $I_{q,r}$ for the arc with centre at r/q. Thus

$$I_{q,r} \subset \left[\frac{r}{q} - \frac{1}{qQ}, \frac{r}{q} + \frac{1}{qQ}\right]$$

for $q \leq Q$, $1 \leq r \leq q$, $(r, q) = 1$. Let

$$I'_{q,r} = \left[\frac{r}{q} - \frac{\mathcal{L}^{4A}}{qY}, \frac{r}{q} + \frac{\mathcal{L}^{4A}}{qY}\right], \quad I''_{q,r} = \begin{cases} I_{q,r}\backslash I'_{q,r} & \text{if } q \leq \mathcal{L}^{2A} \\ I_{q,r} & \text{if } q > \mathcal{L}^{2A}. \end{cases}$$

The major and minor arcs are defined by

$$\mathfrak{M} = \bigcup_{q \leq \mathcal{L}^{2A}} \bigcup_{r=1}^{q} {}^{*} I'_{q,r}, \quad \mathfrak{m} = [1/Q, 1 + 1/Q]\backslash \mathfrak{M}$$

respectively. As usual, an asterisk denotes a restriction to those r coprime to q.

Let us write

$$\mathfrak{S}(2n) = 2 \prod_p \left(1 - \frac{1}{(p-1)^2}\right) \prod_{\substack{p|n \\ p>2}} \left(\frac{p-1}{p-2}\right),$$

$$c_q(m) = \sum_{r=1}^{q}{}^{*} e\left(\frac{mr}{q}\right), \qquad \tau(\chi) = \sum_{r=1}^{q}{}^{*} \chi(r) e\left(\frac{r}{q}\right).$$

For the well-known formula

$$\mathfrak{S}(2n) = \sum_{q=1}^{\infty} \frac{\mu^2(q)}{\phi^2(q)} c_q(-2n)$$

see [31], (3.24). We recall the well-known results ([5], pages 66, 67 for example) for the Gauss sum :

$$|\tau(\chi)| \le q^{\frac{1}{2}}, \quad \tau(\chi_0) = \mu(q). \tag{2.1}$$

To prove Theorem 2 it suffices to show that

$$\sum_{2n \in K} \left| \int_{1/Q}^{1+1/Q} S_1(\alpha) S_2(\alpha) e(-2n\alpha)\, d\alpha - \frac{uvY}{\mathcal{L}\mathcal{L}'} \mathfrak{S}(2n) \right|^2 \ll HY^2 \mathcal{L}^{-A-7}.$$

We accomplish this in two stages by proving

$$\sum_{2n \in K} \left| \int_{\mathfrak{M}} S_1(\alpha) S_2(\alpha) e(-2n\alpha)\, d\alpha - \frac{uvY}{\mathcal{L}\mathcal{L}'} \mathfrak{S}(2n) \right|^2 \ll HY^2 \mathcal{L}^{-A-7}, \tag{2.2}$$

$$\sum_{2n \in K} \left| \int_{\mathfrak{m}} S_1(\alpha) S_2(\alpha) e(-2n\alpha)\, d\alpha \right|^2 \ll HY^2 \mathcal{L}^{-A-7}. \tag{2.3}$$

We begin the proof of (2.2) by replacing the S_j by suitable approximations. For $q \le \mathcal{L}^{2A}$ we have $(k, q) = 1$ whenever $a(k) \ne 0$. Hypothesis (i) suggests the use of the identity

$$S_1\left(\frac{r}{q} + \eta\right) = \sum_k a(k) e(k\eta) \frac{1}{\phi(q)} \sum_\chi \tau(\bar{\chi}) \chi(kr)$$

$$= \frac{1}{\phi(q)} \sum_\chi \tau(\bar{\chi}) \chi(r) \sum_k a(k) \chi(k) e(k\eta)$$

$$= S_1'\left(\frac{r}{q} + \eta\right) + \sum_\chi \frac{\tau(\bar{\chi})}{\phi(q)} \chi(r) \sum_k \left(\left(a(k)\chi(k) - \frac{u\delta_\chi}{\mathcal{L}}\right) e(k\eta)\right).$$

$$\tag{2.4}$$

Here, for $r/q + \eta \in I'_{q,r}$,

$$S'_1\left(\frac{r}{q} + \eta\right) = \frac{\mu(q)}{\phi(q)} \frac{u}{\mathcal{L}} \sum_k e(k\eta).$$

(We have used (2.1) to rewrite the contribution from χ_0.)

To bound the double sum in (2.4) we use partial summation. With

$$v(t) = \sum_{\substack{k \in I \\ k < t}} \left(a(k)\chi(k) - \frac{u\delta_\chi}{\mathcal{L}}\right),$$

hypothesis (i) yields

$$\sum_k \left(a(k)\chi(k) - \frac{u\delta_\chi}{\mathcal{L}}\right) e(k\eta) = \int_I e(t\eta)\,dv(t)$$

$$= \left[e(t\eta)v(t)\right]_{x-2Y}^{x} - 2\pi i \eta \int_I e(t\eta)v(t)\,dt$$

$$\ll (1 + Y|\eta|) \max_{t \in I} |v(t)| \ll \mathcal{L}^{4A} Y \mathcal{L}^{-13A} \ll Y\mathcal{L}^{-9A}.$$

Here and below we suppose that $x > C_2(A')$ where A' is sufficiently large in terms of A. The double sum in (2.4) is thus $\ll Y\mathcal{L}^{-8A}$.

Since $\mathfrak{M}$ has Lebesgue measure $\ll \mathcal{L}^{6A}Y^{-1}$,

$$\int_{\mathfrak{M}} |(S_1 - S'_1)S_2|\, d\alpha \ll \mathcal{L}^{6A}Y^{-1} . Y\mathcal{L}^{-8A} . Y\mathcal{L}^B \ll Y\mathcal{L}^{-A}. \tag{2.5}$$

(We have used hypothesis (ii) to get $|S_2| \ll \sum_m |b(m)| \ll Y\mathcal{L}^B$.)

The bound (2.5) clearly permits us to replace S_1 by S'_1 in proving (2.2). In exactly the same way, we may replace $S'_1 S_2$ by $S'_1 S'_2$, where

$$S'_2\left(\frac{r}{q} + \eta\right) = \frac{\mu(q)}{\phi(q)} \frac{v}{\mathcal{L}'} \sum_m e(m\eta) \quad \text{for} \quad \frac{r}{q} + \eta \in I'_{q,r};$$

that is, we need only prove the analogue of (2.2) for $S'_1 S'_2$.

We may rewrite the integral

$$\int_{\mathfrak{M}} S'_1(\alpha) S'_2(\alpha) e(-2n\alpha)\, d\alpha$$

in the form

$$\frac{uv}{\mathcal{L}\mathcal{L}'} \sum_{q \le \mathcal{L}^{2A}} \sum_{r=1}^{q}{}^{*} \frac{\mu^2(q)}{\phi^2(q)} e\left(-\frac{2nr}{q}\right) \int_{-\eta_0}^{\eta_0} \sum_k \sum_m e((k+m-2n)\eta)\, d\eta, \tag{2.6}$$

where $\eta_0 = \mathcal{L}^{4A}/qY$. If we replace the domain of integration in (2.6) by $[-\frac{1}{2}, \frac{1}{2}]$, we introduce an error

$$\ll \sum_{q \leq \mathcal{L}^{2A}} \frac{1}{\phi(q)} \int_{\eta_0}^{\frac{1}{2}} \frac{d\eta}{\eta^2} \ll \sum_{q \leq \mathcal{L}^{2A}} \frac{qY}{\phi(q)\mathcal{L}^{4A}} \ll Y\mathcal{L}^{-A}.$$

This replaces the left-hand side of (2.6) by

$$\frac{uv}{\mathcal{L}\mathcal{L}'} \sum_{q \leq \mathcal{L}^{2A}} \frac{\mu^2(q)}{\phi^2(q)} c_q(-2n) \sum_{\substack{k \\ k+m=2n}} \sum_m 1 = \frac{uv}{\mathcal{L}\mathcal{L}'} \sum_{q \leq \mathcal{L}^{2A}} \frac{\mu^2(q)}{\phi^2(q)} c_q(-2n)(Y + O(1))$$

for $2n \in K$. Consequently,

$$\sum_{2n \in K} \left| \int_{\mathfrak{M}} S_1'(\alpha) S_2'(\alpha) e(-2n\alpha) d\alpha - \frac{uvY}{L^2} \mathfrak{S}(2n) \right|^2$$

$$\ll HY^2 \mathcal{L}^{-A-7} + \sum_{2n \in K} Y^2 \left| \sum_{q > \mathcal{L}^{2A}} \frac{\mu^2(q)}{\phi^2(q)} c_q(-2n) \right|^2$$

$$\ll HY^2 \mathcal{L}^{-A-7} + Y^2 \sum_{2n \in K} \left\{ \sum_{d|2n} \frac{1}{\phi(d)} \min\left(\frac{d}{\mathcal{L}^{2A}}, 1\right) \right\}^2. \quad (2.7)$$

For the last step we use Vaughan [31, (3.23)]. Now

$$\sum_{2n \in K} \left\{ \sum_{\substack{d|2n \\ d \leq \mathcal{L}^{2A}}} \frac{1}{\phi(d)} \frac{d}{\mathcal{L}^{2A}} \right\}^2 \ll \mathcal{L}^{-4A+1} \sum_{j \in K} \tau^2(j) \ll H\mathcal{L}^{-A-7}, \quad (2.8)$$

$$\sum_{2n \in K} \left\{ \sum_{\substack{d|2n \\ d > \mathcal{L}^{2A}}} \frac{1}{\phi(d)} \right\}^2 \ll \mathcal{L}^{-4A+1} \sum_{j \in K} \tau^2(j) \ll H\mathcal{L}^{-A-7}. \quad (2.9)$$

The analogue of (2.2) for $S_1' S_2'$ now follows from (2.7)–(2.9). This completes the proof of (2.2).

We now turn to the sum in (2.3), which may be rewritten as

$$\sum_{2n \in K} \int_{\mathfrak{m}} S_1(\zeta) S_2(\zeta) e(-2n\zeta) \int_{\mathfrak{m}} \overline{S_1(\alpha)} \, \overline{S_2(\alpha)} \, e(2n\alpha) \, d\alpha \, d\zeta$$

$$\ll \int_{\mathfrak{m}} |S_1(\zeta) S_2(\zeta)| \int_{\mathfrak{m}} |S_1(\alpha) S_2(\alpha)| \min\left(H, \frac{1}{\|2(\alpha - \zeta)\|}\right) d\alpha \, d\zeta$$

$$\ll (Y\mathcal{L}^B)^{\frac{3}{2}} \sup_{\zeta \in [0,1]} \left(\int_{\mathfrak{m}} |S_2(\alpha)|^2 \min\left(H, \frac{1}{\|2(\alpha - \zeta)\|}\right)^2 d\alpha \right)^{\frac{1}{2}}. \quad (2.10)$$

For the last step we have used the Cauchy-Schwarz inequality, the bound

$$\int_0^1 |S_1(\alpha)|^2 \, d\alpha = \sum_k |a(k)|^2 \ll \sum_k \tau(k)^{2B} \ll Y\mathcal{L}^B,$$

and the corresponding bound for S_2.

In view of (2.10), it suffices to show that

$$\sup_\zeta \int_{\mathcal{I}(\zeta)} |S_2(\alpha)|^2 \, d\alpha \ll Y\mathcal{L}^{-3A}, \qquad (2.11)$$

where $\mathcal{I}(\zeta) = (\zeta - 1/H, \, \zeta + 1/H) \cap \mathfrak{m}$. Since $I_{q,r}$ has length at least

$$1/Q^2 > 2/H,$$

there are at most two punctured arcs $I''_{q,r}$ with $q \le Q$ and $(r,q) = 1$, which intersect $(\zeta - 1/H, \, \zeta + 1/H)$. Instead of (2.11), then, we may show that

$$\int_{I''_{q,r}} |S_2(\alpha)|^2 \, d\alpha \ll Y\mathcal{L}^{-3A}$$

for $q \le Q$, $(r,q) = 1$.

By the analogue of (2.4) for S_2, it suffices to show that

$$\int_{\eta + r/q \in I''_{q,r}} \left| S_2' \left(\frac{r}{q} + \eta \right) \right|^2 \, d\eta \ll Y\mathcal{L}^{-3A}, \qquad (2.12)$$

$$\frac{q}{\phi^2(q)} \int_{-1/qQ}^{1/qQ} \left\{ \sum_\chi |W(\chi, \eta)| \right\}^2 \, d\eta \ll Y\mathcal{L}^{-3A}. \qquad (2.13)$$

Here

$$W(\chi, \eta) = \sum_m \left(b(m)\chi(m) - \frac{v\delta_\chi}{\mathcal{L}'} \right) e(m\eta).$$

The left hand side of (2.12) is

$$\ll \frac{1}{\phi^2(q)} \frac{1}{\mathcal{L}^2} \int_{I(\eta)} \min\left(\frac{1}{Y^2}, \frac{1}{\eta^2} \right) d\eta \qquad (2.14)$$

with

$$I(\eta) = \begin{cases} [\mathcal{L}^{4A}/qY, \frac{1}{2}] & \text{if } q \le \mathcal{L}^{2A} \\ [0, \frac{1}{2}] & \text{if } q > \mathcal{L}^{2A}. \end{cases}$$

For $q \leq \mathcal{L}^{2A}$, the expression in (2.14) is $\ll q^{-2}qY\mathcal{L}^{-4A} \ll Y\mathcal{L}^{-3A}$; for $q > \mathcal{L}^{2A}$, we have instead the bound $\ll q^{-2}Y \ll Y\mathcal{L}^{-3A}$. This establishes (2.12).

We split the sum over χ into sums over E_q and

$$E_q^c = \{\chi \ (\mathrm{mod}\ q)\colon \chi \notin E_q\}.$$

Applying Cauchy's inequality to each subsum, it suffices to prove that

$$\frac{q|E|}{\phi^2(q)} \int_{-1/qQ}^{1/qQ} \sum_{\chi \in E} |W(\chi,\eta)|^2 \, d\eta \ll Y\mathcal{L}^{-3A} \qquad (2.15)$$

for $E = E_q$, E_q^c in order to establish (2.12).

By hypothesis (ii), (iv) and Parseval's inequality the left hand side of (2.15) is bounded when $E = E_q$ by

$$\ll \frac{q}{\phi^2(q)}|E_q| \int_{-\frac{1}{2}}^{\frac{1}{2}} \sum_{\chi \in E_q} |W(\chi,\eta)|^2 \, d\eta \ll \frac{q}{\phi^2(q)}|E_q|^2 \sum_m \left(|b(m)|^2 + 1\right)$$

$$\ll q^{-1}\mathcal{L}|E_q|^2 Y\mathcal{L}^B \ll Y\mathcal{L}^{-3A}.$$

For $E = E_q^c$ we appeal to Gallagher's Lemma ([21], Lemma 1.9):

$$\int_{-1/qQ}^{1/qQ} |W(\chi,\eta)|^2 \, d\eta \ll \frac{1}{(qQ)^2} \int_{Y/2}^{3Y} \left\{ \sum_{m \in J_y'} \left(b(m)\chi(m) - \frac{v\delta_\chi}{\mathcal{L}}\right) \right\}^2 dy \quad (2.16)$$

where

$$J_y' = \left[y - \tfrac{1}{2}qQ, \, y + \tfrac{1}{2}qQ\right].$$

Recalling that $J_y = [y, \, y + yz]$, for some $z \in [qQ/6Y, \, 6qQ/Y]$, the right hand side of (2.16) is

$$\ll U_\chi := \frac{1}{(qQ)^2} \int_{Y/2}^{3Y} \left\{ \sum_{m \in J_y} \left(b(m)\chi(m) - \frac{v\delta_\chi}{\mathcal{L}}\right) \right\}^2 dy;$$

this may be demonstrated using a device of Saffari and Vaughan ([29], proof of (6.21)). By hypothesis (iv),

$$\sum_{\chi \in E_q^c} U_\chi \ll Y\mathcal{L}^{-4A}.$$

The bound (2.15) with $E = E_q^c$ follows at once. This completes the proof of Theorem 2.

3. Dirichlet polynomials

In this section we assemble some results about mean and large values of Dirichlet polynomials

$$F(s, \chi) = \sum_{n \asymp F} \frac{a_n \chi(n)}{n^s}.$$

Here χ is a Dirichlet character $(\bmod\, q)$ with $q \leq x$; the symbolism $n \asymp F$ means that $F(s, \chi)$ 'has length F', that is, $c_1 F < n < c_2 F$ where c_1 and c_2 are absolute constants. Sometimes we impose a tighter condition $n \sim F$, where $u \sim U$ means $u \in (U, 2U]$. Note the convention that the same letter is used for $F(s, \chi)$ and its length. We suppose that $2 \leq F \leq x$.

The coefficients of the polynomials we consider will satisfy the bound

$$|a_n| \leq \tau(n)^B. \tag{3.1}$$

Let $2 < T \leq x/4$. Let $\mathcal{S}$ be a set of pairs (t, χ), with $t \in [-T, T]$ and χ a character $(\bmod\, q)$. We say that $\mathcal{S}$ is *well-spaced* if $|t - t'| \geq 1$ whenever $(t, \chi), (t', \chi) \in \mathcal{S}$.

Suppressing dependence on $\mathcal{S}$, we write

$$\|F\|_p = \left\|F(\tfrac{1}{2} + it, \chi)\right\|_p = \left(\sum_{(t,\chi) \in \mathcal{S}} \left|F(\tfrac{1}{2} + it, \chi)\right|^p \right)^{1/p} \quad \text{if} \quad p \geq 1,$$

$$\|F\|_\infty = \sup_{(t,\chi) \in \mathcal{S}} \left|F(\tfrac{1}{2} + it, \chi)\right|.$$

We sometimes need to impose a condition:

$$\|F\|_\infty \ll F^{\frac{1}{2}} \mathcal{L}^{-A} \quad \text{for all} \quad A > 0. \tag{3.2}$$

Lemma 1. *Let $\mathcal{S}$ be well-spaced; then*

$$\|F\|_2^2 \ll \mathcal{L}(qT + F)G \tag{3.3}$$

where

$$G = G(F) = \sum_{n \asymp F} \frac{|a_n|^2}{n}.$$

If

$$\left|F(\tfrac{1}{2} + it, \chi)\right| \geq V$$

for $(t, \chi) \in \mathcal{S}$, then the cardinality of $\mathcal{S}$ is bounded by

$$|\mathcal{S}| \ll \mathcal{L}^2 \left(\frac{GF}{V^2} + \frac{G^3 F q T}{V^6} \right). \tag{3.4}$$

Proof. The inequality (3.3) follows readily on combining Lemma 1.4 and Theorem 6.4 of Montgomery [21]. To obtain (3.4) we begin by proving

$$\|F\|_2^2 \ll \mathcal{L}\big(F + (qT)^{\frac{1}{2}}|\mathcal{S}|\big)G$$

using the same reasoning as in the proof of Theorem 8.3 of [21]. We may then follow the argument of Huxley [14], §2 to complete the proof of (3.4).

In the following, when X occurs it denotes *one* of x, Y^2.

Theorem 4. *Let* $M(s,\chi)$, $N(s,\chi)$, $L(s,\chi)$ *be Dirichlet polynomials with*

$$MNL = X. \tag{3.5}$$

Suppose that M, N, L *satisfy* (3.1) *and* L *satisfies* (3.2). *Let* $qT = X^{1-\theta}$ *where* $\frac{1}{2} + \epsilon < \theta < \frac{7}{12}$. *Suppose that* $M = X^{\alpha_1}$, $N = X^{\alpha_2}$,

$$|\alpha_1 - \alpha_2| < 2\theta - 1 - \epsilon, \tag{3.6}$$
$$1 - (\alpha_1 + \alpha_2) < \gamma(\theta) - \epsilon \tag{3.7}$$

where

$$\gamma(\theta) = \min\big(4\theta - 2, (20\theta - 9)/11, (72\theta - 37)/11\big). \tag{3.8}$$

Then

$$\|MNL\|_1 \ll X^{\frac{1}{2}}\mathcal{L}^{-A} \quad \text{for all} \quad A > 0. \tag{3.9}$$

Proof. By a simple dyadic argument we may suppose that

$$\left|M\big(\tfrac{1}{2} + it, \chi\big)\right| \sim M^{\sigma_1 - \frac{1}{2}}, \quad \left|N\big(\tfrac{1}{2} + it, \chi\big)\right| \sim N^{\sigma_2 - \frac{1}{2}}, \quad \left|L\big(\tfrac{1}{2} + it, \chi\big)\right| \sim L^{\sigma_3 - \frac{1}{2}}$$

for $(t, \chi) \in \mathcal{S}$, where the σ_j are fixed real numbers. It now suffices to show that

$$\mathcal{I} := |\mathcal{S}|M^{\sigma_1 - \frac{1}{2}}N^{\sigma_2 - \frac{1}{2}}L^{\sigma_3 - \frac{1}{2}} \ll X^{\frac{1}{2}}\mathcal{L}^{-A}.$$

We note that, for any positive integer $g \leq B$,

$$|\mathcal{S}|\mathcal{L}^{-B} \ll (L^g)^{2 - 2\sigma_3} + qT(L^g)^{\min(1 - 2\sigma_3, 4 - 6\sigma_3)}. \tag{3.10}$$

To get this inequality we use the Dirichlet polynomial $L(s, \chi)^g$, using (3.1) to bound $G(P^g)$ by $\mathcal{L}^B$, and then combine (3.3) and (3.4). Similarly,

$$|\mathcal{S}|\mathcal{L}^{-B} \ll M^{2 - 2\sigma_1} + qTM^{f(\sigma_1)}, \tag{3.11}$$
$$|\mathcal{S}|\mathcal{L}^{-B} \ll N^{2 - 2\sigma_2} + qTN^{f(\sigma_2)}, \tag{3.12}$$

where $f(\sigma) = \min(1 - 2\sigma, 4 - 6\sigma)$. Let

$$c(\sigma) = \begin{cases} 1 & \text{if } \sigma \leq \frac{3}{4} \\ 4\sigma - 2 & \text{if } \sigma > \frac{3}{4}. \end{cases}$$

Thus the first terms in (3.11) and (3.12) dominate if and only if $M^{c(\sigma_1)} \geq qT$. To prove (3.9) we must consider four cases, some dividing into sub-cases.

Case 1. $M^{c(\sigma_1)} \geq qT$, $N^{c(\sigma_2)} \geq qT$. From (3.11), (3.12),

$$\mathcal{I}\mathcal{L}^{-B} \ll (M^{2-2\sigma_1} N^{2-2\sigma_2}) M^{\sigma_1 - \frac{1}{2}} N^{\sigma_2 - \frac{1}{2}} \|L\|_\infty$$
$$\ll M^{\frac{1}{2}} N^{\frac{1}{2}} L^{\frac{1}{2}} \mathcal{L}^{-A} = X^{\frac{1}{2}} \mathcal{L}^{-A}$$

since L satisfies (3.2). Since A is arbitrary, this yields (3.9).

Case 2 (i). $M^{c(\sigma_1)} < qT$, $N^{c(\sigma_2)} < qT$, $\sigma_3 < \frac{3}{4}$. From (3.11), (3.12)

$$\mathcal{I}\mathcal{L}^{-B} \ll (qTM^{1-2\sigma_1} \cdot qTN^{1-2\sigma_2})^{\frac{1}{2}} M^{\sigma_1 - \frac{1}{2}} N^{\sigma_2 - \frac{1}{2}} L^{\frac{1}{4}}$$
$$= qTL^{\frac{1}{4}} \ll X^{\frac{1}{2}} \mathcal{L}^{-A}$$

since, according to (3.7), $L \ll X^2 (qT)^{-4} \mathcal{L}^{-4A}$.

Case 2 (ii). $M^{c(\sigma_1)} \leq qT$, $N^{c(\sigma_2)} \leq qT$, $\sigma_3 > \frac{3}{4}$. From (3.10) with $g = 3$,

$$\mathcal{I}\mathcal{L}^{-B} \ll (qTM^{f(\sigma_1)})^{\frac{5}{12}} (qTN^{f(\sigma_2)})^{\frac{5}{12}} (L^{6-6\sigma_3})^{\frac{1}{6}} M^{\sigma_1 - \frac{1}{2}} N^{\sigma_2 - \frac{1}{2}} L^{\sigma_3 - \frac{1}{2}}$$
$$+ (qTM^{f(\sigma_1)})^{\frac{17}{36}} (qTN^{f(\sigma_2)})^{\frac{17}{36}} (qTL^{12-18\sigma_3})^{\frac{1}{18}} M^{\sigma_1 - \frac{1}{2}} N^{\sigma_2 - \frac{1}{2}} L^{\sigma_3 - \frac{1}{2}}$$
$$= I_1 + I_2, \text{ say.}$$

For any $\alpha \in \left[\frac{1}{6}, \frac{1}{2}\right]$, and any real σ

$$\alpha f(\sigma) + \sigma - \frac{1}{2} \leq \alpha f(\tfrac{3}{4}) + \frac{3}{4} - \frac{1}{2} = \frac{1}{4}(1 - 2\alpha) \tag{3.13}$$

(the left-hand side is increasing for $\sigma \leq \frac{3}{4}$ and decreasing for $\sigma \geq \frac{3}{4}$). Thus

$$I_1 \ll (qT)^{\frac{5}{6}} (MN)^{\frac{1}{24}} L^{\frac{1}{2}} = (qT)^{\frac{5}{6}} (X/L)^{\frac{1}{24}} L^{\frac{1}{2}} \ll X^{\frac{1}{2}} \mathcal{L}^{-A}$$

since $L \ll X^{(20\theta - 9)/11 - \epsilon}$ from (3.5), (3.6). Similarly

$$I_2 \ll qT(MN)^{\frac{1}{72}} L^{\frac{1}{6}} = qT(X/L)^{\frac{1}{72}} L^{\frac{1}{6}} \ll X^{\frac{1}{2}} \mathcal{L}^{-A}$$

since $L \ll X^{(72\theta - 37)/11 - \epsilon}$. This establishes (3.9) in Case 2 (ii).

Case 3 (i). $M^{c(\sigma_1)} \geq qT > N^{c(\sigma_2)}, \sigma_3 \leq \frac{3}{4}$. Then

$$\mathcal{IL}^{-B} \ll M^{(2-2\sigma_1)/2}\left(qTN^{f(\sigma_2)}\right)^{\frac{1}{2}} M^{\sigma_1 - \frac{1}{2}} N^{\sigma_2 - \frac{1}{2}} L^{\frac{1}{4}}$$
$$\ll M^{\frac{1}{2}}(qT)^{\frac{1}{2}} L^{\frac{1}{4}} \ll X^{\frac{1}{2}} \mathcal{L}^{-A}$$

since, from (3.6),

$$L \ll (X/qTM)^2 X^{-\epsilon} = X^{2\theta - 2\alpha_1 - \epsilon}. \tag{3.14}$$

Case 3 (ii). $M^{c(\sigma_1)} \geq qT > N^{c(\sigma_2)}4, \sigma_3 > \frac{3}{4}$. Let

$$d(g) = 2\theta - \frac{1 + 2g\theta}{2g + 1}.$$

Then $d(g)$ is increasing as a function of g, $d(0) = 2\theta - 1$ and $d(g_0) > \theta - \epsilon/3$ for some $g_0 \leq B$. It easily follows from (3.6), (3.7) that $\alpha_1 \in \left(d(0), d(g_0)\right]$, whence

$$d(g - 1) < \alpha_1 \leq d(g)$$

for some g with $1 \leq g \leq g_0$. From (3.10)–(3.12),

$$\mathcal{IL}^{-C} \ll \left(M^{2-2\sigma_1}\right)^{\frac{1}{2}} \left(qTN^{f(\sigma_2)}\right)^{\frac{1}{2} - \frac{1}{2g}} L^{g\frac{2-2\sigma_3}{2g}} M^{\sigma_1 - \frac{1}{2}} N^{\sigma_2 - \frac{1}{2}} L^{\sigma_3 - \frac{1}{2}}$$
$$+ \left(M^{2-2\sigma_1}\right)^{\frac{1}{2}} \left(qTN^{f(\sigma_2)}\right)^{\frac{1}{2} - \frac{1}{6g}} \left(qTL^{g(4-6\sigma_3)}\right)^{\frac{1}{6g}} M^{\sigma_1 - \frac{1}{2}} N^{\sigma_2 - \frac{1}{2}} L^{\sigma_3 - \frac{1}{2}}$$
$$\ll M^{\frac{1}{2}} L^{\frac{1}{2}}(qT)^{\frac{1}{2} - \frac{1}{2g}} N^{\frac{1}{4g}} + M^{\frac{1}{2}} L^{\frac{1}{6}}(qT)^{\frac{1}{2}} N^{\frac{1}{12g}}.$$

Now

$$M^{\frac{1}{2}} L^{\frac{1}{2}}(qT)^{\frac{1}{2} - \frac{1}{2g}} N^{1/4g} \ll X^{\frac{1}{2}} \mathcal{L}^{-A} \tag{3.15}$$

is equivalent to

$$\frac{X}{ML} = N \gg (qT)^{(2g-2)/(2g-1)} \mathcal{L}^{A_1}$$

or to

$$L \ll X^{(1+(2g-2)\theta)/(2g-1)-\alpha_1} \mathcal{L}^{-A_1}$$

$(A_1, A_2, \ldots$ denote constants depending on A, g). Since $\alpha_1 > d(g-1)$, we have

$$2\theta - 2\alpha_1 < \frac{1 - (2g-2)\theta}{2g - 1} - \alpha_1$$

and (3.15) is a consequence of (3.14). Similarly,

$$M^{\frac{1}{2}} L^{\frac{1}{6}}(qT)^{\frac{1}{2}} N^{\frac{1}{12g}} \ll X^{\frac{1}{2}} \mathcal{L}^{-A} \tag{3.16}$$

is equivalent to $L \ll X^{(6g\theta - 1 - (6g-1)\alpha_1)/(2g-1)} \mathcal{L}^{-A_2}$. Since $\alpha_1 \leq d(g)$, we have

$$2\theta - 2\alpha_1 \leq \frac{6g\theta - 1 - (6g-1)\alpha_1}{2g - 1}.$$

Thus (3.16) is also a consequence of (3.14), and (3.9) is established in Case 3 (ii).

Case 4. As Case 3, with M and N interchanged. In view of the symmetry between M and N, Case 4 follows in the same way as Case 3, and the proof of Theorem 4 is complete.

We note that Theorem 4 was suggested by Lemma 1 (i) of Baker and Harman [1], which it strengthens and extends.

Lemma 2. *Let $M(s, \chi)$ be a Dirichlet polynomial and suppose that*

$$qT \leq M^\gamma \quad where \quad 1 \leq \gamma < B.$$

Suppose that $M(s, \chi)$ satisfies (3.1), (3.2). Then

$$\|M\|_\beta \ll M^{\frac{1}{2}} \mathcal{L}^{-A} \tag{3.17}$$

for every $A > 0$, provided that

$$\beta \geq 4\gamma - 2h + \epsilon, \tag{3.18}$$

where h is an integer satisfying $h \ll 1$ and

$$2h - \epsilon < \beta < 6h - \epsilon. \tag{3.19}$$

Proof. This is similar to Lemma 4 of Harman [8]. Put $Q(s, \chi) = M(s, \chi)^h$; we may suppose that $Q(\frac{1}{2} + it, \chi) \sim V$ for all $(t, \chi) \in \mathcal{S}$. Using (3.4) we therefore obtain

$$\|M\|_\beta^\beta \ll |\mathcal{S}| V^{\beta/h} = |\mathcal{S}| V^{2+(\beta-2h)/h}$$
$$\ll \mathcal{L}^B \left(QV^{(\beta-2h)/h} + qTQV^{-4+(\beta-2h)/h} \right). \tag{3.20}$$

For the first term in the last expression we have

$$QV^{(\beta-2h)/h} \leq Q \left(Q\mathcal{L}^{-A_3} \right)^{(\beta-2h)/2h} \ll Q^{\beta/2h} \mathcal{L}^{-A\beta} \ll M^{\beta/2} \mathcal{L}^{-A\beta}$$

from (3.20) and the hypothesis (3.2). For the remaining term, we get the desired bound if

$$V > \left(qTQ^{(2h-\beta)/2h} \right)^{h/(6h-\beta)} \mathcal{L}^{A_4}. \tag{3.21}$$

Suppose now (3.21) is violated. Then

$$|\mathcal{S}| V^{\beta/h} = V^{(\beta-2h)/h} |\mathcal{S}| V^2$$
$$\ll \left(qTQ^{(2h-\beta)/2h} \right)^{(\beta-2h)/(6h-\beta)} (Q + qT) \mathcal{L}^{A_5}. \tag{3.22}$$

We now find that the right-hand side of (3.22) is $\ll Q^{\beta/2h} \mathcal{L}^{-A\beta}$ provided that $M^{2h+\beta-\epsilon} \geq (qT)^4$ which is a consequence of (3.18). This completes the proof of Lemma 2.

Lemma 3. *Let* $L(s,\chi)$, $M(s,\chi)$, $N(s,\chi)$, $R(s,\chi)$ *be Dirichlet polynomials with*

$$LMNR = X.$$

Suppose L, M, N, R *satisfy* (3.1). *Suppose that*

$$qT \le X^{0.45-\epsilon/2}, \tag{3.23}$$

$$M \ge qT. \tag{3.24}$$

Suppose that L, N *and* R *satisfy* (3.2). *Then each of the following sets of conditions implies* $\|LMNR\|_1 \ll X^{1/2} \log^{-A} X$ *for all* $A > 0$:

(i) $N \gg X^{0.1125}$, $R \gg X^{0.225}$, $L \gg X^{\frac{9}{70}}$,

(ii) $N \gg X^{0.225}$, $R \gg X^{0.15}$, $L \gg X^{0.1}$,

(iii) $N \gg X^{0.15}$, $R^2 L \gg X^{0.45}$, $L \gg X^{0.18}$,

(iv) $N \ll X^{0.15}$, $R \ll X^{0.15}$, $NR \gg X^{\frac{9}{35}}$, $LNR \gg X^{\frac{63}{130}+B\epsilon}$.

Proof. By Hölder's inequality, (3.1), (3.3) and (3.24),

$$\|LMNR\|_1 \le \|M\|_2 \|N\|_{\beta_1} \|L\|_{\beta_2} \|R\|_{\beta_3}$$
$$\ll \mathcal{L}^B M^{\frac{1}{2}} \|N\|_{\beta_1} \|L\|_{\beta_1} \|R\|_{\beta_3} \tag{3.25}$$

whenever β_1, β_2, β_3 are positive and $\sum_j 1/\beta_j = \frac{1}{2}$.

We note that when β_1 is an even integer the inequality

$$\|N\|_{\beta_1} \ll \mathcal{L}^B N^{\frac{1}{2}} \tag{3.26}$$

follows from (3.3) whenever

$$N \gg (qT)^{2/\beta_1}; \tag{3.27}$$

similarly for R.

(i) Take $\beta_1 = 8$, $4\beta_2 = 8$, $\beta_3 = 4$. By (3.23), (3.27) we have (3.26) and its analogue for R. Apply Lemma 2 with $\beta = 8$, $\gamma = \frac{7}{2} - \epsilon$, $h = 3$ in conjunction with (3.25) to get the result.

(ii) Take $\beta_1 = 4$, $\beta_2 = 12$, $\beta_3 = 6$. By (3.23), (3.27) we have (3.26) and its analogue for R. Apply Lemma 2 with $\beta = 12$, $\gamma = \frac{9}{2} - \epsilon$, $h = 3$.

(iii) Instead of (3.25) we have

$$\|LMNR\|_1 \le \|M\|_2 \|N\|_6 \|R|L|^{\frac{1}{2}}\|_4 \||L|^{\frac{1}{2}}\|_{12} .$$

Now (3.26) holds, from (3.27) with $\beta_1 = 6$; similarly

$$\|R|L|^{\frac{1}{2}}\|_4 \ll (RL^{\frac{1}{2}})^{\frac{1}{2}}$$

since $R^2 L \ge qT$. We now get the desired inequality because

$$\||L|^{\frac{1}{2}}\|_{12} = \|L\|_6^{\frac{1}{2}} \ll L^{\frac{1}{4}} \mathcal{L}^{-A}$$

from Lemma 2 with $\beta = 6$, $\gamma = \frac{5}{2} - \epsilon$, $h = 2$.

(iv) Case (a). Suppose that $L \geq X^{0.225}$. If $\min(N,R) \geq X^{0.1125}$ we can use (i) since $\max(N,R) \gg X^{9/70}$. If $\min(N,R) < X^{0.1125}$, say $X^{0.1} < N < X^{0.1125}$, then take $\beta_1 = 10$, $\beta_2 = 4$, $\beta_3 = \frac{20}{3}$. We have (3.26) and its analogue for L. Apply Lemma 2 to R with

$$\beta = \tfrac{20}{3}, \quad \gamma = \frac{0.45}{\frac{9}{35} - 0.1125} = \tfrac{28}{9}, \quad h = 3.$$

Case (b). Let $L = X^{\alpha_1}$, $\alpha_1 < 0.225$; then $\alpha_1 > 0.18$. We may apply Lemma 2 to L with

$$\gamma = \frac{0.45}{\alpha_1}, \quad h = 2, \quad \beta = 4\gamma - 4 + \epsilon.$$

We also apply Lemma 2 to N and R. Write $N = X^{\alpha_2}$, $R = X^{\alpha_3}$. In the case of N,

$$\gamma = \frac{0.45}{\alpha_2}, \quad h = 3, \quad \beta = \frac{1.8}{\alpha_2} - 6 + \epsilon,$$

and in the case of R,

$$\gamma = \frac{0.45}{\alpha_3}, \quad h = 3,$$

with the following equation giving the value of β:

$$\frac{1}{\beta} + \frac{1}{(1.8/\alpha_2) - 6 + \epsilon} + \frac{1}{(1.8/\alpha_1) - 4 + \epsilon} = \tfrac{1}{2}. \qquad (3.28)$$

This will give the desired result provided that $\beta \geq (1.8/\alpha_3) - 6 + \epsilon$ in (3.28). Thus we must show

$$S := \frac{1}{(1.8/\alpha_1) - 4} + \frac{1}{(1.8/\alpha_2) - 6} + \frac{1}{(1.8/\alpha_3) - 6} \geq \tfrac{1}{2} + B\epsilon \qquad (3.29)$$

where $0.18 < \alpha_1 < 0.225$, $0.1 < \alpha_2$, $\alpha_3 < 0.15$, $\alpha_1 + \alpha_2 + \alpha_3 \geq \frac{63}{130} + B\epsilon$. We need only consider a fixed value c of $\alpha_1 + \alpha_2 + \alpha_3$. Now fix α_1. It is easy to show that $S = S(\alpha_2, \alpha_3)$ is least when $\alpha_2 = \alpha_3$. Now

$$S(\alpha, \alpha) = \frac{2\alpha}{1.8 - 6\alpha} + \frac{c - 2\alpha}{1.8 - 4(c - 2\alpha)}$$

has increasing derivative vanishing at $\alpha = 2c/7$. The minimum value of S is thus $c/(1.8 - 12c/7)$, which is increasing in c and takes the value $\frac{1}{2}$ at $\frac{63}{130}$. The desired result follows readily.

Lemma 4. *Let $M(s,\chi)$, $N(s,\chi)$, $L(s,\chi)$ be Dirichlet polynomials with $MNL = X$. Suppose that M, N, L satisfy (3.1), L satisfies (3.2), (3.23) holds, and*

$$\max(M, N) \le X^{0.46+\epsilon/2}, \quad L \le X^{\frac{8}{35}}. \tag{3.30}$$

Then (3.9) holds.

Proof. This is a variant of Lemma 2 of Heath-Brown and Iwaniec [12].

We follow the proof of Theorem 4 as far as Case 1. Now consider

Case 2: $M^{c(\sigma_1)} < qT$, $N^{c(\sigma_2)} < qT$. From (3.10) with $g = 2$, (3.11), (3.12) we have

$$\mathcal{I}\mathcal{L}^{-B} \ll \min\{qTM^{f(\sigma_1)}, qTN^{f(\sigma_2)}, L^{4-4\sigma_3}\} M^{\sigma_1-\frac{1}{2}}N^{\sigma_2-\frac{1}{2}}L^{\sigma_3-\frac{1}{2}}$$

$$+ \min\{qTM^{f(\sigma_1)}, qTN^{f(\sigma_2)}, qTL^{2f(\sigma_3)}\} M^{\sigma_1-\frac{1}{2}}N^{\sigma_2-\frac{1}{2}}L^{\sigma_3-\frac{1}{2}}$$

$$\ll (qTM^{1-2\sigma_1})^{\frac{5}{16}}(qTM^{4-6\sigma_1})^{\frac{1}{16}}(qTN^{1-2\sigma_2})^{\frac{5}{16}}(qTN^{4-6\sigma_2})^{\frac{1}{16}}$$

$$\times (L^{4-4\sigma_3})^{\frac{1}{4}}M^{\sigma_1-\frac{1}{2}}N^{\sigma_2-\frac{1}{2}}L^{\sigma_3-\frac{1}{2}}$$

$$+ \min \left\{ \begin{matrix} (qTM^{1-2\sigma_1})^{\frac{5}{16}}(qTM^{4-6\sigma_1})^{\frac{1}{16}}(qTN^{1-2\sigma_2})^{\frac{5}{16}}(qTN^{4-6\sigma_2})^{\frac{1}{16}} \\ \times (qTL^{2-4\sigma_3})^{\frac{1}{4}}M^{\sigma_1-\frac{1}{2}}N^{\sigma_2-\frac{1}{12}}L^{\sigma_3-\frac{1}{2}}, \\ (qTM^{1-2\sigma_1})^{\frac{7}{16}}(qTM^{4-6\sigma_1})^{\frac{1}{48}}(qTN^{1-2\sigma_2})^{\frac{7}{16}}(qTN^{4-6\sigma_2})^{\frac{1}{48}} \\ \times (qTL^{8-12\sigma_3})^{\frac{1}{12}}M^{\sigma_1-\frac{1}{2}}N^{\sigma_2-\frac{1}{2}}L^{\sigma_3-\frac{1}{2}} \end{matrix} \right\}$$

$$\ll (qT)^{\frac{3}{4}}(MN)^{\frac{1}{16}}L^{\frac{1}{2}} + qT(MN)^{\frac{1}{16}}\min(1, L^{\frac{1}{6}}(MN)^{-\frac{1}{24}})$$

$$\ll (qT)^{\frac{3}{4}}X^{\frac{1}{16}}L^{\frac{7}{16}} + qT(MN)^{\frac{1}{16}}(L^{\frac{1}{6}}(MN)^{-\frac{1}{24}})^{\frac{3}{10}}$$

$$\ll (qT)^{\frac{3}{4}}X^{\frac{13}{80}} + qTX^{\frac{1}{20}} \ll X^{\frac{1}{2}}\mathcal{L}^{-A}$$

by (3.23), (3.30).

Case 3: $M^{c(\sigma_1)} \ge qT$, $N^{c(\sigma_2)} < qT$. As in Case 2,

$$\mathcal{I}\mathcal{L}^{-B} \ll (M^{2-2\sigma_1})^{\frac{1}{2}}(qTN^{1-2\sigma_2})^{\frac{1}{8}}(qTN^{4-6\sigma_2})^{\frac{1}{8}}(L^{4-4\sigma_3})^{\frac{1}{4}}$$

$$\times M^{\sigma_1-\frac{1}{2}}N^{\sigma_2-\frac{1}{2}}L^{\sigma_3-\frac{1}{2}}$$

$$+ \min \left\{ \begin{matrix} (M^{2-2\sigma_1})^{\frac{1}{2}}(qTN^{1-2\sigma_2})^{\frac{1}{8}}(qTN^{4-6\sigma_2})^{\frac{1}{8}}(qTL^{2-4\sigma_3})^{\frac{1}{4}}, \\ (M^{2-2\sigma_1})^{\frac{1}{2}}(qTN^{1-2\sigma_2})^{\frac{3}{8}}(qTN^{4-6\sigma_2})^{\frac{1}{24}}(qTL^{8-12\sigma_3})^{\frac{1}{12}} \end{matrix} \right\}$$

$$\times M^{\sigma_1-\frac{1}{2}}N^{\sigma_2-\frac{1}{2}}L^{\sigma_3-\frac{1}{2}}.$$

We see that

$$
\begin{aligned}
\mathcal{IL}^{-B} &\ll (qT)^{\frac{1}{4}} M^{\frac{1}{2}} N^{\frac{1}{8}} L^{\frac{1}{2}} + (qT)^{\frac{1}{2}} M^{\frac{1}{2}} \min\{N^{\frac{1}{8}}, N^{\frac{1}{24}} L^{\frac{1}{6}}\} \\
&\ll (qT)^{\frac{1}{4}} X^{\frac{1}{8}} (ML)^{\frac{3}{8}} + (qT)^{\frac{1}{2}} M^{\frac{1}{2}} N^{\frac{1}{16}+\frac{1}{48}} L^{\frac{1}{12}} \\
&\ll (qT)^{\frac{1}{4}} X^{\frac{1}{8}} (ML)^{\frac{3}{8}} + (qT)^{\frac{1}{2}} X^{\frac{1}{12}} M^{\frac{5}{12}} \\
&\ll X^{\frac{1}{2}} \mathcal{L}^{-A}
\end{aligned}
\tag{3.31}
$$

by (3.23), (3.30).

Since Case 4 reduces to Case 3 on interchanging M and N, this establishes Lemma 4.

We now need results for the special Dirichlet polynomials

$$
\sum_{n \asymp N} \frac{\chi(n)}{n^s}, \qquad \sum_{\substack{n \asymp N \\ (n, P(z))=1}} \frac{\chi(n)}{n^s}
$$

where

$$
z \geq \exp\left(\mathcal{L}^{\frac{9}{10}}\right).
\tag{3.32}
$$

We first recall some results from Prachar [25]. There is at most one character $\chi \pmod{q}$ in the set $\Gamma = \{\chi \pmod{q}\colon q \leq x\}$ for which $L(s, \chi)$ has a zero $\beta + i\gamma$ with $\beta > 1 - \epsilon/\mathcal{L}$, $|\gamma| \leq x$. If χ exists, say $\chi = \chi_1$, then χ_1 is real and primitive with conductor $q \gg \mathcal{L}^A$ for all A; moreover, for $\chi \in \Gamma \setminus \{\chi_1\}$, we have

$$
\beta < 1 - \frac{\epsilon}{\max(\log q, \mathcal{L}^{\frac{4}{5}})}.
$$

Let E_q consist of those characters $\chi \pmod{q}$ for which $L(s, \chi)$ has a zero $\beta + i\gamma$ with

$$
\beta \geq 1 - \frac{\epsilon}{\mathcal{L}^{\frac{4}{5}}}, \quad |\gamma| \leq x.
$$

Of course, if $\chi \in E_q$, then either $\chi = \chi_1$ or $\log q > \mathcal{L}^{4/5}$. Thus

$$
|E_q| = 0 \quad \text{if} \quad q \leq \mathcal{L}^{2A}, \qquad |E_q| \leq 1 \quad \text{if} \quad \mathcal{L}^{2A} < q \leq \exp\left(\mathcal{L}^{\frac{4}{5}}\right).
$$

For $q > \exp\left(\mathcal{L}^{4/5}\right)$, we appeal to a zero density theorem of Montgomery [21], Theorem 12.1:

$$
\sum_{\chi} N(\sigma, x, \chi) \ll (qx)^{3(1-\sigma)} \mathcal{L}^{14}
$$

where $N(\sigma, x, \chi)$ is the number of zeros of $L(s, \chi)$ in $[1 - \sigma, 1] \times [-x, x]$. Taking $1 - \sigma = \epsilon/\mathcal{L}^{4/5}$ we get $|E_q| \ll \exp\left(7\epsilon\mathcal{L}^{1/5}\right)$. In all cases we have the condition $|E_q| \ll q^{1/2}\mathcal{L}^{-A}$ needed in order to apply Theorem 2.

From now on, any well-spaced set $\mathcal{S}$ referred to will obey the restriction

$$\chi \notin E_q \quad \text{for} \quad (t, \chi) \in \mathcal{S}. \tag{3.33}$$

Lemma 5. *Suppose that* $\min\{|t| : (t, \chi) \in \mathcal{S}\} \gg \mathcal{L}^A \delta_\chi$ *for all* $A > 0$. *Let*

$$M(s, \chi) = \sum_{\substack{m \asymp M \\ (m, P(z))=1}} \frac{\chi(m)}{m^s}$$

where z satisfies (3.31). Then M satisfies (3.2).

Proof. We first give a similar bound for

$$N(s, \chi) = \sum_{N < n \leq N'} \Lambda(n) \frac{\chi(n)}{n^{it}}, \quad N' \in (N, 2N]$$

where $N \geq \mathcal{L}^{9/10}$. By [29], Lemma 3.19, this is

$$\int_{1+\mathcal{L}^{-1}-ix/2}^{1+\mathcal{L}^{-1}+ix/2} -\left(\frac{L'}{L}\right)(w + it, \chi) \frac{(N')^w - N^w}{w} \, dw + O\left(\frac{N\mathcal{L}}{x}\right).$$

By (3.32) we may replace the contour by two short horizontal segments and the vertical segment

$$1 - \frac{\epsilon}{2\mathcal{L}^{\frac{4}{5}}} + \left[-\frac{ix}{2}, \frac{ix}{2}\right].$$

We must then add the residue at $1-it$: $\delta_\chi\{(N')^{1-it} - N^{1-it}\}/(1-it)$. The integrand is bounded using

$$-\left(\frac{L'}{L}\right)(w + it, \chi) = \sum_{|t-\rho|<1} \frac{1}{w + it - \rho} + O(\mathcal{L}) \tag{3.34}$$

where ρ runs over zeros of $L(s, \chi)$ (Davenport [5], p. 102). The right-hand side of (3.34) is clearly $O(\mathcal{L}^2)$ by choice of E_q. Thus the horizontal integrals are $O(N\mathcal{L}^2/x)$ and the vertical integral is

$$O\left(\mathcal{L}^3 N \exp\left(\frac{-\epsilon \log N}{2\mathcal{L}^{\frac{4}{5}}}\right)\right).$$

We conclude that N satisfies $|N(it, \chi)| \ll N\mathcal{L}^{-A}$ for $\chi \in E_q$, $|t| < x/2$, $|t| \gg \mathcal{L}^A \delta_\chi$, $N \geq \mathcal{L}^{9/10}$; and the same bound holds if we replace $N(s, \chi)$ by

$$\sum_{N < p \leq N'} \frac{\chi(p)}{p^{it}}.$$

The lemma follows on decomposing $M(s, \chi)$ into $O(\mathcal{L})$ sums

$$M_r(s, \chi) = \sum_{\substack{p_1 \geq \cdots \geq p_r \geq z \\ p_1 \cdots p_r \asymp M}} \frac{\chi(p_1) \cdots \chi(p_r)}{(p_1 \cdots p_r)^s}.$$

For, by a partial summation,

$$\left| M_r\left(\tfrac{1}{2} + it, \chi\right) \right| \leq \sum_{\substack{p_1 \geq \cdots \geq p_{r-1} \geq z}} \frac{1}{M^{\frac{1}{2}}} \sup_u \left| \sum_{\substack{z \leq p_r \leq u \\ p_1 \cdots p_r \asymp M}} \frac{\chi(p_r)}{p_r^{it}} \right|$$

$$\ll \frac{1}{M^{\frac{1}{2}}} \sum_{\substack{p_1 \geq \cdots \geq p_{r-1} \geq z \\ p_1 \cdots p_{r-1} \ll M}} \left(\frac{M}{p_1 \cdots p_{r-1}} \right)^{\frac{1}{2}} \mathcal{L}^{-A}$$

$$\ll \mathcal{L}^{-A} \sum_{n \ll M} \frac{1}{n^{\frac{1}{2}}} \ll M^{\frac{1}{2}} \mathcal{L}^{-A}.$$

Lemma 6. *Let* $x^\epsilon \leq N \leq x$, $N \leq N_1 \leq 2N$, $q \leq \mathcal{L}^A$. *Then, for* $t \leq x$,

$$\sum_{N < n \leq N_1} \frac{\chi(n)}{n^{\frac{1}{2} + it}} \ll N^{\frac{1}{2} - \eta} + \frac{N^{\frac{1}{2}}}{(1 + |t|)},$$

where $\eta = \eta(\epsilon) > 0$.

Proof. If $N > |t|^{1/2} x^\epsilon$ this is a consequence of Theorem 1 of Fujii, Gallagher and Montgomery [6]. Suppose now that $N \leq |t|^{1/2} x^\epsilon$. It suffices to show that, for a fixed r with $1 \leq r \leq q$,

$$\sum_{N < qk + r \leq N_1} (qk + r)^{\frac{1}{2} - it} \ll N^{\frac{1}{2} - 2\eta}.$$

By a partial summation, this reduces to showing that

$$\sum e(f(k)) \ll N^{1 - 2\eta} \tag{3.35}$$

where k runs through all the integers in the interval $\left((N - r)/q, \, (N_1 - r)/q \right]$ and

$$f(\nu) = t \log(q\nu + r), \qquad |f^{(j)}(\nu)| \asymp |t| q^{2j} N^{-j} \quad \text{for} \quad j \leq B.$$

By Titchmarsh [30], Theorems 5.9 and 5.11, writing $K = 2^{j-1}$,

$$S \ll \frac{N}{q} \left(|t| q^{2j} N^{-j} \right)^{\frac{1}{2K-2}} + \left(\frac{N}{q} \right)^{1 - \frac{2}{K}} \left(|t| q^{2j} N^{-j} \right)^{-\frac{1}{2K-2}}.$$

Choose j so that

$$N^{-\frac{3}{2}} \leq |t| q^{2j} N^{-j} < N^{-\frac{1}{2}}.$$

Then $j \geq 3$, $j = O(1)$, and (3.35) is proved since

$$S \ll N \left(N^{-\frac{1}{2}} \right)^{\frac{1}{2K-2}} + N^{1 - \frac{2}{K} + \frac{3}{4K-4}}.$$

Lemma 7. *For $q \leq x$, $T \leq x$, we have*

$$\sum_{\chi \, (\mathrm{mod}\, q)} \int_{-T}^{T} \left| L(\tfrac{1}{2} + it, \chi) \right|^4 dt \ll qT\mathcal{L}^B.$$

Proof. See Ramachandra [26] (where there is no restriction to primitive characters as in the earlier work of Montgomery [21]).

Lemma 8. *Let χ be a character $(\mathrm{mod}\, q)$, then*

$$L(\sigma + it, \chi) \ll \left(q(|t| + 1) \right)^{1+\epsilon-\sigma} + \frac{1}{|\sigma + it - 1|}.$$

Proof. This is a straightforward bound of Rademacher [28].

Lemma 9. *Let*

$$N(s, \chi) = \sum_{n \leq N} \frac{\chi(n)}{n^s}.$$

Then for $2 \leq T \leq x$, $q \leq x$, $N \leq x$,

$$\|N\|_4^4 \ll qT\mathcal{L}^B + \mathcal{L}^4 |S| \left(\frac{q^2}{T^2} + \frac{N^2}{T^4} \right) + \delta_\chi N^2 \sum_{(t,\chi) \in S} \frac{1}{(1 + |t|)^4}.$$

Proof. By Theorem 3.19 of [30], writing $c = \tfrac{1}{2} + \mathcal{L}^{-1}$,

$$\sum_{n \leq N} \frac{\chi(n)}{n^{it+\frac{1}{2}}} = \frac{1}{2\pi i} \int_{c-iT}^{c-iT_1} L(w + \tfrac{1}{2} + it, \chi) \frac{N^w}{w} \, dw$$

$$+ O\left(\mathcal{L}\frac{N^{\frac{1}{2}}}{T} + \frac{1}{N^{\frac{1}{2}}} \right).$$

We move the contour to $[-iT, iT]$; the horizontal segments contribute

$$\ll \max_{\sigma \in [0,c]} \left((qT)^{c-\sigma} N^\sigma T^{-1} \right) \ll q^{\frac{1}{2}} T^{-\frac{1}{2}} + N^{\frac{1}{2}} T^{-1}.$$

We must also add the residue at $\tfrac{1}{2} - it$: $\delta_\chi N^{\frac{1}{2}-it}/(\tfrac{1}{2} - it)$. Thus

$$\sum_{n \leq N} \frac{\chi(n)}{n^{it+\frac{1}{2}}} = J(t, \chi) + O\left(\frac{q^{\frac{1}{2}}}{T^{\frac{1}{2}}} + \frac{\mathcal{L}N^{\frac{1}{2}}}{T} + \frac{\delta_\chi N^{\frac{1}{2}}}{1 + |t|} \right) \qquad (3.36)$$

where

$$J(t, \chi) = \frac{1}{2\pi i} \int_{-iT}^{iT} L(w + \tfrac{1}{2} + it, \chi) N^w \, dw.$$

Moreover,

$$\sum_{(t,\chi)\in\mathcal{S}} |J(t,\chi)|^4 \ll \sum_{(t,\chi)\in\mathcal{S}} \left\{\int_{-T}^{T} \left|L(\tfrac{1}{2}+iv+it,\chi)\right| \frac{dv}{1+|v|}\right\}^4$$

$$\ll \sum_{(t,\chi)\in\mathcal{S}} \left\{\int_{-2T}^{2T} \left|L(\tfrac{1}{2}+iu,\chi)\right| \frac{du}{1+|t-u|}\right\}^4.$$

Using Hölder's inequality, this is

$$\ll \sum_{(t,\chi)\in\mathcal{S}} \left\{\int_{-2T}^{2T} \left|L(\tfrac{1}{2}+iu,\chi)\right|^4 \frac{du}{1+|t-u|}\right\}\left\{\int_{-2T}^{2T} \frac{du}{1+|t-u|}\right\}^3$$

$$\ll \mathcal{L}^3 \sum_{\chi} \int_{-2T}^{2T} \left|L(\tfrac{1}{2}+iu,\chi)\right|^4 \left\{\sum_{\substack{t \\ (t,\chi)\in\mathcal{S}}} \frac{1}{1+|t-u|}\right\} du$$

$$\ll \mathcal{L}^4 \sum_{\chi} \int_{-2T}^{2T} \left|L(\tfrac{1}{2}+iu,\chi)\right|^4 du \ll \mathcal{L}^B qT.$$

The lemma follows from this estimate and (3.36).

In the remainder of the paper we write

$$T_0 = \exp(\mathcal{L}^{\frac{1}{3}}), \quad T_1 = x^{0.45-2\epsilon/3}.$$

Lemma 10. *Let $L(s,\chi)$, $M(s,\chi)$, $N(s,\chi)$ be Dirichlet polynomials with $LMN = x$. Suppose that M, N satisfy (3.1) and*

$$L(s,\chi) = \sum_{l \asymp L} \frac{\chi(l)}{l^s}.$$

Suppose that $q \leq \mathcal{L}^A$, that $|t| \in \left[\tfrac{1}{2}T, T\right]$ for all $(t,\chi_0) \in \mathcal{S}$, and that

$$T_0 \leq T \leq T_1. \tag{3.37}$$

Suppose further that either

$$\max(M, x^{0.45}) \max(N, x^{0.225}) \leq x^{0.775+\epsilon/8} \tag{3.38}$$

or

$$\max(M, N) \leq x^{0.46+\epsilon/8}. \tag{3.39}$$

Then (3.9) holds.

Proof. It is convenient to begin by eliminating the case $L > qT$. In this case, the reflection principle, as used for example in eq. (21) of [24], yields

$$\|L\|_\infty \ll \mathcal{L}^B,$$
$$\|MNL\|_1 \ll (T_1 + M)^{\frac{1}{2}}(T_1 + N)^{\frac{1}{2}}\mathcal{L}^B$$
$$\ll (T_1 + T_1^{\frac{1}{2}}x^{0.275+\epsilon/16} + x^{\frac{1}{2}}T_0^{-\frac{1}{2}})\mathcal{L}^B$$
$$\ll x^{\frac{1}{2}}\mathcal{L}^{-A}.$$

We may now suppose that $L \leq qT$, so that $\|L\|_4^4 \ll \mathcal{L}^{3A}qT$ from Lemma 9. Thus when (3.39) holds we may replace $L^{4-4\sigma_3}$ by $\mathcal{L}^{3A}qTL^{2-4\sigma_3}$ where it occurs in the proof of Lemma 4. Since $\mathcal{L}^{3A}qT < x^{16/35}$, it may readily be verified that the proof goes through as before. Now suppose that (3.38) holds. With σ_j as in the proof of Theorem 4,

$$\|MNL\|_1 \leq \|M\|_2\|N\|_4\|L\|_4$$
$$\ll \mathcal{L}^{2A}(T + M)^{\frac{1}{2}}(T + N^2)^{\frac{1}{4}}(T^{\frac{1}{4}} + L^{\frac{1}{2}}T^{-\frac{3}{4}}).$$

Since
$$(T + M)^{\frac{1}{2}}(T + N^2)^{\frac{1}{4}} \ll x^{0.775/2+\epsilon/16}$$

the contribution from $T^{1/4}$ in the last bracket is $O(x^{1/2}\mathcal{L}^{-A})$. The remaining term clearly contributes

$$\ll \mathcal{L}^{2A}L^{\frac{1}{2}} + x^{\frac{1}{2}}T_0^{-\frac{1}{2}} \ll x^{\frac{1}{2}}\mathcal{L}^{-A}.$$

4. Sieving the interval I

The next lemma shows that, modulo a mild arithmetic requirement, (1.7) can be reduced to a mean value bound for Dirichlet polynomials.

Lemma 11. *Let*

$$F(s,\chi) = \sum_{k\sim x} c_k \frac{\chi(k)}{k^s}$$

be a Dirichlet polynomial satisfying (3.1). Suppose that

$$\int_{T_0}^{T_1} \left|F(\tfrac{1}{2} + it, \chi)\right| dt \ll x^{\frac{1}{2}}\mathcal{L}^{-A} \quad \text{for all} \quad A > 0. \tag{4.1}$$

Then

$$\sum_{k\in I} c_k\chi(k) = \frac{Y}{y_1}\sum_{k\in I_1} c_k\chi(k) + O(Y\mathcal{L}^{-A}) \quad \text{for all} \quad A > 0, \tag{4.2}$$

where $y_1 = x\exp(-3\log^{1/3}x)$ and $I_1 = (x - y_1, x]$.

Proof. We follow Heath-Brown [10], using Perron's formula

$$\frac{1}{2\pi i}\int_{\frac{1}{2}-iT_1}^{\frac{1}{2}+iT_1} u^s\,\frac{ds}{s} = \mathcal{E}(u) + O\left(\frac{u^{\frac{1}{2}}}{T_1|\log u|}\right)$$

where $\mathcal{E}(u) = 0$ when $0 < u < 1$, $\mathcal{E}(u) = 1$ when $u > 1$. Then it is easy to see that

$$\frac{1}{2\pi i}\int_{\frac{1}{2}-iT_1}^{\frac{1}{2}+iT_1} F(s,\chi)\,\frac{x^s - (x-y)^s}{y}\,ds = \sum_{n\in I} c_n\chi(n) + O\left(\frac{x^{1+\epsilon/4}}{T_1}\right) \quad (4.3)$$

(compare [10], p. 1372). On the vertical line in question, clearly

$$|F(s,\chi)| \ll x^{\frac{1}{2}}\mathcal{L}^B,$$

while

$$\frac{x^s - (x-y)^s}{s} = \begin{cases} x^{s-1}y + O\left(|s|x^{-\frac{3}{2}}y^2\right) & \text{if } |\operatorname{Im} s| \le T_0, \\ O(x^{-\frac{1}{2}}y) & \text{if } |\operatorname{Im} s| \ge T_0. \end{cases}$$

Hence, for any y with $2 \le y \le y_1$ we obtain

$$\frac{1}{2\pi i}\int_{\frac{1}{2}-iT_0}^{\frac{1}{2}+iT_0} F(s,\chi)\,\frac{x^s - (x-y)^s}{s}\,ds = yE(x,\chi) + O\left(T_0^2 x^{-\frac{3}{2}}y^2 x^{\frac{1}{2}}\mathcal{L}^B\right)$$

$$= yE(x,\chi) + O\left(y\mathcal{L}^{-A}\right) \quad (4.4)$$

where

$$E(x,\chi) = \frac{1}{2\pi i}\int_{\frac{1}{2}-iT_0}^{\frac{1}{2}+iT_0} F(s,\chi)x^{s-1}\,ds.$$

Combining (4.3), (4.4),

$$\frac{1}{y}\sum_{x-y<n\le x} c_n\chi(n) = E(x,\chi) + O\left(\mathcal{L}^{-A} + \frac{x^{0.55+11\epsilon/12}}{y}\right). \quad (4.5)$$

The result follows by combining the cases $y = Y$ and $y = y_1$ of (4.5).

In building the sequence $a_0(k)$ we must draw on a supply of Dirichlet polynomials satisfying (4.2). Theorem 4 and Lemmata 3, 4 provide an 'initial supply.' Our aim is to extend this supply and we begin with a sieve fundamental lemma. Let $w = \exp\left(\log^{9/10}x\right)$. For $n \ge 1$, $z \ge 2$ let

$$\psi(n,z) = \begin{cases} 1 & \text{if } (n,P(z)) = 1 \\ 0 & \text{otherwise.} \end{cases} \quad (4.6)$$

Lemma 12. *Let*

$$c_k = \sum_{\substack{mnl=k \\ m\sim M,\, n\sim N}} a_m b_n \psi(l, w). \tag{4.7}$$

Suppose that (3.1) *holds for* a_m, b_n, *and that one of*

$$\max(x^{0.45}, M)\max(x^{0.225}, N) \le x^{0.775} \tag{4.8}$$

$$\max(M, N) \le x^{0.46+\epsilon/8} \tag{4.9}$$

applies. Then (4.2) *holds.*

(The last phrase is our abbreviation for ' (4.2) holds for all $\chi \pmod{q}$ when $q \le \mathcal{L}^A$ '.)

Proof. We combine Lemmata 10, 11 with an argument of Harman ([9], Lemma 2*). We write $\gamma = x^{\epsilon/12}$. Now

$$c_k = \sum_{m,n} \sum_{\substack{d|l,d|P(w) \\ mnl=k}} a_m b_n.$$

Here (and often, in what follows) we suppress the summation conditions $m \sim M$, $n \sim N$.

According to Lemma 15 of Heath-Brown [11],

$$\sum_{d|l,d|P(w)} \mu(d) = \sum_{\substack{d|l,d|P(w) \\ d\le\gamma}} \mu(d) + O\left(\sum_{\substack{d|l,d|P(w) \\ \gamma\le d<\gamma w}} 1 \right),$$

leading to

$$\sum_{x-y<k\le x} c_k \chi(k) = \sum_{\substack{m,n;d|P(w),d\le\gamma \\ l\equiv 0\,(\text{mod } d),\, x-y<mnl\le x}} a_m b_n \mu(d)\chi(mnl) + O\left({\sum}^{\dagger} |a_m b_n| \right)$$

$$= \sum_{x-y<k\le x} c'_k \chi(k) + O\left(\sum_{x-y<k\le x} c''_k \right),$$

say. Here $y \in \{Y, y_1\}$, and $\dagger$ represents the summation conditions

$$m, n : d|P(w), \quad \gamma \le d < w\gamma, \quad l \equiv 0 \pmod{d}, \quad x - y < mnl \le y.$$

It now suffices to prove

$$\sum_{k\in I} c'_k \chi(k) = \frac{Y}{y_1} \sum_{k\in I_1} c'_k \chi(k) + O(Y\mathcal{L}^{-A}), \tag{4.10}$$

$$\sum_{k\in I} c''_k = \frac{Y}{y_1} \sum_{k\in I_1} c''_k + O(Y\mathcal{L}^{-A}), \tag{4.11}$$

$$\sum_{k\in I_1} c''_k = O(y_1\mathcal{L}^{-A}). \tag{4.12}$$

To get (4.10), apply Lemma 10 and 11 to

$$\sum_{k \asymp x} c'_k \frac{\chi(k)}{k^s} = M_1(s,\chi)N(s,\chi)L(s,\chi)$$

where

$$M_1(s,\chi) = \sum_{\substack{m;d \sim D \\ d|P(w)}} \frac{a_m \mu(d)\chi(md)}{(md)^s},$$

$$N(s,\chi) = \sum_n \frac{b_n \chi(n)}{n^s}, \qquad L(s,\chi) = \sum_{nMND \asymp x} \frac{\chi(n)}{n^s},$$

and then sum over $D = \gamma 2^{-j}$ for $j = 0, 1, \ldots$. In each case M_1 has length $MD \le Mx^{\epsilon/8}$, so (3.38) or (3.39) holds. We prove (4.11) in precisely the same way with the polynomial M_1 modified by replacing the condition on d by

$$d|P(w), \quad w \le d < w\gamma, \quad d \sim D.$$

The left-hand side of (4.12) is

$$\ll y_1 \sum_{m,n} \frac{|a_m b_n|}{mn} \sum_{\substack{d|P(w) \\ \gamma \le d < w\gamma}} \frac{1}{d} \ll y_1 \sum_{m,n} \frac{|a_m b_n|}{mn} \exp\left(-\tfrac{1}{8} u\epsilon \log u\epsilon\right) \qquad (4.13)$$

by a straightforward application of Theorem 1 of Hildebrand [13]. Here $w = x^{1/u}$ so that $u = \mathcal{L}^{1/10}$ and (4.12) follows at once.

We would like to 'step up' the size of w in Lemma 12 using the simple identity

$$\psi(j,z) = \psi(j,w) - \sum_{\substack{ph=j \\ w \le p < z}} \psi(h,p) \qquad (4.14)$$

(essentially, Buchstab's identity). This leads us into proving (4.2) with c_k of a form such as

$$c_k = \sum_{mnlr=k} u_m v_n w_l \psi(r,l) \qquad (4.15)$$

where $m \sim M$, $r \sim R$, $n \sim N$, $l \sim L$. The interdependence of r and l prevents several groupings of variables such as m', n, l with $m' = mr$. We overcome this obstacle by using the identity

$$\sum_{d|P(l)} \mu(d)g(d) = g(1) - \sum_{2 \le p < l} \sum_{d|P(p)} \mu(d)g(pd)$$

([9], (10)). In (4.15),

$$c_k = \sum_{mnlr=k} u_m v_n w_l \sum_{\substack{d|r \\ d|P(l)}} \mu(d) = \sum_l w_l \sum_{d|P(l)} \mu(d) \sum_{mnltd=k} u_m v_n$$

$$= \sum_l w_l \sum_{mnlt=k} u_m v_n - \sum_l w_l \sum_{2 \le p < l} \sum_{d|P(p)} \mu(d) \sum_{mnltpd=k} u_m v_n$$

$$= a_k - b_k, \tag{4.16}$$

say. There is no 'interdependence' in the sum a_k. In the sum b_k we group three variables as $tpd \asymp R$. The sole dependence condition $p < l$ can be re-moved, with the loss of a factor $\mathcal{L}$, by Perron's formula (cf. [1], Lemma 1). In what follows we pass over the above procedure without comment. Likewise we leave implicit the frequent appeals to Lemma 5 to provide the bound (3.2) needed for application of Theorem 4 and Lemmata 3, 4.

Lemma 13. *Let*

$$c_k = \sum_{\substack{mnl=k \\ m \sim M, n \sim N}} a_m b_n \psi(l, z) \tag{4.17}$$

where a_m and b_n satisfy (3.1). Suppose that

$$x^{0.4} \le M \le x^{0.46}, \quad N \le x^{0.46}, \quad z \ll x/MN; \tag{4.18}$$

then (4.2) holds.

Proof. If $N > x^{0.375}$ then $x/MN < x^{0.225}$ and the result follows from Lemma 4. Suppose now that $N \le x^{0.375}$. We have

$$c_k = \sum_{mnl=k} a_m b_n \psi(l, w) - \sum_{\substack{w \le p < z \\ mnph=k}} a_m b_n \psi(h, p) = d_k - e_k,$$

say. Lemma 12 covers d_k. Let e'_k denote the contribution to e_k from

$$p \le x^{0.225}. \tag{4.19}$$

In the sum e'_k, consider those p with

$$p \ge \frac{x^{0.9}}{M^2}. \tag{4.20}$$

The contribution to e'_k from p, h with (4.20) and $Nh \le x^{0.46}$ may be dealt with via Lemma 4. For $Nh \ge x^{0.46}$ we have

$$p \ll x^{0.14}, \quad \frac{M}{Nh} \le 1, \quad \frac{Nh}{M} \asymp \frac{x}{M^2 p} < x^{0.1},$$

and we can use Theorem 4. Similarly we can deal with those p with

$$\frac{x^{0.9}}{M^2} \leq \frac{x}{MNp} \leq x^{0.225}, \tag{4.21}$$

by using the same argument with p interchanged with h.

Consider now the part of e'_k where (4.20), (4.21) both fail. Here

$$\frac{Np}{M} < \frac{x^{0.9+0.375}}{M^3} < x^{0.1}$$

and so $MNp < x^{0.775}$. We split up the range of p.

(i) $x^{0.45}/M \leq p < x^{0.9}/M^2$. The contribution to e'_k is

$$\sum_{\substack{mnph=k \\ x^{0.45}/M \leq p < x^{0.9}/M^2}} a_m b_n \psi(h,w) \;-\; \sum_{\substack{w \leq q < p,\, mnpql=k \\ x^{0.45}/M \leq p < x^{0.9}/M^2}} a_m b_n \psi(l,q) \;=\; f_k - g_k,$$

say. Since $np < 2x^{0.775}/M < 2x^{0.375}$ and $M < x^{0.46}$, Lemma 12 covers f_k. For g_k we note that

$$Mp \geq x^{0.45}, \quad q(Mp)^2 \leq p^3 M^2 \leq x^{2.7} M^{-4} \leq x^{1.1}$$

so that

$$\frac{mp}{nl} \asymp \frac{Mp}{x/(Mpq)} \leq x^{0.1},$$

while

$$\frac{nl}{mp} \asymp \frac{x}{(Mp)^2 q} < x^{0.1}, \quad q < x^{0.1}.$$

Theorem 4 then covers g_k.

(ii) $p < x^{0.45}/M$. We can go back to the start of the proof with m replaced by pm, and a_m replaced by

$$\sum_{\substack{pk=m \\ w \leq p < x^{0.45}/M}} a_k.$$

If necessary, iterate this procedure. After $O(\mathcal{L}^{1/10})$ steps, case (ii) can no longer arise. In repeatedly using case (i), we must pay attention to the coefficients which arise; a_m will be replaced by

$$\sum_{p_1 \ldots p_j k=m} a_k \ll \left(\max_{k|m} a_k \right) \tau(m)^B \ll \tau(m)^B.$$

Thus at every stage the argument of case (i) suffices.

It remains to deal with the contribution to e_k from $p > x^{0.225}$. The sum

$$e_k'' = \sum_{\substack{mnph=k \\ p > x^{0.225}}} a_m b_n \psi(h, p)$$

runs over primes h since $mnp^3 \geq mp^3 > x$. Moreover,

$$p \leq h < x^{0.375}, \quad e_k'' = \sum_{\substack{p > x^{0.225} \\ mnph=k}} a_m b_n \psi(h, x^{0.2}), \quad np \ll \frac{x}{Mp} \ll x^{0.375}.$$

Grouping n and p, we deal with e_k'' in the same way as e_k'. This completes the proof.

Lemma 13 covers all sums that arise below where one variable lies in $[x^{0.4}, x^{0.46}]$. Next we give a result for $(x^{0.46}, x^{0.5}]$.

Lemma 14. *Let*

$$c_k = \sum_{\substack{m \sim M, n \sim N, \\ mnl=k}} a_m b_n \psi(l, z)$$

where a_m *and* b_n *satisfy* (3.1). *If*

$$x^{0.46} < M \leq x^{0.5}, \quad MN \leq x^{0.775}, \quad zM^2 \leq x^{1.1}, \tag{4.22}$$

then (4.2) *holds.*

Proof. We have

$$c_k = \sum_{mnl=k} a_m b_n \psi(l, w) - \sum_{\substack{mnph=k \\ w \leq p < z}} a_m b_n \psi(h, p) = h_k - j_k,$$

say. The case (4.6) of Lemma 12 covers h_k. In j_k,

$$p < \frac{x^{1.1}}{M^2} \ll x^{0.18}, \quad \frac{m}{nh} \asymp \frac{M}{x/Mp} < x^{0.1}, \quad \frac{nh}{m} < \frac{x}{M^2} < x^{0.1}.$$

Thus Theorem 4 covers j_k.

Lemma 15. *Let*

$$c_k = \sum_{\substack{m \sim M, n \sim N \\ mnl=k}} a_m b_n \psi(l, x^{0.1})$$

where a_m, b_n *satisfy* (3.1). *If* $M \leq N \leq x^{0.4}$ *then* (4.2) *holds.*

Proof. We use (4.14) repeatedly (at most $\mathcal{L}^{1/10}$ times). Thus

$$c_k = \sum_{mnl=k} a_m b_n \chi(l, w) - \sum_{\substack{mnph=k \\ w \leq p < x^{0.1}}} a_m b_n \chi(h, p) = l_k - r_k,$$

say. We deal with l_k by Lemma 12. The contribution to r_k from those terms with $x^{0.4} < pm \leq x^{0.46}$ is covered by Lemma 13. For $pm > x^{0.46}$,

$$h \asymp \frac{x}{pMN} \leq \frac{x^{1.1}}{p^2 MN} \leq \frac{x^{1.1}}{(pM)^2} \leq x^{0.18},$$
$$\frac{mp}{n} \ll p < x^{0.1}, \qquad \frac{n}{mp} < 1$$

and we may use Theorem 4. For the contribution to r_k from $pm \leq x^{0.4}$ we iterate the procedure, dealing with the coefficients that arise as in the proof of Lemma 13. This completes the proof.

We can improve on Lemma 15 in some ranges as follows.

Lemma 16. *Lemma 14 holds if in place of (4.22) we suppose that*

$$c_k = \sum_{\substack{m \sim M, n \sim N \\ mnl=k}} a_m b_n \psi(l, z),$$

$$M \in [x^{\alpha/10}, x^{(\alpha+1)/10}] \quad \text{where} \quad \alpha \in \{0, 1, 2, 3, 4\},$$

$$x^{0.1} < z \ll \left(\frac{x^{1.1}}{M^2}\right)^{1/(9-2\alpha)}, \qquad MN \leq x^{0.775} z^{\alpha-4}. \tag{4.23}$$

Proof. By downward induction. The case $\alpha = 4$ is covered by Lemmata 13 and 14. In the induction step, we write

$$c_k = \sum_{mnl=k} a_m b_n \psi(l, x^{0.1}) - \sum_{\substack{mnph=k \\ x^{0.1} \leq p < z}} a_m b_n \psi(h, p) = s_k - t_k,$$

say. Then s_k is covered by Lemma 15, since $N \leq x^{0.375}$.
In t_k, we readily verify that $mp < x^{0.5}$, so that

$$x^{\beta/10} \leq mp < x^{(\beta+1)/10}, \qquad \alpha < \beta \leq 4.$$

We now verify that

$$p \ll \left(\frac{x^{1.1}}{M^2 p^2}\right)^{1/(9-2\beta)}, \qquad MpN \leq x^{0.775} p^{\beta-4}.$$

The first inequality above follows from (4.23), while

$$MpN \leq x^{0.775} z^{\alpha-4} p < x^{0.775} p^{\beta-4}.$$

This completes the proof of Lemma 16.

We now construct the sequence $a_0(k)$ by starting with $\rho(k)$, using (4.14) up to six times, and discarding some non-negative terms from the resulting identity. The results we appeal to are : Theorem 4, Lemmata 3, 4, 13, 14, 15, 16.

The first two applications of (4.14) are as follows. Let $k \in I$; we have

$$\rho(k) = \psi(k, x^{\frac{1}{2}}) = \psi(k, x^{\frac{11}{90}}) - \sum_{\substack{x^{11/90} \leq p_1 < x^{1/2} \\ p_1 n_2 = k}} \psi(n_2, p_1)$$

$$= \psi(k, x^{\frac{11}{90}}) + \sum_{\substack{x^{11/90} \leq p_1 < x^{1/2} \\ p_1 n_2 = k}} (-\psi(n_2, z_1)) + \sum_{\substack{p_1 p_2 n_3 = k, \, z_1 \leq p_2 < p_1 \\ x^{11/90} \leq p_1 < x^{1/2}}} \psi(n_3, p_2)$$

$$= c_k(1) + c_k(2) + c_k(3). \tag{4.24}$$

The same idea may be repeated to generate sums of the form

$$\sum_{\substack{(n_1, \ldots, n_j) \in \mathcal{C}_j \\ n_1 \ldots n_{j+1} = k}} (-1)^j \psi(n_1, z_1) \ldots \psi(n_{j+1}, z_{j+1}) \tag{4.25}$$

where $n_j \geq x^{0.1}$ in the Jordan measurable region $\mathcal{C}_j$ and z_i is a function of $\{n_t : t \neq i\}$. Of course, it is not obvious which variable to choose in applying the next step

$$\psi(n_{j+1}, z_{j+1}) = \psi(n_{j+1}, w_{j+1}) - \sum_{\substack{w_{j+1} \leq p_{j+1} < z_{j+1} \\ p_{j+1} n_{j+2} = n_{j+1}}} \psi(n_{j+2}, p_{j+1}),$$

but we must be sure that the *first term* on the right yields a sum over $\mathcal{C}_j$ satisfying (4.2). We refer to this step as 'decomposing n_{j+1} to reach $p_{j+1} n_{j+2}$.' In general (exceptions being described below) we decompose the *largest variable* permitted by Lemmata 14, 15, 16, and there is *no decomposition for any region $\mathcal{C}_j$ where (4.2) can be shown to hold*. Nor do we decompose further when $j = 6$.

It is always possible to decompose one of $n_1, \ldots, n_{j+1}$ for $j \leq 4$, as we shall see below. However, we shall discard certain regions $\mathcal{C}_4$, that is, use only the trivial lower bound 0 for the sum (4.25). We do this in order to avoid generating $\mathcal{C}_5$ which do not allow one more decomposition, since we want a *lower* bound for $\rho(k)$. Naturally we must also discard any $\mathcal{C}_6$ for which (4.2) cannot be shown to hold.

We write down the identity

$$\rho(k) = \psi(k, x^{\frac{11}{90}}) + \cdots + \sum_{\mathcal{C}_6} \sum_{\substack{(n_1, \ldots, n_6) \in \mathcal{C}_6 \\ n_1 \ldots n_7 = k}} \psi(n_1, z_1) \ldots \psi(n_7, z_7)$$

obtained by the above algorithm. Discarding regions C_4 and C_6 as explained above gives an inequality

$$\rho(k) \geq a_0(k),$$

in which $a_0(k)$ satisfies (4.2).

It is a simple consequence of the Siegel-Walfisz theorem that

$$\sum_{k \in I_1} a_0(k)\chi(k) = \frac{\delta_\chi y_1 \left(u_0 + O(\mathcal{L}^{-A})\right)}{\log x}.$$

The constant u_0 is found by subtracting, from 1, integrals in 4 and 6 dimensions (involving Buchstab's function) corresponding to the discarded regions; compare Baker and Harman [2], [3]. Let $(p_1, p_2) = (x^{\alpha_1}, x^{\alpha_2})$. We have drawn the (α_1, α_2) region under consideration on Diagram 1. Numbers indicate which lemma can be applied in a subregion. The rest of this section is devoted to a proof that only regions C_4 for which (α_1, α_2) lie in a very small part of the plane (the regions Γ and Δ_j) need to be discarded.

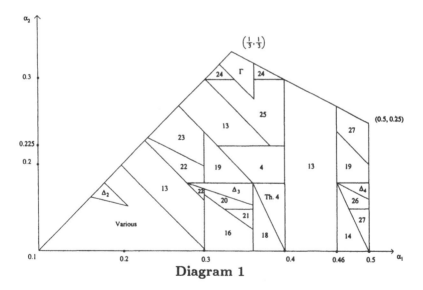

Diagram 1

We begin with a lemma summarizing some general observations about the decomposition procedure.

Lemma 17. *In the decomposition procedure described above,*

(a) *if a subproduct m of $n_1 \ldots n_{j+1}$ satisfies $x^{0.4} \ll m \ll x^{0.46}$, then (4.2) holds for the sum (4.25);*

(b) *if $j \leq 4$, one of the variables may be decomposed;*

(c) *if $j \geq 5$, any variable that is at least $x^{0.225}$ may be decomposed.*

Proof.   (a) Let us say

$$n_1 \ldots n_l \in \left[x^{0.4}, x^{0.46} \right].$$

We can use Lemma 13 if $n_{j+1} > x^{0.14}$ (since then $x/(n_1 \ldots n_l n_{j+1}) < x^{0.46}$) and Theorem 4 if $x^{0.1} \leq n_{j+1} < x^{0.14}$.

(b) If $j = 2$, we have variables p_1, p_2, n_3 with $p_2 \leq p_1$, $p_2 \leq n_3$. By (a) we may exclude the case where p_1 or n_3 is in $\left[x^{0.4}, x^{0.46} \right]$. Both variables cannot exceed $x^{0.46}$, so two of p_1, p_2, n_3 are $\leq x^{0.4}$ and we may use Lemma 15.

If $j = 3$, with $n_1 \leq n_2 \leq n_3 \leq n_4$ then clearly $n_3 \leq x^{0.4}$, $n_1 n_2 \leq x^{0.5}$, $n_1 n_2 n_3 \leq x^{0.75}$. If (4.2) does not hold, one of Lemmata 14, 15 allows us to decompose n_4.

If $j = 4$, $n_1 \leq n_2 \leq n_3 \leq n_4 \leq n_5$ and $n_5 \leq x^{0.225}$, we have

$$x^{0.4} \leq n_4 n_5 \leq x^{0.45}$$

and (a) applies. If $n_5 > x^{0.225}$ we may decompose $n_1 \ldots n_4$ into two products $\leq x^{0.5}$. This is obvious if $n_4 > x^{0.275}$ or $n_1 n_2 > x^{0.275}$. If

$$\max(n_1 n_2, n_4) \leq x^{0.275}$$

then $n_2 \leq x^{0.175}$ and $\max(n_1 n_4, n_2 n_3) \leq x^{0.45}$. We may now finish the proof using case (a) and Lemmata 14, 15.

(c) The general case follows readily from the case $j = 5$, so we consider this case. Plainly $n_1 n_2 n_3 \leq x^{0.775 \times 0.6} < x^{0.5}$, $n_4 n_5 < x^{0.775 - 0.3} < x^{0.5}$ and we may use the preceding argument.

In the rest of this section we decompose, as we may, only regions C_j that are not known to satisfy (4.2) and (in particular) do not have property (a) of Lemma 17.

Lemma 18. *We may decompose the subsum of $c_k(3)$ with*

$$x^{0.36} \leq p_1 < x^{0.4}, \quad p_1 p_2 \leq x^{0.54}$$

to reach C_6.

Proof. We decompose successively to get $p_1 p_2 n_3$, $p_1 p_2 p_3 n_4$, $p_1 p_2 p_3 p_4 n_5$ (note that $\max(p_1, p_2 p_3) \leq x^{0.4}$). Excluding sets with (4.2), we consider two cases for the remaining C_4.

(i) $p_2 \geq x^{0.15}$. Decompose p_1 into $p_5 n_6$ (since $p_3 p_4 n_5 \leq x^{0.49}$). We have

$$n_6 > x^{0.18}, \quad n_6 p_5^2 \geq x^{0.36 + 0.1} > x^{0.45}, \quad p_3 p_4 n_5 \gg x^{0.46}.$$

Lemma 3 (iii) yields (4.2).

(ii) $p_2 < x^{0.15}$. Then $p_2 p_3 p_4 < x^{0.45}$, hence $\max(p_1, p_2 p_3 p_4) < x^{0.4}$. Decompose n_5 to get $p_1 p_2 p_3 p_4 p_5 n_6$. Now use Lemma 17 (c).

Lemma 19. *Lemma 14 holds if in place of (4.22) we suppose that*

$$N \geq x^{0.18}, \quad x^{0.3} \leq M \leq x^{0.36}, \quad MN \leq x^{0.54}, \quad MNz \leq x^{0.775}.$$

Remark. It follows that (4.2) holds for the subsum of $c_k(3)$ with

$$x^{0.3} \leq p_1 \leq x^{0.36}, \quad x^{0.18} \leq p_2 \leq x^{0.54} p_1^{-1}.$$

This is immediate if $p_1 p_2 p_2 \leq x^{0.775}$. In the contrary case,

$$p_1 = \frac{p_1^2 p_2^2}{p_1 p_2^2} \leq x^{0.305}, \quad p_2 \geq x^{0.235}, \quad p_1 p_2^3 \geq x^{1.05}$$

and $p_1 p_2 n_3$ counts only primes n_3. Because $MN(x/p_1 p_2)^{1/2} \ll x^{0.77}$, Lemma 19 gives the desired result.

If $p_3 \leq x^{0.15}$, then

$$n_5 \gg \frac{x}{mnp_3 p_4} \gg \frac{x}{MNp_3^2} \gg x^{0.15}.$$

Hence one of p_3, n_5 always exceeds $x^{0.15}$. Restrict n, p_3, p_4, n_5 to dyadic intervals with left endpoints N_1, N_2, N_3, N_4 arranged so that

$$x^{0.1} \leq N_1 \leq N_2 \leq N_3 \leq N_4,$$
$$N_3 > x^{0.15}, \quad N_4 > x^{0.18}, \quad N_1 N_2 N_3 N_4 \asymp x/M.$$

We have $MN_1 \geq x^{0.4}$, $MN_1 \gg x^{0.46}$; then $N_3 N_4 \ll x^{0.44}$, indeed

$$N_3 N_4 < x^{0.4}.$$

We cannot then have $N_1 N_2 N_4 < x^{0.4}$, since this forces $N_3 > x^{0.24}$ which is absurd. Thus $N_1 N_2 N_4 \geq x^{0.46}$. It remains to deal with

$$N_4 \geq x^{0.18}, \quad N_3 > x^{0.15}, \quad MN_1 \gg x^{0.46}, \quad N_2^2 N_4 \geq N_1 N_2 N_4 > x^{0.46}.$$

Now Lemma 3 (iii) applies.

Lemma 20. *Lemma 14 holds if in place of (4.22) we suppose that*

$$N \geq x^{0.15}, \quad MNz^2 \leq x^{0.82}, \quad x^{0.3} \leq M \leq x^{0.36}, \quad MN \leq x^{0.54}.$$

Proof. Decompose successively to reach $mnp_3 p_4 n_5$, noting here that

$$N \leq x^{0.24}, \quad N^2 z^2 \leq x^{1.06-0.3}, \quad np_3 < x^{0.4}.$$

Now

$$n_5 \gg x/(MNz^2) \gg x^{0.18}, \quad p_3^2 n_5 \geq p_3 p_4 n_5 \gg x^{0.46}, \quad p_4 m \geq x^{0.4},$$

hence $p_4 m \geq x^{0.46}$. Now apply Lemma 3 (iii).

Lemma 21. *We may decompose the subsum of $c_k(3)$ for which*

$$x^{0.3} \leq p_1 \leq x^{0.36}, \quad p_2 \leq x^{0.15},$$

to reach $\mathcal{C}_6$.

Proof. We repeat the argument of case (ii) of the proof of Lemma 18.

Lemma 22. *Let*

$$c_k = \sum_{\substack{p_1 \sim M, \ p_2 \sim M \\ p_1 p_2 n_3 = k}} \psi(n_3, p_2).$$

Then c_k satisfies (4.2) *if either*

$$x^{0.23} \leq M \leq x^{0.3}, \quad MN \geq x^{0.46}, \quad MN^2 \leq x^{0.7}, \quad N \geq x^{0.18} \quad (4.26)$$

or

$$x^{0.23} \leq M \leq x^{0.3}, \quad MN \geq x^{0.46}, \quad MN^3 \leq x^{0.82}. \quad (4.27)$$

Proof. When (4.26) holds, decompose successively to reach $p_1 p_2 p_3 p_4 n_5$. We have

$$p_1 p_2 = \left(p_1 p_1 p_2^2\right)^{\frac{1}{2}} \leq x^{0.5}, \quad p_1 p_2 p_3 \leq p_1 p_2^2 < x^{0.775}.$$

By Lemma 14, we may therefore assume that $p_4 \geq x^{1.1}/(p_1 p_2)^2$. Thus

$$p_1 p_4 \geq x^{1.1}/p_1 p_2^2 \geq x^{0.4},$$

indeed $p_1 p_4 \geq x^{0.46}$, $p_4 \geq x^{0.16}$, so $p_3 \geq x^{0.16}$, $n_5 \geq x^{0.16}$, and Lemma 3 (iii) applies.

When (4.27) holds, we again consider $p_1 p_2 p_3 p_4 n_5$. Since $p_1 p_2^3 \leq x^{0.82}$ we have $n_5 \geq x^{0.18}$. Now

$$n_5 p_4^2 = \frac{p_1 p_2 p_3 p_4 n_5}{p_1 p_2 p_3} p_4 \gg \frac{p_4 x}{p_1 p_2^2} \gg \frac{x^{1.1}}{p_1^{1/3} (p_1 p_2^3)^{2/3}} \gg x^{1.1 - 0.1 - 1.64/3} > x^{0.45}.$$

Lemma 3 (iii) applies unless $p_3 < x^{0.15}$. In the latter case, $p_1 p_3 < x^{0.45}$, indeed $p_1 p_3 < x^{0.4}$. In this case,

$$n_5 p_3 = \frac{p_1 p_2 p_3 p_4 n_5}{p_1 p_2 p_4} \gg \frac{x}{p_1 p_2 p_3} \gg \frac{x^{0.6}}{p_2} > x^{0.4},$$

indeed, $n_5 p_3 \gg x^{0.46}$. However, $n_5 p_3 \leq x^{0.54}/p_4 \leq x^{0.44}$ so this case does not occur.

Lemma 23. *We may decompose the subsum of $c_k(3)$ with*

$$x^{0.23} \le p_1 \le x^{0.3}, \quad x^{0.7} \le p_1 p_2^2, \quad x^{0.46} \le p_1 p_2 \le x^{0.54},$$

to reach C_6.

Proof. Decompose successively, obtaining $p_1 p_2 p_3 n_4$, $p_1 p_2 p_3 p_4 n_5$. (Note that $p_1 p_2 p_3 \le x^{0.77}$ and $p_2 p_3 \le x^{0.5}$.) We have $p_3 n_5 \le x^{0.44}$, hence $p_3 n_5 \le x^{0.4}$. If $p_1 p_4 \le x^{0.46}$, we find that $p_1 p_4 \le x^{0.4}$, decompose p_2 and then use Lemma 17 (c) to decompose p_1. If $p_1 p_4 > x^{0.46}$ then

$$p_2 \ge x^{0.2}, \quad p_3 \ge x^{0.16}, \quad p_2 p_4 n_5 \ge x^{0.4},$$

hence $p_2 p_4 n_5 \ge x^{0.46}$. Now $p_2 n_5^2 \ge x^{0.46}$ and Lemma 3 (iii) applies.

As above, write $(p_1, p_2) = (x^{\alpha_1}, x^{\alpha_2})$ and consider the decomposition procedure for $\alpha_1 \le 0.3$. Lemma 13 covers $\alpha_1 + \alpha_2 \ge 0.54$ since then $x^{0.4} \ll x/p_1 p_2 \ll x^{0.46}$; and also covers $\alpha_1 + \alpha_2 \in [0.4, 0.46]$.

Within the region $\{(\alpha_1, \alpha_2) : \alpha_1 \le 0.3, \quad 0.46 \le \alpha_1 + \alpha_2 \le 0.54, \}$ no C_4 is discarded, except when (α_1, α_2) lies in the small triangle

$$\triangle_1 : \quad \alpha_1 \le 0.3, \quad \alpha_1 + 3\alpha_2 \ge 0.82, \quad \alpha_2 \le 0.18.$$

contained in $\triangle_3$ in Diagram 1. Here we appeal to Lemmata 22, 23.

(For part of the region covered by Lemma 23, (4.2) holds:

$$p_2 \ge x^{0.225}, \quad p_1^4 p_2^3 \ge x^{1.8}, \quad p_1 p_2 \le x^{0.54}.$$

Here $p_1 n_3 \le x^{0.775}, x^{0.46} < n_3 \le x^{0.5}$, permitting us to take

$$z = x^{1.1}/n_3^2 > x^{-0.9} p_1^2 p_2^2 > p_2^{\frac{1}{2}}$$

in Lemma 14.)

For $\alpha_1 \le 0.3$ we are left to consider $\alpha_1 + \alpha_2 \le 0.4$. If $\alpha_1 + \alpha_2 \ge 0.36$, we may reach C_6. (We easily reach $p_1 p_2 p_3 p_4 p_5 n_6$. If $p_3 \ge x^{0.15}$, we have

$$p_1 \ge x^{0.18}, \quad p_1 p_2^2 \ge x^{0.46}, \quad p_4 p_5 n_6 \ge x^{0.4},$$

indeed $p_4 p_5 n_6 \ge x^{0.46}$; so Lemma 3 (iii) applies. If $p_3 < x^{0.15}$ we argue as in case (ii) of Lemma 18).

If $0.3 \le \alpha_1 + \alpha_2 \le 0.36$ and $\alpha_2 \le 0.15$, we repeat the argument just used. We also reach C_6 for

$$0.3 \le \alpha_1 + \alpha_2 \le 0.36, \quad \alpha_2 > 0.15, \quad \alpha_1 + 4\alpha_2 \le 0.82.$$

To see this, we easily reach $p_1p_2p_3p_4p_5n_6$, with $n_6 \geq 0.18$. If $p_3p_4p_5 > x^{0.4}$, then $p_3p_4p_5 > x^{0.46}$ and Lemma 3 (iii) applies. If $p_3p_4p_5 \leq x^{0.4}$ then we decompose n_6. Thus, within the strip $0.3 \leq \alpha_1 + \alpha_2 \leq 0.4$, we are left with the triangle

$$\triangle_2: \quad \alpha_1 + 4\alpha_2 \geq 0.82, \quad \alpha_2 \leq \alpha_1, \quad \alpha_1 + \alpha_2 \leq 0.36$$

for which certain C_4 are discarded.

As for $\alpha_1 + \alpha_2 \leq 0.3$, we can reach C_6 by the argument of Lemma 18 (ii). We turn our attention to $0.3 \leq \alpha_1 \leq 0.4$. First take

$$0.3 \leq \alpha_1 \leq 0.4, \quad \alpha_1 + \alpha_2 \leq 0.54. \tag{4.28}$$

If $\alpha_1 \geq 0.36$ or $\alpha_2 \leq 0.15$, we reach C_6 (Lemmata 18, 21). If

$$\alpha_1 < 0.36, \quad \alpha_1 + \alpha_2 \leq 0.54, \quad \alpha_1 + 3\alpha_2 \leq 0.82, \quad \alpha_2 \geq 0.15$$

then Lemma 20 yields (4.2); while if $\alpha_2 \geq 0.18$ we may appeal to the remark following Lemma 19. Within (4.28), we are left with a region $\triangle_3 \backslash \triangle_1$, where $\triangle_3$ is the triangle

$$\triangle_3: \quad \alpha_2 \leq 0.18, \quad \alpha_1 \leq 0.36, \quad \alpha_1 + 3\alpha_2 \geq 0.82,$$

for which regions C_4 with (4.2) unavailable are discarded.

Next we consider

$$0.3 \leq \alpha_1 \leq 0.4, \quad \alpha_1 + \alpha_2 \geq 0.6, \quad \alpha_1 + 2\alpha_2 \leq 1 \tag{4.29}$$

We can dismiss the case $\alpha_2 \leq 0.225$ immediately, using Lemma 4.

Lemma 24. *Lemma 22 remains true if* (4.26), (4.27) *are replaced by either*

$$M \geq x^{0.3}, \quad N \geq x^{0.3}, \quad MN \leq x^{0.64} \tag{4.30}$$

or

$$x^{0.36} \leq M \leq x^{0.4}, \quad N \geq x^{0.3}. \tag{4.31}$$

Remark. Note that in Lemma 24, k is a product of three primes any of which can be decomposed.

Proof. When (4.30) holds, decompose to get $p_1p_2p_3n_4$ and then decompose p_2 to get $p_1p_3n_4p_4n_5$. (Note that $p_3n_4 \leq x^{0.4}$). We have

$$n_4 \geq (p_3n_4)^{\frac{1}{2}} \geq x^{0.18},$$

while $p_1p_4 \geq x^{0.4}$ and indeed $p_1p_4 \geq x^{0.46}$. Moreover,

$$n_5 \geq N^{\frac{1}{2}} \geq x^{0.15}, \quad n_4p_3^2 \gg p_3\left(\frac{x}{MN}\right) \gg x^{0.46}.$$

Hence Lemma 3 (iii) is applicable.

When (4.31) holds we proceed as in (4.30) with the roles of p_1 and n_3 reversed.

Lemma 25. *For the subsums of $c_k(3)$ for which either*

$$x^{0.3} \ll p_1 \ll x^{0.4}, \quad x^{0.225} \ll p_2 \ll x^{0.3}, \quad x^{0.6} \ll p_1 p_2 \ll x^{0.64} \quad (4.32)$$

or

$$x^{0.36} \ll p_1 \ll x^{0.4}, \quad x^{0.225} \ll p_2 \ll x^{0.3}, \quad p_1 p_2 \gg x^{0.6}, \quad (4.33)$$

we may decompose to reach C_6.

Proof. Suppose first that (4.32) holds. Decompose n_3 into $p_3 n_4$ and then p_1 to reach $p_2 p_3 n_4 p_4 n_5$ (Note that $n_4 \leq x^{0.5}$ and $n_4 p_2 p_3 \leq x^{0.7}$).

If $p_2 p_4 < x^{0.4}$, then $p_3 n_4 \ll x/p_1 p_2 \ll x^{0.4}$ so we decompose n_5 to reach $p_2 p_4 p_3 n_4 p_5 n_6$, and complete the proof using Lemma 17 (c). We may thus suppose $p_2 p_4 > x^{0.46}$ and $p_2 p_1^{1/2} > x^{0.46}$.

If $p_2 p_3 < x^{0.4}$, then

$$p_4 n_4 < p_1^{\frac{1}{2}} \left(\frac{x^{0.9}}{p_1 p_2} \right) < x^{0.44}.$$

Arguing as before, we may suppose $p_2 p_3 > x^{0.46}$, so that $p_3 > x^{0.16}$. Since

$$n_4 \geq n_3^{\frac{1}{2}} \geq x^{0.18}, \quad n_5 \geq x^{0.15}, \quad p_2 p_4 \geq x^{0.46}$$

we may apply Lemma 3 (iii).

Now suppose that (4.33) holds. Since

$$x^{0.3} \ll n_3 \ll x^{0.4}, \quad x^{0.6} \ll n_3 p_2 \ll x^{0.64},$$

we may argue as in (4.32) with the roles of p_1 and n_3 reversed.

Returning to the region (4.29), we are left with a quadrilateral

$$\text{T}: \quad \alpha_2 \leq \alpha_1 \leq 0.36, \quad \alpha_1 + 2\alpha_2 \leq 1, \quad \alpha_1 + \alpha_2 \geq 0.64,$$

for which regions C_4 with (4.2) unavailable are discarded.

It remains to consider $0.46 \leq \alpha_1 \leq 0.5$. Lemma 19 with a role reversal gives a region

$$0.46 \leq \alpha_1 \leq 0.5, \quad 0.64 \leq \alpha_1 + \alpha_2 \leq 0.7, \quad \alpha_2 \geq 0.18 \quad (4.34)$$

where (4.2) holds. For $0.46 \leq \alpha_1 \leq 0.54$, $\alpha_2 \geq 0.225$, $4\alpha_1 + \alpha_2 \leq 2.2$, we may deduce (4.2) from Lemma 14, since

$$n_3 p_1 \ll x^{0.775}, \quad x^{1.1}/p_1^2 > p_2^{\frac{1}{2}}.$$

Lemma 26. *Lemma* 22 *remains true if* (4.26) *is replaced by*

$$x^{0.46} \leq M \leq x^{0.5}, \quad N \geq x^{0.15}, \quad MN^2 \leq x^{0.82}.$$

Proof. Decompose using Lemma 14 to reach $p_1 p_2 p_3 n_4$. Since $MN^2 \leq x^{0.82}$ we have $n_4 \geq x^{0.18}$. If $n_4 \geq x^{0.225}$ then Lemma 3 (ii) applies. Otherwise,

$$n_4 p_3^2 = \frac{(p_1 p_2 p_3 n_4)^2}{p_1^2 p_2^2 p_3} \gg \frac{x^2}{p_1 (p_1 p_2^2) p_3} \gg x^{2-0.5-0.82-0.225} \gg x^{0.45}$$

and Lemma 3 (iii) applies.

Lemma 27. *For the subsum of $c_k(3)$ with either $\alpha_2 \geq 0.18$ or $\alpha_2 \leq 0.15$ we may decompose to reach C_6.*

Proof. Suppose that $\alpha_2 \geq 0.18$. Decompose p_1 to reach $p_2 n_3 p_3 n_4$. Since (4.34) may be excluded, we have $p_2 \geq x^{0.2}$ and $n_3 \leq x^{0.3}$. We decompose n_3 to reach $p_2 p_3 n_4 p_4 n_5$. (Note that $p_3 n_4 \leq x^{0.5}$, $p_2 \leq x^{0.27}$.) We have

$$p_3 n_5 \leq (x^{0.5})^{\frac{1}{2}} x^{0.3-0.1} = x^{0.45},$$

indeed $p_3 n_5 \leq x^{0.4}$. If $p_4 n_4 \leq x^{0.4}$ we may decompose p_2, and finally decompose n_4 by Lemma 17 (c).

Suppose, then, that $p_4 n_4 > x^{0.46}$. Now Lemma 3 (iii) applies since

$$p_3 \geq p_2 \geq x^{0.18}, \quad n_5^2 p_2 \geq n_5 p_4 p_2 \geq x^{0.5}, \quad p_4 n_4 \geq x^{0.46}.$$

On the other hand, if $\alpha_2 \leq 0.15$ we decompose to reach $p_1 p_2 p_3 n_4$. If $p_2 p_3 \leq x^{0.275}$ we decompose n_4 via Lemma 14, reaching $p_1 p_2 p_3 p_4 n_5$. We may then decompose p_1 to reach $p_2 p_3 p_4 n_5 p_5 n_6$, since

$$p_4 n_5 \leq x^{1-0.46-0.1-0.1} < x^{0.4}.$$

Finally, Lemma 17 (c) permits us to decompose n_6.

There remains the case $p_2 p_3 \geq x^{0.275}$. Now Lemma 3 (iv) applies since

$$p_2 p_3 \geq x^{\frac{9}{35}}, \quad p_3 \leq p_2 \leq x^{0.15}, \quad p_2 p_3 n_4 \geq x^{0.5}, \quad p_1 \geq x^{0.46}.$$

In our discussion of the region

$$0.46 \leq \alpha_1 \leq 0.5, \quad \alpha_1 + 2\alpha_2 \leq 1, \quad \alpha_2 \geq 0.1,$$

there remains only the triangle

$$\triangle_4: \quad \alpha_2 \leq 0.18, \quad \alpha_1 \leq 0.5, \quad \alpha_1 + 2\alpha_2 \geq 0.82$$

for which regions C_4 with (4.2) unavailable are discarded.

A calculation in BASIC on a personal computer shows that the four dimensional integrals corresponding to $\triangle_2$, $\triangle_3$, $\triangle_4$ and Γ are less than 0.0003, 0.002, 0.002 and 0.0003 respectively. The total of all six dimensional integrals is less than 0.0001, and so

$$u_0 > 0.9953.$$

We now turn to the construction of $c_1(k)$. This is done by an analogous decomposition in which the final identity is

$$\rho(k) = \psi\left(k, x^{\frac{11}{90}}\right) + \cdots - \sum_{\substack{C_7(n_1,\dots,n_7) \in C_7 \\ n_1 \dots n_8 = k}} \sum \psi(n_1, z_1) \dots \psi(n_8, z_8).$$

Lemma 17 (c) shows that we always reach C_5. We then discard certain C_5 in order to avoid generating C_6 which do not allow one more decomposition, and any C_7 for which (4.2) is unavailable, to get an inequality

$$\rho(k) \leq b_0(k).$$

The constant u_1 is found by adding, to 1, integrals in 5 and 7 dimensions corresponding to the discarded regions. We can give a neat description of the discarded C_5 as follows.

Lemma 28. *In* (4.25) *suppose that*

$$j = 5, \quad n_1 \leq \cdots \leq n_6, \quad n_1 n_2 n_3 \leq x^{0.46};$$

then we may decompose to reach C_7.

Proof. We may suppose $n_1 n_2 n_3 < x^{0.4}$, so $n_5 n_6 \geq x^{0.4}$ and indeed

$$n_5 n_6 \geq x^{0.46}.$$

If both n_5, n_6 are $\geq x^{0.225}$ then we use Lemma 17 (c) to decompose both in turn. Assume now that $n_5 < x^{0.225} < n_6$. Plainly we may suppose also that $n_1 n_2 n_3 n_4 \geq x^{0.46}$. Note that

$$n_2 n_3 n_4 \leq x^{0.44}.$$

If $n_6 < x^{0.3}$ then we group variables as $n_1 n_6$, $n_2 n_3 n_4$, n_5 and apply Lemma 4. Thus we may suppose that $n_6 \geq x^{0.3}$ and easily reach the inequality $n_1 n_6 \geq x^{0.46}$. It follows that $n_3 n_4 n_5 \leq x^{0.44}$ and indeed

$$n_3 n_4 n_5 \leq x^{0.4}.$$

We may now decompose n_6 to reach $n_1 n_2 n_3 n_4 n_5 p_6 n_7$. If $n_7 \geq x^{0.225}$ we may decompose again. Otherwise, Lemma 4 applies to $n_3 n_4 n_5$, $n_1 n_2 p_6$, n_7. (To see this we observe that $n_1 n_2 n_3 n_4 n_5 \geq x^{0.57}$ and indeed

$$n_1 n_2 n_3 n_4 n_5 \geq x^{0.6},$$

so that $p_6 \leq x^{0.2}$. If $n_1 n_2 \leq x^{0.26}$ we see that

$$\max(n_3 n_4 n_5, n_1 n_2 p_6) \leq x^{0.4}.$$

If $n_1 n_2 > x^{0.26}$ then

$$n_1 n_2 n_3 n_4 n_5 > x^{0.65}, \quad p_6 < x^{0.175} \quad n_1 n_2 p_6 \leq x^{0.4 \times \frac{2}{3} + 0.175},$$

leading to the same conclusion.) This completes the proof.

Note that in a region $\mathcal{C}_5$ with $n_1 n_2 n_3 \geq x^{0.46}$ we have $n_4 \geq x^{0.15}$. We can therefore give an asymptotic formula if $n_6 \geq x^{0.18}$. Thus the integrals corresponding to discarded regions take the form

$$\int \frac{d\alpha_1 \dots d\alpha_5}{\alpha_1 \dots \alpha_5 (1 - \alpha_1 - \alpha_2 - \alpha_3 - \alpha_4 - \alpha_5)}$$

where

$$\alpha_1 \leq \alpha_2 \leq \alpha_3 \leq \alpha_4 \leq \alpha_5 \leq 1 - \alpha_1 - \dots - \alpha_5 < 0.18$$

with

$$0.46 \leq \alpha_1 + \alpha_2 + \alpha_3 \leq 0.5.$$

This integral is less than 10^{-6}. Allowing for the number of possible decompositions starting with $p_1 p_2 n_3$ (certainly less than a dozen) we will have to count this integral a few times. Nevertheless, the 7-dimensional integrals are so small that we may readily conclude that

$$u_1 < 1.0001.$$

5. Sieving the interval J

We begin with an analogue of Lemma 11. Let $F = E_q^c \backslash \{\chi_0\}$. Through-out this section we have $1 \leq q \leq Q$.

Lemma 29. *Let* $T_2 = T_2(q) = Y/qQ$; *let*

$$G(s, \chi) = \sum_{m \asymp Y} b_m \frac{\chi(m)}{m^s}$$

be a Dirichlet polynomial satisfying (3.1). *Suppose that*

$$M_0(G) = \sum_{\chi \in F} \int_{-Y}^{Y} \frac{|G(\frac{1}{2} + it, \chi)|^2}{T_2^2 + t^2} \, dt \ll \frac{Y}{T_2^2} \mathcal{L}^{-A} \quad \text{for all } A > 0 \quad (5.1)$$

and

$$M_1(G) = \int_{[-Y,Y] \backslash [-T_0, T_0]} \frac{|G(\frac{1}{2} + it, \chi_0)|^2}{T_2^2 + t^2} \, dt \ll \frac{Y}{T_2^2} \mathcal{L}^{-A} \quad \text{for all } A > 0. \quad (5.2)$$

Then

$$\left\| \sum_{m \in J_y} b(m)\chi(m) - \delta_\chi W(y, z) \right\|_{E_q^c}^2 \ll (qQ)^2 Y \mathcal{L}^{-A} \quad \text{for all } A > 0$$

with the abbreviations

$$\|H\|_E = \left(\sum_{\chi \in E} \int_{Y/2}^{3Y/2} |H(y, \chi)|^2 \, dy \right)^{\frac{1}{2}},$$

$$W(u, v) = \frac{1}{2\pi i} \int_{\frac{1}{2} - iT_0}^{\frac{1}{2} + iT_0} G(s, \chi_0) \frac{(y + yz)^s - y^s}{s} \, ds.$$

Proof. By [29], Lemma 3.19,

$$\sum_{m \in J_y} b(m)\chi(m) = \frac{1}{2\pi i} \int_{\frac{1}{2} - iY}^{\frac{1}{2} + iY} G(s, \chi) \frac{(y + yz)^s - y^s}{s} \, ds + O(Y^\epsilon).$$

Consequently it is enough to show that

$$\left\| \frac{1}{2\pi i} \int_{\frac{1}{2} - iY}^{\frac{1}{2} + iY} G(s, \chi) \frac{(y + yz)^s - y^s}{s} \, ds \right\|_F^2 \ll (qQ)^2 Y \mathcal{L}^{-A} \quad (5.3)$$

and

$$\left\|\left(\int_{c+iT_0}^{c+iY} + \int_{c-iY}^{c-iT_0}\right) G(s,\chi) \frac{(y+yz)^s - y^s}{s} ds\right\|_{y,\{\chi_0\}}^2 \ll (qQ)^2 Y\mathcal{L}^{-A}.$$

(5.4)

By a variant of [8], Lemma 2,

$$\int_{Y/2}^{3Y} \left|\int_T^{T'} g(t) y^{it} dt\right|^2 dy \ll Y\mathcal{L} \int_T^{T'} |g(t)|^2 dt$$

whenever g is continuous on $[T, T'] \subset [Y/2, 3Y]$. Thus the left hand side of (5.3) is

$$\ll Y^2\mathcal{L} \sum_{\chi \in F} \int_{-Y}^Y \frac{|G(\frac{1}{2}+it,\chi)|^2}{T_2^2 + t^2} dt \ll \frac{Y^2}{T_2^2} Y\mathcal{L}^{-A} \ll (qQ)^2 Y\mathcal{L}^{-A},$$

where we used

$$\left|\frac{(1+z)^{\frac{1}{2}+it} - 1}{\frac{1}{2}+it}\right|^2 \ll \frac{1}{T_2^2 + t^2}$$

in the first step, and (5.1) in the second step. Similarly (5.2) yields (5.4).

Our next step is analogous to Lemma 12.

Lemma 30. *Let*

$$b_m = \sum_{n \sim N, nl=m} a_n \psi(l,w)$$

where a_n satisfies (3.1) and

$$N \ll Y^{\frac{1}{2}+\frac{1}{4}\theta_2 - \epsilon};$$

(5.5)

then (5.1) and (5.2) hold.

Proof. As in the proof of Lemma 12, suppressing the summation range for n,

$$b_m = \sum_n \sum_{\substack{d \leq \gamma \\ ndr=m}} a_n \mu(d) + \sum_{nl=m} a_n t_\gamma(l)$$

where $\gamma = x^{\epsilon/2}$ and

$$|t_\gamma(l)| \leq \sum_{\substack{d|l,d|P(w) \\ \gamma \leq d < w\gamma}} 1.$$

Thus $G = G_1 + G_2$ where

$$G_1(s,\chi) = \sum_{\substack{ndr \asymp Y \\ d|P(w),d\leq\gamma}} \frac{a_n\mu(d)\chi(ndr)}{(ndr)^s}, \qquad G_2(s,\chi) = \sum_{nl\asymp Y} \frac{a_n t_\gamma(l)\chi(nl)}{(nl)^s}.$$

The contribution to $M_j(G_1)$ from $n \sim N$, $d \sim D$, $r \sim R = Y/ND$ and $|t| \in [T-1, 2T]$ is

$$\ll \frac{1}{T_2^2 + T^2} \left\| V(s, \chi) R(s, \chi) \right\|^2 \tag{5.6}$$

where $V(s, \chi)$ has length $\asymp ND$ and satisfies (3.1), while $R(s, \chi)$ has length R and coefficients 1, and the well-spaced set S has $|t| \geq \delta_\chi T_0$ for $(\chi, t) \in S$. The expression in (5.6) is

$$\ll \frac{1}{T_2^2 + T^2} \left\| V^2 \right\| \left\| R^2 \right\|.$$

In the case $j = 0$, the expression in (5.6) is

$$\ll \frac{\mathcal{L}^B}{T_2^2 + T^2} (qT + N^2 D^2)^{\frac{1}{2}} \left(q(T + T_2) + \frac{q^3}{T_2} + \frac{qR^2}{T_2^3} \right)^{\frac{1}{2}}$$

from (3.3) and Lemma 9 (with T replaced by $T + T_2$). It suffices to prove this is $\ll Y T_2^{-2} \mathcal{L}^{-A}$ for $T = T_2$. Since it may be readily verified that

$$(qT_2 + N^2 D^2)^{\frac{1}{2}} \ll Y^{\frac{1}{2} - \epsilon/2} Q^{\frac{1}{2}}, \quad \left(qT_2 + \frac{q^3}{T_2} + \frac{qY^2}{T_2^3} \right)^{\frac{1}{2}} \ll (qT_2)^{\frac{1}{2}} = \frac{Y^{\frac{1}{2}}}{Q^{\frac{1}{2}}},$$

we obtain (5.1) with G_1 in place of G.

In the case $j = 1$, we replace q by 1 in (3.3), taking the Dirichlet polynomial there to have coefficients $a_n \chi_0(n)$. The expression in (5.5) is

$$\ll \frac{\mathcal{L}^B}{T_2^2 + T^2} (T + N^2 D^2)^{\frac{1}{2}} \left(qT + \frac{q^2}{T} + \frac{R^2}{T^3} \right)^{\frac{1}{2}}$$

$$\ll \frac{\mathcal{L}^B}{T_2^2 + T^2} \left(qT^2 + q^3 + \frac{R^2}{T_0^2} + qTN^2 D^2 + \frac{q^2 N^2 D^2}{T_0} + \frac{Y^2}{T_0^3} \right)^{\frac{1}{2}}. \tag{5.7}$$

All the terms inside the bracket except qT^2, qTN^2D^2 clearly contribute $\ll YT_2^{-2}\mathcal{L}^{-A}$ to the expression in (5.7), while

$$\frac{\mathcal{L}^B}{T_2^2 + T^2} \left\{ (qT^2)^{\frac{1}{2}} + (qTR^2D^2)^{\frac{1}{2}} \right\} \ll \frac{Y\mathcal{L}^{-A}}{T_2^2}$$

(we need only consider $T = T_2$). This establishes (5.2) with G_1 in place of G.

As for G_2, the analogous bounds are a straightforward consequence of (3.3) and the estimate for the sum of the moduli of the coefficients of G_2 implicit in (4.13).

In order to exploit Lemma 2 we need the following simple result.

Lemma 31. *Let ϕ be a constant with*

$$\frac{1}{2j} \geq \phi > \frac{1}{2j+2},$$

where j is an integer ≥ 4. Then if

$$0 < \lambda \leq \frac{\phi}{1-\phi} + \frac{1}{2j-1+2\epsilon} \qquad (5.8)$$

we have

$$\frac{2}{2h+\alpha} \leq \lambda \leq \frac{\phi}{1-\phi} + \frac{1}{2h-1+2\alpha} \qquad (5.9)$$

for some $\alpha \in [\epsilon, 1]$ and integer h with $j \leq h \ll 1/\lambda$.

Proof. First, we observe that

$$\frac{2}{2j+1} - \frac{1}{2j+1+\epsilon} \leq \frac{\phi}{1-\phi} \leq \frac{1}{2j-1} \quad \text{so that} \quad \lambda \leq \frac{1}{2j-1} + \frac{1}{2j-1+2\epsilon}.$$

Case 1. $2/(2h+\epsilon) < \lambda \leq 2/(2h-1)$ for some integer h; necessarily $j \leq h \ll \lambda^{-1}$. Take $\alpha = \epsilon$. Then

$$\lambda \leq \frac{2}{2h-1} \leq \frac{\phi}{1-\phi} + \frac{1}{2h-1+2\epsilon} \quad \text{if} \quad h > k.$$

If $h = k$, (5.8) yields (5.9).

Case 2. $2/(2h+1) < \lambda \leq 2/(2h+\epsilon)$ for some h; necessarily $j \leq h \ll \lambda^{-1}$. Define α in $[\epsilon, 1]$ by $\lambda = 2/(2h+\alpha)$. It remains to check that

$$\frac{2}{2h+\alpha} \leq \frac{1}{2h+1} + \frac{1}{2h-1+2\alpha}$$

by a straightforward calculation.

Lemma 32. *Let ϕ, j be as in Lemma 31. Suppose T satisfies*

$$qTY^\epsilon = Y^{1-\phi}.$$

Let $Y = MN$ with

$$Y^\epsilon \leq N \leq Y^{\phi + \frac{1-\phi}{2j-1} - \epsilon}.$$

Suppose that $M(s,\chi)$, $N(s,\chi)$ satisfy (3.1) and $M(s,\chi)$ satisfies (3.2). Then

$$\|MN\|_2^2 \ll Y\mathcal{L}^{-A}.$$

Proof. Define λ by $N = (qTY^\epsilon)^\lambda$; we know that (5.9) holds. Then (5.9) holds for some h with $j \leq h \ll 1$, and $\alpha \in [\epsilon, 1]$. Let

$$\beta = 2(h+\alpha), \quad \tau = \frac{2\beta}{\beta-2} = \frac{2(h+\alpha)}{h-1+\alpha}.$$

Then (5.9) yields

$$M^{2+\tau} = \left((qTY^\epsilon)^{1/(1-\phi)} N^{-1} \right)^{2+\tau} \geq (qTY^\epsilon)^4$$

since

$$\frac{1}{1-\phi} - \frac{4}{2+\tau} = \frac{\phi}{1-\phi} + \frac{1}{2h-1+2\alpha} \geq \lambda.$$

By Hölder's inequality

$$\|MN\|_2^2 \leq \|N\|_\beta^2 \|M\|_\tau^2. \tag{5.10}$$

We apply Lemma 2 to M with τ, 1, $(\tau+2)/4 - \epsilon$ in place of β, h, γ so that (3.18) holds, while (3.19) follows from $\beta \ll 1$. Thus, for every $A > 0$,

$$\|M\|_\tau \ll M\mathcal{L}^{-A}. \tag{5.11}$$

It is easy to adapt the proof of Lemma 2 to obtain the weaker conclusion

$$\|M\|_\beta \ll M^{\frac{1}{2}} \mathcal{L}^B$$

if the hypothesis (3.2) is deleted. We apply this bound to $N(s, \chi)$. We have $qT \leq N^\gamma$ where $\gamma = 1/(\lambda + \epsilon)$. Now

$$\beta - 4\gamma + 2h = 4h + 2\alpha - \frac{4}{\lambda + \epsilon} \geq \frac{4}{\lambda} - \frac{4}{\lambda + \epsilon},$$

and we deduce that

$$\|N\|_\beta \ll \mathcal{L}^B N^{\frac{1}{2}}. \tag{5.12}$$

The lemma follows on combining (5.10)–(5.12).

Now we take θ_2 as in Section 1, so that

$$qT_2 Y^\epsilon \asymp Y^{1-\theta_2/2+\epsilon}.$$

We apply Lemma 32 with $j = 8$. Let $c_1 = (1 + 7\theta_2)/15 - \epsilon$, and suppose M satisfies (3.2) and

$$Y^\epsilon \leq N \leq Y^{c_1}, \quad G(s, \chi) = N(s, \chi)M(s, \chi), \quad MN = Y.$$

The contribution to $M_0(G)$ or $M_1(G)$ from any interval $[kT_2, (k+1)T_2]$ is $YT_2^{-2}(k^2+1)^{-1}\mathcal{L}^{-A}$. Accordingly, (5.1) and (5.2) hold.

Let

$$c_2 = \frac{36\theta_2 - 2}{11} - \epsilon.$$

Lemma 33. *Let $M(s, \chi)$ and $N(s, \chi)$ satisfy* (3.1) *and* (3.2) *with*

$$LMN = Y, \quad M = X^{\beta_1}, \quad N = X^{\beta_2} \quad 0 \le \beta_2 \le \beta_1 \le \tfrac{1}{2}.$$

If either

$$\epsilon \le \beta_2 \le c_1 \tag{5.13}$$

or (β_1, β_2) *lies in the triangle R, defined by*

$$R: \quad \beta_2 > c_1, \quad 2\beta_2 - \theta_2 + \epsilon \le \beta_1 \le c_2$$

or (β_1, β_2) *lies in either of the parallelograms P_1, P_2 defined by*

$$P_1: \quad |4\beta_1 + 3\beta_2 - 2| \le \theta_2 - \epsilon, \; c_1 \le \beta_2 \le c_2,$$
$$P_2: \quad |4\beta_1 + \beta_2 - 2| \le \theta_2 - \epsilon, \; c_1 \le \beta_2 \le c_2$$

then (5.1), (5.2) *hold with $G = LMN$.*

Proof. The case (5.13) follows from the remarks preceding Lemma 33. For $(\beta_1, \beta_2) \in R$, we apply Theorem 4 with $N^2 L$, LM, M, T_2 in place of M, N, L, T. Thus (3.5) holds with $X = Y^2$,

$$\alpha_1 = \tfrac{1}{2}(2\beta_2 + 1 - \beta_1 - \beta_2) = \tfrac{1}{2}(1 + \beta_2 - \beta_1), \qquad \alpha_2 = \tfrac{1}{2}(1 - \beta_2).$$

Moreover,

$$qT_2 \asymp Y^{1 - \frac{1}{2}\theta_2} \asymp X^{1-\theta} \quad \text{with} \quad \theta = \tfrac{1}{2} + \tfrac{1}{4}\theta_2.$$

Thus (3.7) follows from $\beta_1 \le c_2$, and (3.6) follows from

$$-\theta_2 + \epsilon \le 2\beta_2 - \beta_1 \le \theta_2 - \epsilon. \tag{5.14}$$

Note that in R, $2\beta_2 \ge 2c_1 \ge c_2 - \theta_2 + \epsilon \ge \beta_1 - \theta_2 + \epsilon$, so (5.14) holds. Thus $\|LMN\|_2^2 \ll Y\mathcal{L}^{-A}$ if $t \in [T, T + T_2]$ for $(\chi, t) \in \mathcal{S}$, which leads directly to (5.1) and (5.2).

Similarly, to get P_1 we use $M^2 N$, L^2, N in place of M, N, L, so that

$$\alpha_1 = \tfrac{1}{2}(2\beta_1 + \beta_2), \qquad \alpha_2 = 1 - \beta_1 - \beta_2.$$

Thus (3.6) becomes

$$|4\beta_1 + 3\beta_2 - 2| \le \theta_2 - \epsilon$$

and (3.7) becomes

$$\beta_2 \le c_2.$$

We get P_2 by replacing β_1 by $1 - \beta_1 - \beta_2$ in P_1.

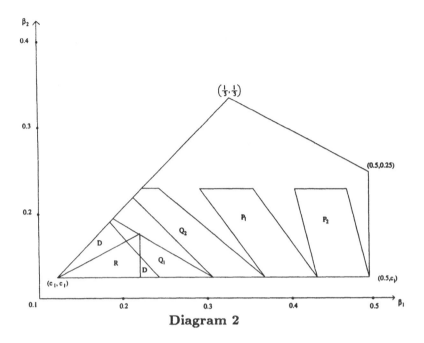

Diagram 2

Remark. The case $\theta_2 = \frac{1}{8} - \epsilon$ is illustrated in Diagram 2.

Our next lemma is analogous to Lemma 15.

Lemma 34. *Let*

$$b_m = \sum_{\substack{nl=k \\ n \le Y^{1/2+\theta_2/4-2\epsilon}}} a_n \psi\big(l, Y^{c_1}\big)$$

where (a_n) *satisfies* (3.1). *Then* (5.1) *and* (5.2) *hold.*

Proof. By Buchstab's identity,

$$b_m = \sum_{\substack{nl=k \\ n \le Y^{1/2+\theta_2/4-2\epsilon}}} a_n \psi\big(l, w\big) - \sum_{\substack{np_1 n_1=k, w \le p_1 < Y^{c_1} \\ n \le Y^{1/2+\theta_2/4-2\epsilon}}} a_n \psi(n_1, p_1)$$

$$= f_m - g_m, \quad \text{say.}$$

Here f_m satisfies (5.1) and (5.2) by Lemma 30. Let h_m be the contribution to g_m from $n \sim N$, $p_1 \sim P$. With

$$N(s, \chi) = \sum_{p_1 \sim P} \frac{\chi(p_1)}{p_1^s},$$

we know from Lemma 5 that (3.2) holds, and of course if $P \geq Y^\epsilon$ we may obtain (5.1) and (5.2) from Lemma 33. For $w \leq P \leq Y^\epsilon$, we iterate the decomposition in the same fashion as case (ii) of Lemma 13. After j steps we have reached a decomposition $np_1 \cdots p_j n_{j+1}$ where $p_1 \cdots p_{j-1} \leq Y^\epsilon$, and need only decompose again if $p_1 \cdots p_j \leq Y^\epsilon$ (since $p_1 \cdots p_j \leq Y^{2\epsilon}$). In this case,

$$np_1 \cdots p_j \leq Y^{\frac{1}{2}+\frac{1}{4}\theta_2-\epsilon}$$

and a further decomposition is possible. The process terminates after $O(\mathcal{L}^{1/10})$ steps.

Lemma 35. *Let*
$$b_m = \sum_{p_1 p_2 n_3 = m} \psi(n_3, p_2)$$

with $p_1 \sim x^{\beta_1}$, $p_2 \sim x^{\beta_2}$ implicit in the summation. Suppose that (β_1, β_2) lies in one of the quadrilaterals Q_1, Q_2 defined by

$$Q_1 : c_1 \leq \beta_2 \leq \beta_1, \quad 4\beta_1+7\beta_2 \leq 2+\theta_2-\epsilon, \quad \beta_1+\beta_2 \geq \tfrac{1}{2} - \frac{(\theta_2+3c_1)}{4}+\epsilon;$$

$$Q_2 : c_1 \leq \beta_2 \leq \min(\beta_1, c_2), \quad 4\beta_1+5\beta_2 \leq 2+c_1, \quad \beta_1+\beta_2 \geq \tfrac{1}{2} - \frac{(\theta_2+c_1)}{4}+\epsilon.$$

Then (5.1) and (5.2) hold.

Proof. Using Lemma 34, we decompose n_3 to reach $p_1 p_2 p_3 n_4$ with

$$p_2 > p_3 \geq x^{c_1}.$$

(Note that $p_1 p_2 \ll x^{\beta_1+\beta_2} \leq x^{1/2+\theta_2/4-\epsilon}$ in Q_1 or Q_2.) Let $p_3 = x^{\beta_3}$. Then for $(\beta_1, \beta_2) \in Q_1$, we have $c_1 < \beta_3 \leq \beta_2 \leq c_2$,

$$4(\beta_1 + \beta_2) + 3\beta_3 - 2 \leq 4\beta_1 + 7\beta_2 - 2 \leq \theta_2 - \epsilon,$$
$$4(\beta_1 + \beta_2) + 3\beta_3 - 2 \geq 4\beta_1 + 4\beta_2 + 3c_1 - 2 \geq -\theta_2 + \epsilon.$$

Thus we may apply Lemma 33 to obtain (5.1) and (5.2). A similar argument, with P_2 in place of P_1, yields (5.1) and (5.2) for $(\beta_1, \beta_2) \in Q_2$.

We now carry out a final decomposition. As in (4.24),

$$\rho(m) = \psi(m, Y^{c_1}) - \sum_{\substack{p_1 n_2 = m \\ Y^{c_1} \leq p_1 < Y^{1/2}}} \psi(n_2, z_1) + \sum_{\substack{p_1 p_2 n_3 = m \\ Y^{c_1} \leq p_2 < p_1 < Y^{1/2}}} \psi(n_3, p_2). \quad (5.15)$$

Let $c_3 = (1 + 22\theta_2)/60 - 2\epsilon$. We may decompose twice, to reach $p_1 p_2 p_3 p_4 n_5$, that part of the last sum corresponding to the plane region D defined by

$$D: \quad c_1 \leq \beta_2 < \beta_1, \quad \beta_1 + 2\beta_2 \leq \tfrac{1}{2} + c_3.$$

To see this, we may decompose once to get $p_1p_2p_3n_4$ since

$$p_1p_2 \leq Y^{\frac{1}{2}+c_3-c_1} \leq Y^{\frac{1}{2}}.$$

We may decompose once more if $p_1p_2p_3 \leq Y^{\frac{1}{2}+\theta_2/4-2\epsilon}$. Suppose this is not the case, so that

$$\frac{1}{2} - c_3 \leq \beta_4 \leq \frac{1}{2} - \frac{1}{4}\theta_2 + 2\epsilon, \quad c_1 \leq \beta_3 \leq \beta_2 \leq \beta_1 \qquad (5.16)$$

where $p_3 = Y^{\beta_3}$, $n_4 = Y^{\beta_4}$. We shall show that (5.1) and (5.2) hold for

$$b_m = \sum_{p_1p_2p_3n_4=m} \psi(n_4, p_3)$$

where the variables are subject to (5.16). We need only note that

$$-\theta_2 + \epsilon \leq 2 - 4c_3 + c_1 - 2 \leq 4\beta_4 + \beta_3 - 2$$
$$\leq 2 - \theta_2 + 8\epsilon + \frac{1}{6} + \frac{1}{3}c_3 - 2 < \theta_2 - \epsilon.$$

Thus no further decomposition is required when (5.15) holds.

We now obtain $b_0(m)$ from $\rho(m)$ by subtracting, in (5.15), the sum over $p_1p_2n_3$ corresponding to $(\beta_1, \beta_2) \in E_2$, where E_2 is defined by

$$c_1 \leq \beta_2 \leq \beta_1 \leq \frac{1}{2}, \quad \beta_1 + 2\beta_2 \leq 1,$$
$$(\beta_1, \beta_2) \notin D \cup Q_1 \cup Q_2 \cup P_1 \cup P_2 \cup R;$$

and further subtracting the sum over $p_1p_2p_3p_4n_5$ corresponding to

$$\beta = (\beta_1, \beta_2, \beta_3, \beta_4) \in E_4,$$

for which $(\beta_1, \beta_2) \in D$ and Lemma 33 applies to no grouping of

$$\beta_1, \quad \beta_2, \quad \beta_3, \quad \beta_4, \quad 1 - \beta_1 - \beta_2 - \beta_3 - \beta_4.$$

It is a straightforward consequence of the prime number theorem, with error term $Y \exp(-\epsilon \mathcal{L}^{1/2})$, that for $y \sim Y$, $z' \leq \mathcal{L}^{-1}$,

$$\sum_{y \leq m \leq y+yz'} b_0(m) = \frac{v_0 yz'}{\mathcal{L}'} + O\left(Y(T_0^{-5} + z'^2)\right). \qquad (5.17)$$

Here v_0 is obtained by subtracting from 1 the usual integrals over E_2 and E_4 involving Buchstab's function. A computer calculation yields the desired lower bound for (1.8):

$$v_0 > 0.05.$$

A little care is now needed to obtain (1.9) with $\beta(m) = b_0(m)$ and $v = v_0$. We have proved (5.1) and (5.2) for $b_0(m)$. Lemma 29 yields

$$\left\| \sum_{y \leq n \leq y + yz} b_0(m)\chi(m) - \delta_\chi W(y,z) \right\|_{E_q^c}^2 \ll z^2 Y^3 \mathcal{L}^{-A}. \tag{5.18}$$

Now (5.1) and (5.2) hold *a fortiori* if T_2 is replaced by a smaller positive number, so the same argument yields

$$\left\| \sum_{y \leq n \leq y + yz'} b_0(m)\chi(m) - \delta_\chi W(y,z) \right\|_{E_q^c}^2 \ll z'^2 Y^3 \mathcal{L}^{-A} \tag{5.19}$$

where

$$z \leq z' \leq z_0 = 1/T_0^2.$$

Combining (5.19), (5.17),

$$\left\| \delta_\chi \left(W(y,z_0) - \frac{v_0 y z_0}{\mathcal{L}'} \right) \right\|_{E_q^c}^2 \ll z_0^2 Y^3 \mathcal{L}^{-A}. \tag{5.20}$$

The proof of (4.4) yields

$$\left| W(y,z') - z'H(y) \right| \ll z'y\mathcal{L}^{-A} \quad \text{if} \quad z \leq z' \leq z_0 \tag{5.21}$$

where

$$H(y) = \frac{1}{2\pi i} \int_{\frac{1}{2} - iT_0}^{\frac{1}{2} + iT_0} G(s, \chi_0) y^{s-1} ds.$$

Combining (5.20) with (5.21) for $z = z_0$,

$$\left\| \delta_\chi \left(z_0 H(y) - \frac{v_0 y z_0}{\mathcal{L}'} \right) \right\|_{E_q^c}^2 \ll z_0^2 Y^3 \mathcal{L}^{-A}$$

and consequently

$$\left\| \delta_\chi \left(z H(y) - \frac{v_0 y z}{\mathcal{L}'} \right) \right\|_{E_q^c}^2 \ll z^2 Y^3 \mathcal{L}^{-A}. \tag{5.22}$$

Another consequence of (5.21) is

$$\left\| \delta_\chi \left(W(y,z) - z H(y) \right) \right\|_{E_q^c}^2 \ll z^2 Y^3 \mathcal{L}^{-A}. \tag{5.23}$$

Now (1.9) follows from (5.18), (5.23) and (5.22).

We require a simpler variant of this procedure to construct $b_1(m)$. Let $c_4 = \frac{1}{4} + \frac{1}{8}\theta_2 - \epsilon$. Then

$$
\begin{aligned}
\rho(m) &\le \psi(m, Y^{c_4}) \\
&= \psi(m, Y^{c_1}) - \sum_{\substack{p_1 n_2 = m \\ Y^{c_1} \le p_1 < Y^{c_4}}} \psi(n_2, Y^{c_1}) + \sum_{\substack{p_1 p_2 n_3 = m \\ Y^{c_1} \le p_2 < p_1 < Y^{c_4}}} \psi(n_3, Y^{c_1}) \\
&\quad - \sum_{\substack{p_1 p_2 p_3 n_4 = m \\ Y^{c_1} \le p_3 < p_2 < p_1 < Y^{c_4}}} \psi(n_4, p_3) \\
&= g_m - h_m + j_m - k_m,
\end{aligned}
$$

say. We use Lemma 34 to obtain (5.1) and (5.2) for g_m, h_m and j_m. We obtain $b_1(m)$ from $\psi(m, Y^{c_4})$ by deleting from k_m the terms for which Lemma 33 does not apply to any arrangement of the variables, so that $\rho(m) \le b_1(m)$. Let E denote the corresponding 'deleted subset' of

$$
\{\beta : c_1 \le \beta_3 < \beta_2 < \beta_1 \le c_4\},
$$

then the constant $v = v_1$ in (1.8) is seen to be

$$
v_1 = \frac{w(1/c_4)}{c_4} + \int_E w\left(\frac{1 - \beta_1 - \beta_2 - \beta_3}{\beta_3}\right) \frac{d\beta_1}{\beta_1} \frac{d\beta_2}{\beta_2} \frac{d\beta_3}{\beta_3^2}
$$

from the prime number theorem. A computer calculation yields

$$
v_1 < 2.2.
$$

We may now repeat for $b_1(m)$ the argument that we have already used to obtain (1.9) for $b_0(m)$. We have established all the properties that we need for the sequences $b_0(m)$ and $b_1(m)$, and the proof of Theorem 1 is complete.

References

1 R.C. Baker, G. Harman and J. Rivat: Primes of the form $[n^c]$. *J. Number Theory* **50** (1995), 261–277.

2 R.C. Baker and G. Harman: The difference between consecutive primes. *Proc. London Math. Soc.* (3) **72** (1996), 261-280.

3 R.C. Baker and G. Harman: The Brun-Titchmarsh theorem on average. In *Analytic Number Theory, Proceedings of a Conference in Honor of Heini Halberstam* Vol. I., 39–103 (B.C. Berndt, H.G. Diamond and A.J. Hildebrand, eds.). Birkhäuser (1996).

4 J. Brüdern and E. Fouvry: Le crible á vecteurs. *Compositio Math.*, (to appear).

5 H. Davenport: *Multiplicative Number Theory* (2nd edn., revised by H.L Montgomery). Springer (1980).

6 A. Fujii, P.X. Gallagher and H.L. Montgomery: Some hybrid bounds for character sums and Dirichlet L-series. In *Topics in Number Theory* (Colloquia Math. Soc. János Bolyai **13**). North-Holland (1976).

7 G. Harman: On the distribution of αp modulo one. *J. London Math. Soc.* (2) **27** (1983), 9–18.

8 G. Harman: Primes in short intervals. *Math. Zeitschr.* **180** (1982), 335–348.

9 G. Harman: On the distribution of αp modulo one II. *Proc. London Math. Soc.* (3) **72** (1996), 241–260.

10 D.R. Heath-Brown: Prime numbers in short intervals and a generalized Vaughan identity. *Can. J. Math.* **34** (1982), 1365–1377.

11 D.R. Heath-Brown: The number of primes in a short interval. *J. Reine Angew. Math.* **389** (1988), 22–63.

12 D.R. Heath-Brown and H. Iwaniec: On the difference between consecutive primes. *Invent. Math.* **55** (1979), 49–69.

13 A. Hildebrand: On the number of positive integers $\leq x$ and free of prime factors $> y$. *J. Number Theory* **22** (1986), 265–290.

14 M.N. Huxley: On the difference between consecutive primes. *Invent. Math.* **15** (1972), 155–164.

15 H. Iwaniec: On sums of two norms from cubic fields. In *Journées de théorie additive des nombres*, 71–89. Université de Bordeaux I (1977).

16 Jia Chaohua: Goldbach numbers in a short interval (I). *Science in China* **38** (1995), 385–406.

17 Jia Chaohua: Goldbach numbers in a short interval (II). *Science in China* **38** (1995), 513–523.

18 Li Hongze: Goldbach numbers in short intervals. *Science in China* **38** (1995), 641–652.

19 Lou Shituo and Yao Qi: A Chebyshev's type of prime number theorem in a short interval II. *Hardy-Ramanujan J.* **15** (1992), 1–33.

20 H. Mikawa: On the exceptional set in Goldbach's problem. *Tsukuba J. Math.* **16** (1992), 513–543.

21 H.L. Montgomery: *Topics in Multiplicative Number Theory* (Lecture Notes in Math. **227**). Springer (1971).

22 H.L. Montgomery and R.C. Vaughan: The exceptional set in Goldbach's problem. *Acta Arith.* **27** (1975), 353–370.

23 A. Perelli and J. Pintz: On the exceptional set for Goldbach's problem in short intervals. *J. London Math. Soc.* (2) **47** (1993), 41–49.

24 A. Perelli, J. Pintz and S. Salerno: Bombieri's theorem in short intervals. *Ann. Scuola Norm. Sup. Pisa* **11** (1984), 529–538.

25 K. Prachar: *Primzahlverteilung*. Springer (1957).

26 K. Ramachandra: Two remarks in prime number theory. *Bull. Soc. Math. France* **105** (1977), 433–437.

27 K. Ramachandra: A simple proof of the mean fourth power estimate for $\zeta(1/2+it)$ and $L(1/2 + it, \chi)$. *Ann. Scuola Norm. Sup. Pisa* **1** (1974), 81–97.

28 H. Rademacher: On the Phragmen-Lindelöf theorem and some applications. *Math. Zeitschr.* **72** (1959/60), 192–204.

29 B. Saffari and R.C. Vaughan: On the fractional parts of x/n and related sequences II. *Ann. Inst. Fourier* **27** (1977), 1–30.

30 E.C. Titchmarsh: *The Theory of the Riemann Zeta-Function* (2nd edn, revised by D.R. Heath-Brown). Oxford University Press (1986).

31 R.C. Vaughan: *The Hardy-Littlewood Method*. Cambridge University Press (1981).

32 N. Watt: Kloosterman sums and a mean value for Dirichlet polynomials. *J. Number Theory* **53** (1995), 179–210.

33 N. Watt: Short intervals almost all containing primes. *Acta Arith.* **72** (1995), 131–167.

34 K.C. Wong: Primes in almost all short intervals (preprint). (1995).

2. On an Additive Property of Stable Sets

Antal Balog and Imre Z. Ruzsa

1. Introduction

To derive additive properties of a set of integers defined by multiplicative constraints is one of the most difficult problems in number theory. The prime twin conjecture, for example, asserts that p and $p + 2$ are simultaneously prime infinitely often. In light of this fact it is perhaps surprising that there is a rather general family of sets for which additive properties are within the scope of known methods.

A set $\mathcal{A}$ of positive integers is called "stable" if for every fixed positive integer d the relation

$$n \in \mathcal{A} \iff dn \in \mathcal{A} \qquad (1)$$

holds for almost all positive integers n, i.e. the set of positive integers n violating (1) has zero (asymptotic) density. In other words, a stable set is one that is almost invariant with respect to multiplication and division by any fixed d. Typical examples of stable sets are the sets

$$\mathcal{Q}_\gamma = \{n : P(n) > n^\gamma\},$$

where $P(n)$ denotes the greatest prime factor of n and $0 < \gamma < 1$. The concept of stability was introduced by the first named author exclusively to study the behaviour of these sets.

Providing the affirmative answer to a conjecture of the first named author, Hildebrand [4] proved that if $\mathcal{A}$ is stable and has positive lower density, then so has $\mathcal{A} \cap (\mathcal{A} + 1)$, i.e. there are plenty of pairs $n, n + 1$ in $\mathcal{A}$. The lower (asymptotic) density is defined by

$$\underline{d}(\mathcal{A}) = \liminf_{x \to \infty} \frac{1}{x} \#\{n \le x : n \in \mathcal{A}\},$$

and similarly the upper (asymptotic) density is defined by

$$\overline{d}(\mathcal{A}) = \limsup_{x \to \infty} \frac{1}{x} \#\{n \le x : n \in \mathcal{A}\}.$$

Inspired by a question of Prof. Etienne Fouvry we make a modest step ahead by proving that a general binary linear equation is solvable in a stable set and only positive upper density is required.

Written while A. Balog was visiting the University of Illinois at Urbana-Champaign. Research of both authors is supported by HNFSR grant #T017433.

Sieve Methods, Exponential Sums, and their Applications in Number Theory
Greaves, G.R.H., Harman, G., Huxley, M.N., Eds. ©Cambridge University Press, 1996

Theorem. *Let $a > 0$, $b > 0$ and $c \neq 0$ be integers satisfying $(a, b) \mid c$. If $\mathcal{A}$ is a stable set and $\overline{d}(\mathcal{A}) > 0$ then the linear equation $am - bn + c = 0$ has infinitely many solutions in $m \in \mathcal{A}$, $n \in \mathcal{A}$. Moreover $\overline{d}\big(b\mathcal{A} \cap (a\mathcal{A} + c)\big) > 0$.*

It will be clear from the proof that if $\underline{d}(\mathcal{A}) > 0$ then the conclusion of the theorem is valid with lower density in place of upper density.

Writing $b = 1$ we get the special case

Corollary 1. *Let $a > 0$ and $c \neq 0$ be integers. If $\mathcal{A}$ is a stable set and $\overline{d}(\mathcal{A}) > 0$ (or $\underline{d}(\mathcal{A}) > 0$) then also $\overline{d}(\mathcal{A} \cap (a\mathcal{A} + c)) > 0$ (or $\underline{d}(\mathcal{A} \cap (a\mathcal{A} + c)) > 0$ respectively).*

Applying this result to the sets $\mathcal{Q}_\alpha \setminus \mathcal{Q}_\beta$, $0 \leq \alpha < \beta \leq 1$ (having positive density) we obtain

Corollary 2. *Let $0 \leq \alpha < \beta \leq 1$ and $a > 0$, $c \neq 0$ be fixed. The set of integers n for which*

$$n^\alpha < P(n) \leq n^\beta, \quad (an + c)^\alpha < P(an + c) \leq (an + c)^\beta$$

holds has positive lower density.

The proof is built on Hildebrand's method. One of the crucial points is that a stable set is equidistributed in residue classes. This is expressed in Lemma 1 of [5]: If $\mathcal{A}$ is stable, then

$$\underline{d}\big(\mathcal{A} \cap (q\mathbb{N} + r)\big) = \frac{1}{q}\,\underline{d}(\mathcal{A})$$

holds for all positive integers q and r. We, however, need a subtler study of the structure of a stable set and a more precise version of the above statement.

Lemma 1. *Let $\mathcal{A}$ be a stable set and define*

$$\mathcal{A}(x) = \#\{n \leq x, n \in \mathcal{A}\},$$
$$\mathcal{A}(x; q) = \#\{n \leq x, n \in \mathcal{A}, (n, q) = 1\},$$
$$\mathcal{A}(x; q, r) = \#\{n \leq x, n \in \mathcal{A}, n \equiv r \pmod{q}\}.$$

We have the following relations.

$$\lim_{x \to \infty} \frac{1}{x}\big(\mathcal{A}(cx) - c\mathcal{A}(x)\big) = 0 \quad \text{for any} \quad c > 0, \tag{2}$$

$$\lim_{x \to \infty} \frac{1}{x}\Big(\mathcal{A}(x; q) - \frac{\phi(q)}{q}\mathcal{A}(x)\Big) = 0 \quad \text{for any integer} \quad q \geq 1, \tag{3}$$

$$\lim_{x \to \infty} \frac{1}{x}\Big(\mathcal{A}(x; q, r) - \frac{1}{q}\mathcal{A}(x)\Big) = 0 \quad \text{for any integers} \quad q \geq 1, r. \tag{4}$$

Another crucial point is to construct a certain auxiliary set of integers (Lemma 1 of [4] or Lemma 3 of [5]). As a second step of the proof of our theorem we construct such an auxiliary set, which is similar to that of [4] or [3], but satisfies an additional condition. This kind of argument originates from Heath-Brown's paper [2], where he used such a set with a different additional condition.

Lemma 2. *Let $(a, b) = 1$ be positive integers. For any $l \geq 2$ there is a set of positive integers $q_1 < q_2 < \cdots < q_l$ with the properties*

$$q_j - q_i = (q_i, q_j) \quad for \quad 1 \leq i < j \leq l, \tag{5}$$
$$q_i = a^{l+1-i} b^i s_i, \quad (ab, s_i) = 1 \quad for \quad 1 \leq i \leq l. \tag{6}$$

2. Proof of Lemma 1

Let $c > 0$ be fixed, $z \geq 3$ be a parameter and $f(n)$ be the characteristic function of the stable set $\mathcal{A}$. We consider the weighted sum

$$S_z(x) = \sum_{n \leq x} f(n) \omega_z(n), \quad \text{where} \quad \omega_z(n) = \#\{p \mid n, p < z\}.$$

From the elementary estimate

$$\sum_{p < z} \frac{1}{p} = \log \log z + O(1)$$

we have that the additive function $\omega_z(n)$ has an average of $\log \log z$. We can apply the Turán-Kubilius inequality (see for example [1], Lemma 4.1) to $\omega_z(n)$ and get

$$\sum_{n \leq x} \left(\omega_z(n) - \log \log z \right)^2 = O(x \log \log z).$$

On the one hand this implies that

$$S_z(x) = \log \log z \sum_{n \leq x} f(n) + O\left(\sum_{n \leq x} |\omega_z(n) - \log \log z| \right)$$
$$= \log \log z \, \mathcal{A}(x) + O\left(x (\log \log z)^{\frac{1}{2}} \right). \tag{7}$$

On the other hand we have

$$S_z(x) = \sum_{n \leq x} f(n) \sum_{\substack{p \mid n \\ p < z}} 1 = \sum_{p < z} \sum_{mp \leq x} f(mp).$$

The stability of $\mathcal{A}$ implies that $f(mp) = f(m)$ holds for every prime $p < z$ and all m outside a set of density zero, i.e.

$$S_z(x) = \sum_{p<z} \sum_{m \le x/p} f(m) + o_z(x)$$

$$= \sum_{\log z \le p < z} \sum_{m \le x/p} f(m) + o_z(x) + O(x \log\log\log z).$$

For every prime satisfying $\log z \le p < z$ we define a set of primes

$$\mathcal{P}_p = \{(c - \delta)p < p' \le cp\},$$

where $\delta = 1/\log\log z$. The Prime Number Theorem implies that for sufficiently large z the size of $\mathcal{P}_p$ is

$$|\mathcal{P}_p| = \frac{\delta p}{\log p} + O\left(\frac{\delta p}{\log^2 p}\right) \ge 1$$

or equivalently

$$\frac{1}{|\mathcal{P}_p|} = \frac{\log p}{\delta p} + O\left(\frac{1}{\delta p}\right).$$

From elementary estimates

$$S_z(x) = \sum_{\log z \le p < z} \frac{1}{|\mathcal{P}_p|} \sum_{p' \in \mathcal{P}_p} \sum_{m \le x/p} f(m) + o_z(x) + O(x \log\log\log z)$$

$$= \sum_{\log z \le p < z} \frac{1}{|\mathcal{P}_p|} \sum_{p' \in \mathcal{P}_p} \left(\sum_{mp' \le cx} f(m) + O\left(\sum_{x/p < m \le cx/p'} 1 \right) \right)$$
$$+ o_z(x) + O(x \log\log\log z)$$

$$= \sum_{\log z \le p < z} \frac{\log p}{\delta p} \sum_{p' \in \mathcal{P}_p} \sum_{mp' \le cx} f(m) + o_z(x) + O(x \log\log\log z)$$

$$= \sum_{p < z} \frac{\log p}{\delta p} \sum_{p' \in \mathcal{P}_p} \sum_{mp' \le cx} f(m) + o_z(x) + O(x \log\log\log z)$$

$$= \sum_{p' < cx} \sum_{mp' \le cx} f(m) \sum_{p'/c \le p < \min(z, p'/(c-\delta))} \frac{\log p}{\delta p}$$
$$+ o_z(x) + O(x \log\log\log z).$$

Using the Prime Number Theorem we have

$$\sum_{p'/c \le p < p'/(c-\delta)} \frac{\log p}{\delta p} = \frac{1}{\delta} \log \frac{c}{c - \delta} + O\left(\frac{1}{\log p'}\right) = \frac{1}{c} + O(\delta) + O\left(\frac{1}{\log p'}\right),$$

which with the stability of $\mathcal{A}$ gives

$$S_z(x) = \sum_{p' < cz} \sum_{mp' \leq cx} f(mp')\frac{1}{c} + o_z(x) + O(x \log\log\log z)$$

$$= \frac{1}{c} \sum_{n \leq cx} f(n) \sum_{p'|n;\, p' < cz} 1 + o_z(x) + O(x \log\log\log z)$$

$$= \frac{1}{c} \sum_{n \leq cx} f(n)\omega_{cz}(n) + o_z(x) + O(x \log\log\log z).$$

Finally we refer to the Turán-Kubilius inequality again and get

$$S_z(x) = \frac{1}{c} \log\log z \, \mathcal{A}(cx) + o_z(x) + O\big(x(\log\log z)^{\frac{1}{2}}\big). \tag{8}$$

Equations (7) and (8) clearly imply (2) by choosing z sufficiently large.
The proof of (3) is immediate from (2) and the stability of $\mathcal{A}$:

$$\mathcal{A}(x; q) = \sum_{d|q} \mu(d) \sum_{d|n;\, n \leq x} f(n) = \sum_{d|q} \mu(d)\mathcal{A}(x/d) + o(x)$$

$$= \sum_{d|q} \frac{\mu(d)}{d} \mathcal{A}(x) + o(x).$$

The proof of (4) is similar to the proof of (2). Suppose for simplicity that $(q, r) = 1$ and consider the weighted sum

$$T_z(x) = \sum_{\substack{n \leq x \\ n \equiv r \,(\mathrm{mod}\, q)}} f(n)\omega_z(n).$$

We have by the Turán-Kubilius inequality that

$$T_z(x) = \log\log z \, \mathcal{A}(x; q, r) + O\big(x(\log\log z)^{\frac{1}{2}}\big). \tag{9}$$

On the other hand using the stability of $\mathcal{A}$ we get

$$T_z(x) = \sum_{p < z} \sum_{\substack{n \leq x \\ p|n;\, n \equiv r \,(\mathrm{mod}\, q)}} f(n) = \sum_{p < z} \sum_{\substack{m \leq x/p \\ mp \equiv r \,(\mathrm{mod}\, q)}} f(m) + o_z(x).$$

For every prime $p < z$ we define a set of primes $\mathcal{P}_p = \{(1 - \delta)p < p' \leq p\}$, where $\delta = 1/\log\log z$. The Prime Number Theorem implies that the size of $\mathcal{P}_p$ is

$$|\mathcal{P}_p| = \frac{\delta p}{\log p} + O\left(\frac{\delta p}{\log^2 p}\right) \geq 1.$$

Using the Prime Number Theorem for Arithmetic Progressions and the stability of $\mathcal{A}$ again we get

$$T_z(x) = \sum_{p<z} \frac{1}{|\mathcal{P}_p|} \sum_{p'\in\mathcal{P}_p} \sum_{\substack{m\leq x/p \\ mp\equiv r(\mathrm{mod}\ q)}} f(m) + o_z(x) + O(x\log\log\log z)$$

$$= \sum_{p<z} \frac{1}{|\mathcal{P}_p|} \sum_{p'\in\mathcal{P}_p} \sum_{\substack{mp'\leq x \\ mp\equiv r(\mathrm{mod}\ q)}} f(m) + o_z(x) + O(x\log\log\log z)$$

$$= \sum_{p<z} \frac{\log p}{\delta p} \sum_{p'\in\mathcal{P}_p} \sum_{\substack{mp'\leq x \\ mp\equiv r(\mathrm{mod}\ q)}} f(m) + o_z(x) + O(x\log\log\log z)$$

$$= \sum_{\substack{p'<z \\ (m,q)=1}} \sum_{\substack{mp'\leq x}} f(m) \sum_{\substack{p'\leq p<\min(z,p'/(1-\delta)) \\ mp\equiv r(\mathrm{mod}\ q)}} \frac{\log p}{\delta p}$$

$$\qquad + o_z(x) + O(x\log\log\log z)$$

$$= \sum_{\substack{p'<z \\ p'\nmid q}} \sum_{\substack{mp'\leq x \\ (m,q)=1}} f(mp')\frac{1}{\phi(q)} + o_z(x) + O(x\log\log\log z)$$

$$= \frac{1}{\phi(q)} \sum_{\substack{n\leq x \\ (n,q)=1}} f(n) \sum_{\substack{p'|n \\ p'<z}} 1 + o_z(x) + O(x\log\log\log z)$$

$$= \frac{1}{\phi(q)} \sum_{\substack{n\leq x \\ (n,q)=1}} f(n)\omega_z(n) + o_z(x) + O(x\log\log\log z).$$

Finally we refer to the Turán–Kubilius inequality again and get

$$T_z(x) = \frac{1}{\phi(q)} \log\log z\,\mathcal{A}(x;q) + o_z(x) + O\big(x(\log\log z)^{\frac{1}{2}}\big). \qquad (10)$$

Equations (3), (9) and (10) clearly imply (4) for $(q,r)=1$. When $(q,r)=d$ we can write $q=q'd$ and $r=r'd$, where $(q',r')=1$. By the stability of $\mathcal{A}$ we have trivially $\mathcal{A}(x;q,r) = \mathcal{A}(x/d;q',r') + o(x)$ and (4) follows from (2) and the former case. The proof of Lemma 1 is completed.

3. Proof of Lemma 2

We follow an inductive process to build up our set. For $l = 2$ we can choose positive integers x and y such that $b^2 y - a^2 x = 1 + a - b$. The set $q_1 = a^2 b(ax + 1)$ and $q_2 = ab^2(by + 1)$ satisfies (5) and (6) as their difference is $q_2 - q_1 = ab(b^2 y + b - a^2 x - a) = ab \mid q_1$ and $(ax + 1, ab) = (by + 1, ab) = 1$ follows from $b(by + 1) - a(ax + 1) = 1$. Suppose $l \geq 2$ and $q_1 < \cdots < q_l$ have been constructed to satisfy (5) and (6). We construct a new set

$$Q_0 < Q_1 < \cdots < Q_l$$

with the required properties, that is

$$Q_j - Q_i = (Q_i, Q_j) \quad \text{for} \quad 0 \leq i < j \leq l, \tag{5'}$$
$$Q_i = a^{l+1-i} b^{i+1} S_i, \quad (ab, S_i) = 1 \quad \text{for} \quad 0 \leq i \leq l. \tag{6'}$$

This slight change of the subscripts will pay later. Let us define

$$\tau = \operatorname*{lcm}_{1 \leq i < j \leq l} (b^{j-i} s_j - a^{j-i} s_i), \quad T = \operatorname*{lcm}_{1 \leq i \leq l} (a^i \tau + b^i s_i).$$

Note that $(a, b) = (ab, s_1 \ldots s_l) = 1$ implies that $(ab, \tau) = (ab, T) = 1$. We can choose positive integers x and y such that $ab^{l+1} x - Ty = 1$. We have $(ab, y) = 1$. We define

$$
\begin{aligned}
Q_0 &= a^{l+1} b\tau Ty = a^{l+1} b S_0, & S_0 &= \tau Ty, \\
Q_1 &= a^{l+2} b^{l+2} \tau x + b q_1 = a^l b^2 S_1, & S_1 &= a^2 b^l \tau x + s_1, \\
Q_2 &= a^{l+2} b^{l+2} \tau x + b q_2 = a^{l-1} b^3 S_2, & S_2 &= a^3 b^{l-1} \tau x + s_2,
\end{aligned}
$$

$$\vdots \qquad\qquad\qquad\qquad \vdots$$

$$Q_l = a^{l+2} b^{l+2} \tau x + b q_l = ab^{l+1} S_l, \qquad S_l = a^{l+1} b\tau x + s_l.$$

Condition (6') follows immediately. To prove condition (5') we observe that for $1 \leq i < j \leq l$ we have

$$Q_j - Q_i = b q_j - b q_i = a^{l+1-j} b^{i+1} (b^{j-i} s_j - a^{j-i} s_i) \mid a^{l+2} b^{l+2} \tau$$

by definition and

$$Q_j - Q_i = b q_j - b q_i \mid b q_i$$

by (5). These imply that $Q_j - Q_i \mid Q_i$, while $(Q_i, Q_j) \mid Q_j - Q_i$ is trivial. For the case $i = 0$ we observe that for $1 \leq j \leq l$ we have

$$Q_j - Q_0 = a^{l+2} b^{l+2} \tau x + b q_j - a^{l+1} b\tau Ty = a^{l+1} b\tau + b q_j$$
$$= a^{l+1-j} b(a^j \tau + b^j s_j) \mid a^{l+1} bT \mid Q_0,$$

while $(Q_0, Q_j) \mid Q_j - Q_0$ is trivial.

4. Proof of the theorem

Let a, b, c, $\mathcal{A}$ be as in the theorem and let $\bar{d}(\mathcal{A}) = \epsilon > 0$. We can suppose that $(a, b) = 1$, as otherwise we take $d = (a, b)$ and prove our theorem with the new triplet given by $a = a'd$, $b = b'd$, $c = c'd$. We can also suppose that $c > 0$, as otherwise we change the role of a and b.

From Lemma 2 there is a system of integers $q_1 < \cdots < q_l$ satisfying (5), (6) with $l \geq 3/\epsilon$, say. For $i = 1, \ldots, l$ we write

$$\mathcal{K}_i = \{k : kq_i + 1 \in \mathcal{A}\}.$$

For all $i = 1, \ldots, l$ we have from (2) and (4) of Lemma 1 that

$$\mathcal{K}_i(x) = \#\{k \leq x : k \in \mathcal{K}_i\} = \mathcal{A}(q_i x + 1; q_i, 1) = \mathcal{A}(x) + o(x).$$

Using that $\mathcal{A}$ is a stable set, all but $o(x)$ elements $n \leq q_l x + 1$ of $\mathcal{A}$ have the property that the product of n with any of the numbers

$$\frac{cq_i}{a(q_i, q_j)}, \quad \frac{cq_j}{b(q_i, q_j)}, \quad (1 \leq i < j \leq l)$$

is again in $\mathcal{A}$. Note that as q_i is divisible by a higher power of a than q_j is, by condition (6), and similarly q_j is divisible by a higher power of b than q_i is, the factors $cq_i/a(q_i, q_j)$ and $cq_j/b(q_i, q_j)$ are integers.

By the inclusion–exclusion principle we have (with obvious notation)

$$x \geq \left(\bigcup_{i=1}^{l} \mathcal{K}_i\right)(x) \geq \sum_{1 \leq i \leq l} \mathcal{K}_i(x) - \sum_{1 \leq i < j \leq l} (\mathcal{K}_i \cap \mathcal{K}_j)(x),$$

and this implies that for at least one pair of integers $i < j$ we have

$$(\mathcal{K}_i \cap \mathcal{K}_j)(x) \geq \frac{2}{l(l-1)}\left(l\mathcal{A}(x) - x - o(x)\right).$$

In other words, if $x \geq x_0(a, b, c, l, q_1, \ldots, q_l)$ is chosen such that

$$\mathcal{A}(x) \geq \tfrac{1}{2}\bar{d}(\mathcal{A})x = \tfrac{1}{2}\epsilon x$$

then there are at least x/l^2 integers $k \leq x$ such that

$$kq_i + 1 \in \mathcal{A}, \quad kq_j + 1 \in \mathcal{A}.$$

Thus there are at least $x/(2l^2)$ integers $k \leq x$ such that

$$\frac{cq_j}{b(q_i, q_j)}(kq_i + 1) \in \mathcal{A}, \quad \frac{cq_i}{a(q_i, q_j)}(kq_j + 1) \in \mathcal{A}. \tag{11}$$

Denoting the first expression of (11) by n and the second expression of (11) by m we see by a simple calculation using (5) that $bn - am = c$. The proof is, therefore, completed.

References

1 P.D.T.A. Elliott: *Probabilistic Number Theory I.* Springer (1980).
2 D.R. Heath-Brown: The divisor function at consecutive integers. *Mathematika* **31** (1984), 141–149.
3 D.R. Heath-Brown: Consecutive almost–primes. *J. Indian Math. Soc.* **52** (1987), 39–49.
4 A. Hildebrand: On a conjecture of Balog. *Proc. Amer. Math. Soc.* **95** (1985), 517–523.
5 A. Hildebrand: On integer sets containing strings of consecutive integers. *Mathematika* **36** (1989), 60–70.

3. Squarefree Values of Polynomials and the abc-Conjecture

J. Browkin, M. Filaseta, G. Greaves and A. Schinzel

1. Introduction

The well-known *abc*-conjecture of Masser and Oesterlé states:

Given $\epsilon > 0$, there is a number $C(\epsilon)$ such that for any relatively prime positive integers a, b, c with $a + b = c$, we have

$$c < C(\epsilon)\big(Q(abc)\big)^{1+\epsilon}, \tag{1.1}$$

where $Q(k) = \prod_{p|k} p$ denotes the product of the distinct primes dividing k.

See e.g. [11] for a discussion of the history and implications of the conjecture. It can of course be expressed in terms of not necessarily positive integers a, b, c (which may be taken to satisfy $a + b + c = 0$ if we prefer) in which case the bound is asserted for $\max\{|a|, |b|, |c|\}$.

The inequality (1.1) can be rewritten in the form

$$\log(a + b) < (1 + \epsilon)\log Q\big(ab(a + b)\big) + \log C(\epsilon).$$

Denote

$$L_{a,b} = \frac{\log(a + b)}{\log Q\big(ab(a + b)\big)}, \tag{1.2}$$

and let $(L_{a,b})$ denote a sequence whose values are these numbers $L_{a,b}$, taken in some fixed order. In this notation the conjecture asserts

$(L_{a,b})$ *is a bounded sequence whose greatest limit point does not exceed 1.*

On the other hand, there exist infinitely many examples of $L_{a,b}$ which are larger than 1 (see [13]). Currently, the greatest known, discovered by E. Reyssat (see [14] or [2]) is $1.62991\ldots$, arising from taking $a = 2$, $b = 3^{10} \times 109$, $c = 23^5$.

M. Filaseta is supported in part by NSF Grant DMS-9400937.
A. Schinzel is supported in part by CEC Grant CIPA-CT92-4022.

Sieve Methods, Exponential Sums, and their Applications in Number Theory
Greaves, G.R.H., Harman, G., Huxley, M.N., Eds. ©Cambridge University Press, 1996

If the abc-conjecture holds, the greatest limit point of $(L_{a,b})$ would in fact equal 1. To see this, take $a = 1$, $b = 2^n - 1$. Then

$$Q\big(ab(a+b)\big) = 2 \prod_{p|2^n-1} p < 2^{n+1},$$

so $L_{a,b} > n/(n+1) \to 1$ as $n \to \infty$. So we can reformulate the abc-conjecture as

$(L_{a,b})$ *is a bounded sequence with its greatest limit point equal to* 1.

Now let

$$\mathcal{L} = \big\{L_{a,b} : a \geq 1,\, b \geq 1,\, (a,b) = 1)\big\}$$

be the set of (distinct) values of $(L_{a,b})$, and let $\mathcal{L}'$ be the "derived" set of limit points of $\mathcal{L}$. We are interested in seeing what can be actually proved about $\mathcal{L}'$. By use of a sieve method we obtain the following theorem.

Theorem 1. *The set $\mathcal{L}'$ of limit points of $\mathcal{L}$ contains the interval $\left[\frac{1}{3}, \frac{15}{16}\right]$.*

The entry $\frac{1}{3}$ here is the best possible because

$$\log(a+b) \geq \tfrac{1}{3}\log\big(ab(a+b)\big) \geq \tfrac{1}{3}\log Q\big(ab(a+b)\big)$$

in (1.2). On the other hand, we will show in Theorem 4 that if the abc-conjecture holds then $\frac{15}{16}$ may be replaced by 1 in Theorem 1.

One might also enquire about the limit points of

$$\mathcal{M} = \left\{\frac{\log a}{\log Q\big(ab(a+b)\big)} : 1 \leq a \leq b,\, (a,b) = 1\right\}$$

in addition to those of $\mathcal{L}$. As will be seen, our methods show that

$$\left[0, \tfrac{3}{4}\right] \subseteq \mathcal{M}', \tag{1.3}$$

so that the set of combined limit points of $\mathcal{L}$ and $\mathcal{M}$ contains $\left[0, \frac{15}{16}\right]$.

Our proof of Theorem 1 rests on the following result about squarefree values of binary forms.

Theorem 2. (a) *Let $1 \leq Y \leq X$, where X is sufficiently large, and let $f(x,y) \in \mathbb{Z}[x,y]$ be a binary form whose irreducible factors f_i are distinct and all have degrees not exceeding μ. Let D denote the largest fixed divisor of $f(x,y)$, and let $S = D/(\prod_{p|D} p)$. Let $N(X,Y)$ denote the number of pairs $\langle x,y \rangle$ with*

$$X < x \leq 2X, \quad Y < y \leq 2Y \tag{1.4}$$

for which $f(x,y)/S$ is squarefree. Suppose for some $\epsilon > 0$

$$X^\mu < (XY)^{3-\epsilon}, \quad Y > X^\epsilon. \tag{1.5}$$

Then

$$N(X,Y) = C_f XY \left\{ 1 + O\left(\frac{1}{\log X}\right) \right\}, \tag{1.6}$$

where the constant $C_f > 0$ depends only on f as in (3.6) below, and the O-constant depends only on ϵ.

(b) *In part* (a), *set $X = Y^\alpha$, where $\alpha > 1$ is fixed. Then* (1.6) *holds*

$$\text{for any} \quad \alpha \qquad \text{when} \quad \mu \le 3$$
$$\text{if} \quad \alpha < 3 \quad \text{when} \quad \mu = 4$$
$$\text{if} \quad \alpha < \tfrac{3}{2} \quad \text{when} \quad \mu = 5.$$

In this theorem, the first inequality in (1.5) implies $Y > X^{\epsilon/3}$ when $\mu \ge 3$; for all μ we assume $Y > X^\epsilon$, for our convenience. In particular we will take advantage of the implication $1/\log Y \ll 1/\log X$. As a consequence, when f is quadratic our theorem does not embody even the result of Nagell [12] in which $Y = 1$. It would be possible to refine Theorem 2 into a result holding uniformly for $1 < Y \le X$, but at the cost of introducing technicalities into the treatment that are not necessary for our purpose here.

Observe that Theorem 2 also marginally fails to include the results of Hooley [9], where $Y = 1$, $\mu = 3$, and of one of the present authors [8], where $X = Y$, $\mu = 6$, and may thus be regarded as a somewhat imperfect bridge between them. The imperfection stems from our use in the proof of Lemma 2 of the estimate $d(n) \ll n^\epsilon$ for the divisor function. There is therefore no need to take extra care about other factors that also contribute no worse than X^ϵ to our requirements in (1.5).

The treatment in [8] was arranged on the assumption $X = Y$, and it is necessary to modify the treatment to obtain part (a) of Theorem 2. For part (b), which is what we actually use, we have only to observe that the given conditions are sufficient for (1.5).

Denote the nth cyclotomic polynomial by $\Phi_n(x)$. In §6 we establish the following theorem.

Theorem 3. *Assume the abc-conjecture holds. Then, for every positive integer n, there exist infinitely many integers m for which $\Phi_n(m)$ is square-free.*

As we indicate in §6, the same method suffices to show that (assuming the abc-conjecture) for all positive integers n the polynomial $(x^n - 1)/(x-1)$ takes infinitely many squarefree values.

Unconditionally, it has not been shown that there exists an irreducible $f(x) \in \mathbb{Z}[x]$ of degree ≥ 4 having the property that $f(m)$ is squarefree for infinitely many integers m. Before the stronger result of Hooley [9] mentioned earlier Erdős [4] had established that all irreducible cubics with largest fixed divisor equal to 1 possess this property.

Finally, and also in §6, we point out what is conjectured to be true in the direction of our Theorem 1.

Theorem 4. *Assume the abc-conjecture holds. Then the set of limit points of $\mathcal{L}$ is precisely the interval $\left[\frac{1}{3}, 1\right]$.*

By the same method an analogous result can be obtained about $\mathcal{M}'$, as described in (1.3):

Assume the abc-conjecture holds. Then the set of limit points of $\mathcal{M}$ is precisely the interval $[0, 1]$.

Throughout this article the symbol p denotes a prime. The arbitrarily small real number $\epsilon > 0$ need not be the same on each occurrence.

2. The treatment of Theorem 1

We describe here how Theorem 1 follows once Theorem 2 has been established.

First we use the following construction, which in conjunction with a simpler version of Theorem 2 leads to the weaker form of our Theorem 1 in which the right-hand end-point $\frac{15}{16}$ is replaced by $\frac{6}{7}$. Take

$$a = y^\nu, \quad b = x^\nu - y^\nu.$$

Thus $c = a + b = x^\nu$. We will specify $\nu = 1, 2, 4,$ or 6. Fix an exponent $\alpha > 1$ and consider

$$X = Y^\alpha, \quad X < x \le 2X, \quad Y < y \le 2Y, \tag{2.1}$$

as in Theorem 2. For the specified values of ν the irreducible factors of $x^\nu - y^\nu$ are of degree at most 2. It is then a relatively easy case of Theorem 2 that we can, for any $\alpha \ge 1$, find x, y satisfying (2.1) so that $xy(x^\nu - y^\nu)$ is squarefree.

For $L_{a,b}$, as defined in (1.2), this construction gives

$$Q\big(ab(a + b)\big) = xy(x^\nu - y^\nu), \quad (a, b) = 1,$$

$$L_{a,b} = \frac{\log x^\nu}{\log\big(xy(x^\nu - y^\nu)\big)} = \frac{\nu\alpha}{(\nu + 1)\alpha + 1} + O\left(\frac{1}{\log X}\right).$$

The limit points of these $L_{a,b}$ cover the interval $\big(\nu/(\nu + 2), \nu/(\nu + 1)\big)$. On taking the union of these for $\nu = 1, 2, 4, 6$ we cover $(\frac{1}{3}, \frac{6}{7}) \setminus \{\frac{1}{2}, \frac{2}{3}\}$, whence

$$\left[\tfrac{1}{3}, \tfrac{6}{7}\right] \subseteq \mathcal{L}', \tag{2.2}$$

since the set $\mathcal{L}'$ contains the derived set of the set of limit points of the sequence $(L_{a,b})$.

The construction with $\nu = 6$ also gives

$$\frac{\log a}{\log Q(ab(a+b))} = \frac{\nu}{(\nu+1)\alpha+1} + O\left(\frac{1}{\log X}\right),$$

which leads in a similar way to the assertion (1.3) about $\mathcal{M}'$.

Next, we show how Theorem 2 can further be used to obtain that $\left[\frac{6}{7}, \frac{12}{13}\right] \subseteq \mathcal{L}'$. We invoke the polynomial identity

$$y^3(2x+y) + (x+y)^3(x-y) = x^3(x+2y).$$

We replace the pair $\langle x, y \rangle$ by $\langle x^3, y^3 \rangle$ and set

$$a = y^9(2x^3+y^3), \quad b = (x^3+y^3)^3(x^3-y^3).$$

Thus $a + b = x^9(x^3 + 2y^3)$. We apply Theorem 2 to

$$f(x,y) = xy(x^3+2y^3)(2x^3+y^3)(x^3+y^3)(x^3-y^3).$$

One easily checks that $S = 2$. We deduce that $Q(ab(a+b)) = f(a,b)/2$ provided this last expression is squarefree. Since the irreducible factors of $f(x,y)$ have degree at most 3, we obtain from Theorem 2 that we can arrange this for any α in (2.1). Proceeding as earlier we obtain

$$L_{a,b} = \frac{12\alpha}{13\alpha+1} + O\left(\frac{1}{\log X}\right),$$

whence $\left[\frac{6}{7}, \frac{12}{13}\right] \subseteq \mathcal{L}'$.

To complete the inference of Theorem 1 from Theorem 2, we describe now how to obtain $\left[\frac{12}{13}, \frac{15}{16}\right] \subseteq \mathcal{L}'$. As above, we make use of a certain polynomial identity, namely

$$(x+y)^7(x-y)(x^2-xy+y^2) + y^7(2x+y)(3x^2+3xy+y^2)$$
$$= x^7(x+2y)(x^2+3xy+3y^2).$$

These polynomial identities were obtained by considering special cases of the identities described by Lemma 3 in [5]. For example, this last identity follows from considering $k = 7$, $s = 3$ there. Such identities were explicitly given earlier by Huxley and Nair in [10], but they also go back even further as part of the general theory of Padé approximants (cf. [1]).

In the above identity, we replace $\langle x, y \rangle$ with $\langle x^2, y^2 \rangle$. Set

$$a = (x^2+y^2)^7(x^2-y^2)(x^4-x^2y^2+y^4), \quad b = y^{14}(2x^2+y^2)(3x^4+3x^2y^2+y^4).$$

We apply Theorem 2 with $\mu = 4$ and

$$
f(x,y) = xy(x+y)(x-y)(x^2+y^2)(2x^2+y^2)(x^2+2y^2)
$$
$$
\times (x^4 - x^2y^2 + y^4)(3x^4 + 3x^2y^2 + y^4)(x^4 + 3x^2y^2 + 3y^4).
$$

Here, $S = 6$. We deduce that $Q\big(ab(a+b)\big) = f(a,b)/6$ for infinitely many pairs $\langle a, b \rangle$ defined as above with $x \in (X, 2X]$, $y \in (Y, 2Y]$, and $X = Y^\alpha$, where α can be an arbitrary number from the interval $(1, 3)$. Proceeding as before, the result $\left[\frac{12}{13}, \frac{15}{16} \right] \subseteq \mathcal{L}'$ easily follows, completing the proof that Theorem 1 is a consequence of Theorem 2.

3. The treatment of Theorem 2

In this section we reduce the proof of Theorem 2 to that of Lemmas 1 and 2 below. The general structure of the argument resembles that in [9], [7] and [8], where some points are discussed at greater length than below.

Let $N'(X,Y)$ denote the number of pairs $\langle x, y \rangle$ of integers satisfying (1.4) such that

$$
f(x,y)/S \not\equiv 0, \bmod p^2 \quad \text{for all} \quad p \leq \xi = \tfrac{1}{3} \log Y, \tag{3.1}
$$

this providing the definition of ξ. Then

$$
N(X,Y) = N'(X,Y) + O\big(E(X,Y)\big), \tag{3.2}
$$

where

$$
E(X,Y) = \sum_{\substack{X < x \leq 2X, Y < y \leq 2Y \\ p^2 | f(x,y) \text{ for some } p > \xi}} 1. \tag{3.3}
$$

The Inclusion-Exclusion Principle suffices to estimate N'. We obtain

$$
N'(X,Y) = \sum_{\substack{X < x \leq 2X, Y < y \leq 2Y}} \sum_{\substack{l^2 | f(x,y)/S \\ p|l \Rightarrow p \leq \xi}} \mu(l)
$$

$$
= \sum_{\substack{l \\ p|l \Rightarrow p \leq \xi}} \mu(l) \sum_{\substack{1 \leq u \leq Sl^2,\ 1 \leq v \leq Sl^2 \\ f(u,v) \equiv 0,\, \bmod Sl^2}} \sum_{\substack{X < x \leq 2X, Y < y \leq 2Y \\ x \equiv u, y \equiv v,\, \bmod Sl^2}} 1
$$

$$
= \sum_{\substack{l \\ p|l \Rightarrow p \leq \xi}} \mu(l) \sum_{\substack{1 \leq u \leq Sl^2,\ 1 \leq v \leq Sl^2 \\ f(u,v) \equiv 0,\, \bmod Sl^2}} \left\{ \frac{X}{Sl^2} + O(1) \right\} \left\{ \frac{Y}{Sl^2} + O(1) \right\}.
$$

Denote

$$
\rho(r) = \sum_{\substack{1 \leq u \leq r,\ 1 \leq v \leq r \\ f(u,v) \equiv 0,\, \bmod r}} 1. \tag{3.4}
$$

When l is squarefree and Y sufficiently large the condition $p \mid l \Rightarrow p \leq \xi$ gives $l \leq \prod_{p \leq \xi} p = \exp(\sum_{p \leq \xi} \log p) \leq e^{9\xi/8} = Y^{3/8}$, since ξ is as defined in (3.1). Thus $Y/Sl^2 \gg 1$ and it follows that

$$N'(X,Y) = XY \sum_{\substack{l \\ p \mid l \Rightarrow p \leq \xi}} \frac{\mu(l)\rho(Sl^2)}{S^2 l^4} + O\left(X \sum_{l \leq Y^{3/8}} \frac{\rho(Sl^2)}{l^2}\right) \qquad (3.5)$$

$$= C_f XY \left\{ 1 + O\left(\frac{1}{\log Y}\right) \right\},$$

where

$$C_f = \prod_p \left\{ 1 - \frac{\rho(p^{\sigma+2})}{p^{2\sigma+4}} \right\} > 0, \qquad (3.6)$$

the exponent $\sigma = \sigma(p)$ being defined by $p^\sigma \parallel S$. The properties of the function ρ needed to support (3.5), in particular $\rho(p^2) \ll p^2$, can be obtained as in [7]. The definition of S given in Theorem 2 implies that $\rho(p^{\sigma+2}) < p^{2\sigma+4}$ for all p, so that $C_f > 0$.

Because of (3.2), (3.5) and (3.6) the proof of the main assertion (1.6) of Theorem 2 will be complete when we have shown

$$E(X,Y) \ll \frac{XY}{\log Y}. \qquad (3.7)$$

The irreducible factors f_i of the form f are all distinct, so the discriminant Δ of f is non-zero. If Y is large enough, as the theorem allows, then $p > \xi$ implies that p does not divide Δ or any non-zero coefficient of any f_i. If $p \nmid (x,y)$ in (3.3) then $p^2 \mid f_i(x,y)$ for some i (since $p \nmid \Delta$). Except when $f_i(x,y)$ is x or y this implies $p \nmid xy$. So we can write

$$E(X,Y) \ll E^{(0)}(X,Y) + \sum_i E_i^{(1)}(X,Y) + \sum_i E_i^{(2)}(X,Y), \qquad (3.8)$$

where

$$E^{(0)}(X,Y) = \sum_{\substack{X < x \leq 2X, Y < y \leq 2Y, p > \xi \\ p^2 \mid xy}} 1,$$

$$E_i^{(1)}(X,Y) = \sum_{\substack{X < x \leq 2X, Y < y \leq 2Y, \xi < p \leq XY\eta \\ f_i(x,y)=kp^2, p \nmid xy}} 1, \qquad (3.9)$$

$$E_i^{(2)}(X,Y) = \sum_{\substack{X < x \leq 2X, Y < y \leq 2Y, p > XY\eta \\ f_i(x,y)=kp^2 \neq 0, p \nmid xy}} 1, \qquad (3.10)$$

We may suppose $f_i(x,y) \neq 0$ in (3.10) since otherwise the contribution to $E(X,Y)$ of the pair $\langle x,y \rangle$ would already have been counted in (3.9). We will specify

$$\eta = 1/\log X, \qquad (3.11)$$

although a smaller choice, even $1/X^\epsilon$, would be satisfactory for our purposes.

When $Z = X$ or $Z = Y$ the number of multiples of d in $(Z, 2Z]$ does not exceed the number in $(0, 2Z]$, which is $\ll Z/d$. Hence

$$E^{(0)}(X,Y) \ll \sum_{p > \xi} \frac{XY}{p^2} \ll \frac{XY}{\xi} \ll \frac{XY}{\log Y}, \qquad (3.12)$$

since we specified $\xi = \frac{1}{3} \log Y$ in (3.1).

The proof of Theorem 2 will be completed via the following two lemmas, which we give in a form that is uniform in X, Y, ξ, η.

Lemma 1. *Suppose* $1 \leq Y \leq X$ *and* $\eta \leq \log X$. *Let* f_i *be any one of the irreducible factors of the form* f, *as described in Theorem 2. Then the expression* $E_i^{(1)}(X,Y)$ *given in (3.9) satisfies*

$$E_i^{(1)}(X,Y) \ll \frac{XY\eta}{\sqrt{\log X}} + \frac{XY}{\xi}.$$

When ξ, η are as in (3.1), (3.11), and $Y > X^\epsilon$ as in Theorem 2, this estimate reduces to $E_i^{(1)}(X,Y) \ll XY/\log X$, which is what is needed for Theorem 2. Observe in particular that the corresponding estimate obtained in [8], namely $X^2/\log X$, is not adequate in the present context.

Lemma 2. *Suppose* $1 < Y \leq X$, *and let* f *be a form with no linear factors over* $\mathbb{Z}$. *Define*

$$S(X,Y,K) = \sum_{|k| \leq K} \sum_{\substack{X < x \leq 2X, \, Y < y \leq 2Y \\ f(x,y)=kp^2, \, p \nmid xy}} 1.$$

Then

$$S(X,Y,K) \ll X^{\epsilon/5}\left(X\sqrt{Y} + K^{\frac{1}{4}}(XY)^{\frac{3}{4}}\right).$$

In (3.10) the form f_i is irreducible. We can suppose it is not linear, for if so then the sum (3.10) would be empty. In this sum, the variable k satisfies $|k| \leq K$, where $K \ll X^\mu/p^2$ and $p > XY\eta$. Under the condition (1.5) of Theorem 2 this gives $K \ll (XY)^{1-\epsilon}/\eta^2$, where η is as in (3.11). Thus Lemma 2 gives

$$E_i^{(2)}(X,Y) \ll X^{\epsilon/5} \left(X\sqrt{Y} + \frac{(XY)^{1-\epsilon/4}}{\sqrt{\eta}} \right) \ll \frac{XY}{\log Y}.$$

With (3.12) and (3.8) these estimates for $E_i^{(1)}(X,Y)$ and $E_i^{(2)}(X,Y)$ establish (3.7) and hence Theorem 2.

In [8] a weaker form of Lemma 2 was obtained (in the case $X = Y$) using Selberg's sieve method. The procedure could be adapted to the context $X \neq Y$ to yield a result useful for the same range of K as is Lemma 2. We use the two-dimensional Large Sieve to obtain the sharper version stated.

4. Lemmas on linear congruences

In the proofs of Lemmas 1 and 2 we will need the following results relating to the solutions of a congruence $ax + by \equiv 0$, mod m, that lie in a box $X < x \leq 2X$, $Y < y \leq 2Y$. It is here that we introduce the rescaling of the procedures used in [8] that is essential in the context $X \neq Y$. In the case $X = Y$ Lemma 3 and part (a) of Lemma 4 reduce to Lemmas 1 and 2 in [8].

We will consider the points $\langle x, y \rangle$ for which

$$x \equiv \omega y, \bmod r, \tag{4.1}$$

where $r > 0$. In the language of the Geometry of Numbers, the solutions of (4.1) form a lattice Λ_ω, given by

$$\langle x, y \rangle = l\langle \omega, 1 \rangle + m\langle r, 0 \rangle.$$

This lattice has (positive, by convention) determinant r.

We use the maximum norm $|\langle x, y \rangle| = \max\{|x|, |y|\}$, although the Euclidean norm, or any other equivalent one, could be employed.

Consider the modified lattice λ_ω consisting of non-integral vectors

$$\mathbf{z} = \langle z_1, z_2 \rangle = \left\langle \frac{x}{X}, \frac{y}{Y} \right\rangle, \tag{4.2}$$

where x, y satisfy (4.1). This has determinant

$$\Delta = \frac{r}{XY}. \tag{4.3}$$

Choose a basis $\mathbf{a}, \mathbf{b}$ for this modified lattice (Minkowski-reduced with respect to the norm $|\cdot|$) as follows: let $|\mathbf{a}|$ be minimal so that $\mathbf{a} \neq \mathbf{0}$, and let $|\mathbf{b}|$ be minimal so that $\mathbf{b}$ is independent of $\mathbf{a}$. Then

$$\Delta \ll |\mathbf{a}||\mathbf{b}| \ll \Delta,$$

where Δ is the determinant described in (4.3), the implied constants being absolute. These inequalities go back at least to Minkowski (see, for example, chapter 8 of [3]), but were derived *ab initio* (for the case $X = Y$) in [8]. (They are also immediately accessible to geometrical intuition; one can see, when $|\cdot|$ is the Euclidean norm, that the angle between $\mathbf{a}$ and $\mathbf{b}$ is not less than $\pi/3$.)

The region $X < x \leq 2X, \quad Y < y \leq 2Y$ becomes the square

$$1 < z_1 \leq 2, \quad 1 < z_2 \leq 2$$

in $\mathbf{z}$-space. Write $\mathbf{z} = l\mathbf{a} + m\mathbf{b}$. Then

$$|l| = |b_2 z_1 - b_1 z_2|/\Delta, \quad |m| = |a_2 z_1 - a_1 z_2|/\Delta.$$

Since $|z_i| \leq 2$ in (4.2), this gives

$$|l| \ll L = \frac{|\mathbf{b}|}{\Delta}, \quad |m| \ll M = \frac{|\mathbf{a}|}{\Delta}, \quad M \leq L, \quad LM = \frac{|\mathbf{a}||\mathbf{b}|}{\Delta^2} \ll \frac{1}{\Delta}. \quad (4.4)$$

Thus the integer vector $\langle l, m \rangle$ is confined to a parallelogram with area $\ll XY/r$ and perimeter $\ll L \ll 1/\Delta M = 1/|\mathbf{a}|$. This proves Lemma 3, as follows.

Lemma 3. *The number $N_\omega(X, Y, r)$ of solutions of (4.1) for which*

$$X < x \leq 2X, \quad Y < y \leq 2Y$$

satisfies

$$N_\omega(X, Y, r) \ll \frac{XY}{r} + \frac{1}{|\mathbf{a}|},$$

where $\mathbf{a} = \mathbf{a}(\omega, r)$ is, as above, the shortest non-zero vector in the lattice $\lambda_\omega = \lambda_\omega(r)$ given by (4.2).

We will need to average the error term $1/|\mathbf{a}|$ appearing in Lemma 3 over certain values of r and of ω, mod r. Since it causes little extra trouble, we establish Lemma 4 in a form uniform in η.

Lemma 4. *Suppose ω, $\bmod\, r$ runs over the roots of a congruence*

$$g(\omega, 1) \equiv 0, \bmod r,$$

where g is a form with no linear factors over $\mathbb{Z}[x, y]$. Let $\mathbf{a}$ be as in Lemma 3. Suppose $1 \leq Y \leq X$, $0 < \eta \leq \log X$, and that X is sufficiently large.

(a) *Denote*

$$\Sigma^{(2)}(X, Y, \eta) = \sum_{X < p \leq XY\eta} \sum_{\omega, \bmod p^2} \frac{1}{|\mathbf{a}(\omega, p)|},$$

where the sum is over primes p. Then

$$\Sigma^{(2)}(X, Y, \eta) \ll \frac{XY\sqrt{\eta}}{\sqrt{\log X}}.$$

(b) *Denote*

$$\Sigma^{(3)}(R, X, Y) = \sum_{1 \leq r \leq R} \sum_{\omega, \bmod r} \frac{1}{\sqrt{r|\mathbf{a}|}}.$$

Then

$$\Sigma^{(3)}(R, X, Y) \ll R^{\epsilon}(\sqrt{X} + (XYR)^{\frac{1}{4}}).$$

From (4.2) we can express $\mathbf{a}$ as $\mathbf{a} = \langle u_1/X, u_2/Y \rangle$, where u_1, u_2 are integers, not both 0. Then

$$|\mathbf{a}| = \max\{|u_1|/X, |u_2|/Y\} > 0.$$

Here $\mathbf{a}$ is in the lattice λ_ω, so

$$u_1 \equiv \omega u_2, \bmod r, \tag{4.5}$$

so that in Lemma 4 we have

$$g(u_1, u_2) \equiv 0, \bmod r. \tag{4.6}$$

In part (a) of the Lemma, where $r = p^2$ and $p > X$, consider first those p, ω for which $u_1 u_2 = 0$. If $u_2 = 0$, so that $u_1 \neq 0$, then (4.5) gives $p^2 \,|\, u_1$, so that in particular $|u_1| \geq p^2$. Hence

$$p^2 \leq \frac{p^4}{X^2} \leq \frac{|u_1|^2}{X^2} \leq |\mathbf{a}|^2 \leq |\mathbf{a}||\mathbf{b}| \ll \Delta = \frac{p^2}{XY},$$

which is impossible for large X. Hence $u_2 \neq 0$. Similarly $u_1 \neq 0$ follows, unless $\omega \equiv 0$, mod p, which does not arise as soon as X exceeds the coefficients of g. So in establishing part (a) we may now suppose $u_1 u_2 \neq 0$.

The terms where $|a| \geq \sqrt{\eta}/\sqrt{\log X}$, i.e. where

$$\min \left\{ \frac{X}{|u_1|}, \frac{Y}{|u_2|} \right\} \leq \frac{\sqrt{\log X}}{\sqrt{\eta}}, \tag{4.7}$$

contribute to $\Sigma^{(2)}$ an amount

$$\ll \sum_{X < p \leq XY\eta} \sum_{\omega, \, \text{mod} \, p^2} \frac{\sqrt{\log X}}{\sqrt{\eta}} \ll \frac{XY\sqrt{\eta}}{\sqrt{\log X}},$$

since there are only $O(1)$ roots ω, mod p^2 .

The remaining contribution to $\Sigma^{(2)}$ is from p, ω with

$$p > X, \quad u_1 \neq 0, \quad u_2 \neq 0$$

and where (4.7) is false. The minimality of $|a|$ and X sufficiently large imply that $|u_i| \leq p^2 \leq (XY\eta)^2 \ll X^5$. The terms with $X/|u_1| < Y/|u_2|$ contribute

$$\ll \sideset{}{'}\sum_{u_1, u_2} \sum_{\substack{p > X \\ p^2 | u_1 - \omega u_2}} \sum_{\omega, \, \text{mod} \, p^2} \frac{X}{|u_1|},$$

where $\sum'$ denotes a sum over all u_1 and u_2 satisfying $0 < |u_i| \ll X^5$ and $\sqrt{\log X}/\sqrt{\eta} < X/|u_1| < Y/|u_2|$. The condition $p^2 | u_1 - \omega u_2$ is (4.5), which implies $p^2 | g(u_1, u_2)$, as in (4.6). Here $g(u_1, u_2) \neq 0$ because the form g has no linear factors over $\mathbb{Z}$. Since $|u_i| \ll X^5$, this divisibility condition occurs for only finitely many primes exceeding X. Since the value $u_2 = 0$ does not occur the number of values of u_2 for given u_1 is at most $2Y|u_1|/X$. Thus this contribution to $\Sigma^{(2)}$ is at most

$$\ll \sum_{|u_1| < \frac{X\sqrt{\eta}}{\sqrt{\log X}}} \frac{X}{|u_1|} \frac{Y}{X} |u_1| \ll \frac{XY\sqrt{\eta}}{\sqrt{\log X}}.$$

The contribution from terms with $Y/|u_2| \leq X/|u_1|$ is estimated similarly. This completes the proof of part (a) of Lemma 4.

We argue along the same general lines for part (b). Deal first with

$$\sum_{P < r \leq 2P} \sum_{\omega, \, \text{mod} \, r} \frac{1}{\sqrt{r}} \min \left\{ \sqrt{X/|u_1|}, \sqrt{Y/|u_2|} \right\}, \tag{4.8}$$

where if u_1 or u_2 equals 0 then the minimum is the expression involving the non-zero u_i. We will take

$$P = \frac{R}{2^\alpha} : \quad 1 \leq \alpha \ll \log R. \tag{4.9}$$

First consider those pairs r, ω for which

$$\min\left\{\frac{X}{|u_1|}, \frac{Y}{|u_2|}\right\} \leq \frac{1}{\psi}, \tag{4.10}$$

where ψ will be specified below. The contribution to (4.8) from these pairs is

$$\sum_{P < r \leq 2P} \sum_{\omega, \bmod r} \frac{1}{\sqrt{r\psi}} \ll \frac{P^{\frac{1}{2}+\epsilon}}{\sqrt{\psi}}, \tag{4.11}$$

since there are $\ll r^\epsilon$ values of ω for each r.

There remains the contribution to (4.8) from those pairs r, ω for which (4.10) is false. It is not necessary to attempt to deal with the terms where $u_1 u_2 = 0$ with the same care as in part (a). The terms where

$$X/|u_1| < Y/|u_2|$$

(possibly with $u_2 = 0$) contribute

$$\sum_{\frac{1}{\psi} < \frac{X}{|u_1|} < \frac{Y}{|u_2|}} \sum_{P \leq r \leq 2P} \sum_{\substack{\omega, \bmod r \\ r | u_1 - \omega u_2}} \frac{1}{\sqrt{r}} \sqrt{\frac{X}{u_1}}. \tag{4.12}$$

As in part (a) the divisibility condition implies $r \mid g(u_1, u_2)$, where again $g(u_1, u_2) \neq 0$, whence the number of such r is

$$\ll g(u_1, u_2)^\epsilon \ll (X\psi)^\epsilon \ll P^\epsilon,$$

when $\psi \leq P/X$ as specified below. The number of $\omega, \bmod r$ is also $\ll P^\epsilon$, so the contribution (4.12) is at most

$$\sum_{|u_1| < X\psi} \sum_{|u_2| < Y|u_1|/X} P^\epsilon \sqrt{X/|u_1|P} \ll \sum_{|u_1| < X\psi} P^\epsilon \sqrt{X/|u_1|P} \left\{1 + \frac{Y|u_1|}{X}\right\}$$

$$\ll \frac{1}{P^{\frac{1}{2}-\epsilon}} \left\{X\sqrt{\psi} + XY\psi^{\frac{3}{2}}\right\}, \tag{4.13}$$

and since $Y < X$ the similar contribution from those terms for which $X/|u_1| > Y/|u_2|$ is not larger.

This contribution (4.13) has to be added to the entry (4.11). Of these, (4.13) will not be larger than (4.11) if we choose

$$\psi = \min\{P/X, \sqrt{P/XY}\},$$

in which case (4.11), and therefore (4.8), is $\ll P^\epsilon(\sqrt{X} + (XYP)^{1/4})$.

On summing over the values of P indicated in (4.9) we obtain the estimate for $\Sigma^{(3)}(R, X, Y)$ stated in Lemma 4.

5. The sifting argument

We complete the proof of Theorem 2 by establishing Lemmas 1 and 2, as enunciated earlier.

In proving Lemma 1 we may suppose that X is as large as desired, since otherwise the Lemma is trivial.

In the notations of Lemmas 3 and 4 we have from (3.9)

$$E_i^{(1)}(X,Y) \leq \sum_{\substack{\xi < p \leq XY\eta}} \sum_{\substack{1 \leq \omega \leq p^2 \\ f_i(\omega,1) \equiv 0, \bmod p^2}} N_\omega(X,Y,p^2). \qquad (5.1)$$

When $p \ll X$ we do not use Lemma 3, but make a separate estimate for each y. The contribution of these p to $E_i^{(1)}(X,Y)$, as defined in (3.9), is

$$\ll \sum_{\substack{\xi < p \ll X \\ p \nmid y}} \sum_{Y < y \leq 2Y} \sum_{\substack{1 \leq \psi \leq p^2 \\ f_i(\psi,y) \equiv 0, \bmod p^2}} \sum_{\substack{X < x \leq 2X \\ x \equiv \psi, \bmod p^2}} 1$$

$$\ll \sum_{\substack{\xi < p \ll X}} \sum_{Y < y \leq 2Y} \left\{ \frac{X}{p^2} + O(1) \right\} \ll \frac{XY}{\xi},$$

since there are at most $O(1)$ values of ψ for each p when $p \nmid y$.

In particular, if the irreducible form f_i is of degree 1 or 2 then all primes p in (3.9) for which $E_i^{(1)}(X,Y) \neq 0$ satisfy $p \ll X$, so that the proof of Lemma 1 is already complete. Accordingly to prove Lemma 1 we may suppose in particular that f_i has no linear factor over $\mathbb{Z}$, so that Lemma 4 may be applied.

Lemma 3 gives

$$N_\omega(X,Y,p^2) \leq \frac{XY}{p^2} + O\left(\frac{1}{|\mathbf{a}|}\right).$$

The contribution to (5.1) from the summand XY/p^2 is

$$\sum_{\xi < p \leq XY\eta} \sum_{\omega, \bmod p} \frac{XY}{p^2} \ll \frac{XY}{\xi}.$$

The contribution from the other summand $1/|\mathbf{a}|$ is estimated in part (a) of Lemma 4. This completes the proof of Lemma 1.

Proceed to the proof of Lemma 2, in which we may suppose $K \leq X^\mu$, where μ is the degree of f. A certain amount of removal of common factors is necessary, as in [8]. Thus we first set

$$(x,y) = d, \quad x = dx_1, \quad y = dy_1, \quad X = dX_1, \quad Y = dY_1. \qquad (5.2)$$

Then $(d, p) = 1$, so we obtain $d^\mu \mid k$ and

$$k = d^\mu k_1, \quad f(x_1, y_1) = p^2 k_1, \quad (x_1, y_1) = 1,$$
$$X_1 < x_1 \leq 2X_1, \quad Y_1 < y_1 \leq 2Y_1, \quad |k_1| \leq K/d^\mu.$$

Second, set $\delta = (y_1, k_1)$, so that

$$\delta \mid c_0 \neq 0, \tag{5.3}$$

where c_0 is the coefficient of x^μ in f; $c_0 \neq 0$ because f has no linear factors over $\mathbb{Z}$. Write

$$y_1 = \delta y_2, \quad x_1 = x_2, \quad k_1 = \delta r, \quad X_2 = X_1, \quad Y_2 = Y_1/\delta. \tag{5.4}$$

Then

$$f(x_1, y_1) = \delta g(x_2, y_2),$$

where $g(x, y)$ is again a binary form having no linear factors over $\mathbb{Z}$. Thus, in Lemma 2, $S(X, Y, K)$ does not exceed the number of solutions of

$$\left. \begin{array}{l} g(x_2, y_2) = p^2 r, \quad (y_2, x_2 r) = 1, \quad \delta \mid c_0, \\ X_2 < x_2 \leq 2X_2, \quad Y_2 < y_2 \leq 2Y_2, \quad r \leq R, \end{array} \right\} \tag{5.5}$$

where

$$R = \frac{K}{d^\mu \delta} \ll X^\mu. \tag{5.6}$$

Since $(y_2, r) = 1$ we can define ω, $\mod r$ by

$$x_2 \equiv \omega y_2, \mod r. \tag{5.7}$$

Then $g(\omega, 1) \equiv 0$, $\mod r$. We deal with this using the lattice structure described in §4, where X, Y have to be replaced by X_2, Y_2. Thus we write $\mathbf{z} = l\mathbf{a} + m\mathbf{b}$, where L, M satisfy (4.4), with

$$\Delta = \frac{r}{X_2 Y_2}. \tag{5.8}$$

Write $\mathbf{a} = \langle u_1/X_2, u_2/Y_2 \rangle$, $\mathbf{b} = \langle v_1/X_2, v_2/Y_2 \rangle$, the notation for $\mathbf{a}$ being as in Lemma 4. In this situation we obtain

$$p^2 r = g(x_2, y_2) = g(lu_1 + mv_1, lu_2 + mv_2) = G(l, m),$$

say, divisible by r for all $\langle l, m \rangle$.

Now let q denote a "sifting" prime, which will satisfy $q \leq Q$, $q \nmid r$, where Q is to be specified. The sifting condition in the ensuing argument is that $G(l, m)/r$ is either a quadratic residue ϖ or is 0, mod q. Denote by $\psi(q)$ the number of pairs $\langle l, m \rangle$, mod q that satisfy this condition. Thus $\psi(q) \leq q^2$.

We can estimate $\psi(q)$ as on p. 56 of [8]. Set $l \equiv \beta m$, mod q when $(m, q) = 1$. Then the congruence $\varpi r \equiv m^\mu G(\beta, 1)$, mod q has $\frac{1}{2}q + O(\sqrt{q})$ solutions for $\langle \beta, \varpi \rangle$, mod q (in fact, half as many as there are solutions $\langle \beta, \gamma \rangle$ of $m^\mu G(\beta, 1) \equiv r\gamma^2$, mod q). In this way we obtain $\psi(q) = \frac{1}{2}q^2 + O(q^{3/2})$.

Define $\psi^*(q) = q^2 - \psi(q)$, so that $\psi^*(q) \geq 0$, and make ψ, ψ^* multiplicative. We need an estimate for

$$\Sigma_Q = \sum_{1 \leq n \leq Q, (n,r)=1} |\mu(n)| \frac{\psi(n)}{\psi^*(n)},$$

with the trivial interpretation $\Sigma_Q = +\infty$ if some $\psi^*(n)$ were 0, in which case the quantity to be estimated in Lemma 5 below is also 0. Estimating Σ_Q can be accomplished, with sufficient accuracy for our purposes, much more simply than in [8]. For our convenience define $\psi(q) = \frac{1}{2}q^2$ when $q \mid r$. Then

$$\Sigma_Q \prod_{p \mid r} \left\{ 1 + \frac{\psi(p)}{\psi^*(p)} \right\} \geq \sum_{n \leq Q} |\mu(n)| \frac{\psi(n)}{\psi^*(n)} \geq \sum_{p \leq Q} \frac{\psi(p)}{\psi^*(p)} \gg \frac{Q}{\log Q},$$

so that

$$\Sigma_Q \gg \frac{Q}{(Qr)^\epsilon}.$$

The sum Σ_Q occurs both in Selberg's sieve and in the Large Sieve method, either of which we might employ at this point. We will use the following two-dimensional version of the arithmetic formulation of the Large Sieve inequality. Lemma 5 was established by Gallagher [6] (whose account is restricted to the case where the numbers N_i are all equal).

Lemma 5. *Suppose $c(\mathbf{n}) = 0$ if $\mathbf{n} = \langle l, m \rangle$ is in any one of a certain set of $\psi(q)$ "forbidden" residue classes, mod q, for each prime $q \leq Q$. Let $\mathcal{B}$ denote the box given by the inequalities*

$$M_1 \leq l \leq M_1 + N_1, \quad M_2 \leq m \leq M_2 + N_2.$$

Then

$$\left| \sum_{\mathbf{n} \in \mathcal{B}} c(\mathbf{n}) \right|^2 \ll \frac{\Pi}{\Sigma_Q} \sum_{\mathbf{n} \in \mathcal{B}} |c(\mathbf{n})|^2$$

where $\Pi = (N_1 + Q^2)(N_2 + Q^2)$ and Σ_Q is as above.

Now let $\Sigma(d,\delta,r,\omega)$ denote the number of solutions of (5.5), for given d, δ, r, X_2, Y_2, ω satisfying (5.2), (5.3), (5.4), (5.7), so that in Lemma 2

$$S(X,Y,K) = \sum_{1\le d\le X, \delta|c_0} \sum_{1\le r\le R} \sum_{g(\omega,1)\equiv 0,\ \text{mod } r} \Sigma(d,\delta,r,\omega),$$

R being as in (5.6). Then Lemma 5 gives the estimate

$$\Sigma(d,\delta,r,\omega) \ll \frac{(L+Q^2)(M+Q^2)(rQ)^\epsilon}{Q}.$$

Here $M = |\mathbf{a}|/\Delta \le L$, $LM \ll 1/\Delta = X_2 Y_2/r$, as in (4.4) and (5.8). So we choose $Q = \sqrt{M}$, and we obtain

$$\Sigma(d,\delta,r,\omega) \ll \frac{LMX^\epsilon}{\sqrt{M}} \ll \frac{X^\epsilon}{\Delta\sqrt{M}} \ll \frac{X^\epsilon}{\sqrt{\Delta|\mathbf{a}|}} = \frac{X^\epsilon\sqrt{X_2 Y_2}}{\sqrt{r|\mathbf{a}|}}.$$

We can apply part (b) of Lemma 4, in which we replace X, Y by X_2, Y_2. Using (5.6) we obtain

$$\sum_{1\le r\le R} \sum_{g(\omega,1)\equiv 0,\ \text{mod } r} \Sigma(d,\delta,r,\omega) \ll X^\epsilon\left(X_2\sqrt{Y_2} + (X_2 Y_2)^{\frac{3}{4}} R^{\frac{1}{4}}\right),$$

where we may replace ϵ by $\epsilon/5$. Since (5.4), (5.2), (5.3) give $X_2 = X/d$, $Y_2 = Y/d\delta$, $\delta|c_0$, where $c_0 \ne 0$ is a constant, summing this estimate over d, δ gives the estimate stated in Lemma 2.

This completes the proof of Theorem 2.

6. The proofs of Theorem 3 and Theorem 4

We begin with Theorem 3. We make use of our approach to establishing (2.2), where we took $a = y^v$, $b = x^v - y^v$. We will simplify matters here by considering $y = 1$.

We suppose, as we may, that $n > 1$. We consider $F(x) = x(x^{n^2} - 1)$. Let X be sufficiently large. Fix N so that if $p > N$ then p does not divide the discriminant of $F(x)$, and so that

$$\sum_{p>N} \frac{n^2 + 1}{p^2} < \tfrac{1}{2}.$$

Let

$$P = \prod_{p\le N} p. \tag{6.1}$$

Set $G(s) = F(P^2 s - P)$. Observe that $G(s)$ has no roots modulo p^2 if $p \leq N$. Also, if $p > N$, then $G(s)$ has at most $n^2 + 1$ roots modulo p^2. Therefore, the number of integers $s \in (X/P^2, 2X/P^2]$ for which $G(s)$ is divisible by p^2 for some prime $p \leq 2X$ is bounded above by

$$\sum_{N < p \leq 2X} (n^2 + 1) \left(\frac{X/P^2}{p^2} + 1 \right) \leq \left(\sum_{p > N} \frac{n^2 + 1}{p^2} \right) \frac{X}{P^2} + (n^2 + 1)\pi(2X) \leq \frac{3X}{4P^2}.$$

By letting X vary, we deduce that there are infinitely many integers t having the property that if p is a prime $\leq t$ then $p^2 \nmid F(t)$. We show now that if t is a sufficiently large integer with this property, then either $\Phi_n(t)$ or $\Phi_{n^2}(t)$ must be squarefree, so that Theorem 3 will then follow since $\Phi_{n^2}(x) = \Phi_n(x^n)$. Suppose that neither are squarefree. Let $a = 1$, $b = t^{n^2} - 1$. Note that there is a $g(x) \in \mathbb{Z}[x]$ such that $F(x) = \Phi_n(x)\Phi_{n^2}(x)g(x)$. By our assumption, we deduce that there are primes p_1, p_2, not necessarily distinct, satisfying $p_1^2 p_2^2 \mid F(t)$. Since $p^2 \nmid F(t)$ for every prime $p \leq t$, we deduce that p_1 and p_2 are $> t$. Hence

$$Q(ab(a + b)) = Q(F(t)) < F(t)/t^2 < t^{n^2 - 1},$$

and we obtain

$$L_{a,b} > \frac{n^2}{n^2 - 1}.$$

By our assumption that the abc-conjecture holds, there can be only finitely many such pairs (a, b). This implies as desired that if t is sufficiently large then either $\Phi_n(t)$ or $\Phi_{n^2}(t)$ is squarefree. This completes the proof of Theorem 3.

As was remarked in §1, a similar approach leads to the slightly stronger result that the polynomial

$$f_n(x) = \frac{x^n - 1}{x - 1}$$

is infinitely often squarefree. To show this one would observe

$$F(x) = x(x - 1)f_n(x)f_n(x^n),$$

where $F(x)$ is as before, and argue as above to show that either $f_n(t)$ or $f_n(t^n)$ is squarefree when t is large.

We follow a similar approach in proving Theorem 4, but do not take $y = 1$. Set

$$F(x, y) = xy(x^{2n} - y^{2n}) \quad \text{for} \quad n = 1, 2, \ldots.$$

With N sufficiently large and P as in (6.1) take $G(s,t) = F(x,y)$, where

$$x = P^2 s - P, \quad y = Pt - 1. \tag{6.2}$$

If $p^2 \mid G(s,t)$ then $p > N$, because $G(s,t) \equiv -P$, mod p^2 when $p \leq N$.

Let $C(X,Y,p^2)$ denote the number of $\langle x, y \rangle$ such that (6.2) holds and

$$F(x,y) \equiv 0, \text{ mod } p^2, \quad X < x \leq 2X, \quad Y < y \leq 2Y, \tag{6.3}$$

where we will take $X = Y^\alpha$ with $\alpha > 1$, as elsewhere in this paper. We wish to estimate

$$\sum_{N < p \leq 2X} C(X,Y,p^2) \tag{6.4}$$

from above. It will not be necessary to use the arguments from §4.

For the terms with $p \nmid y$ observe that t takes at most $Y/P + 1$ values and for each of these the integer s lies in at most $2n + 1$ residue classes mod p^2, in each of which there are at most $X/(Pp)^2 + 1$ values of s for which $X < x \leq 2X$. Thus the contribution to (6.4) from such terms does not exceed

$$\sum_{N < p \leq 2X} (2n + 1) \left(\frac{Y}{P} + 1 \right) \left(\frac{X/P^2}{p^2} + 1 \right).$$

When $p \| y$ the congruence in (6.3) gives $p \mid x$, so the contribution from these terms does not exceed

$$\sum_{N < p \leq 2X} \left(\frac{Y/P}{p} + 1 \right) \left(\frac{X/P^2}{p} + 1 \right).$$

When $p^2 \mid y$ (6.3) gives $p \leq \sqrt{2Y}$, so the contribution from these terms does not exceed

$$\sum_{N < p \leq \sqrt{2Y}} \left(\frac{Y/P}{p^2} + 1 \right) \left(\frac{X}{P^2} + 1 \right).$$

We will choose N (and hence P) sufficiently large, and then take X, Y sufficiently large (in terms of N). Since $X^\epsilon < Y < X$ all the contributions to (6.4) are

$$\ll \frac{XY}{NP^3} + \frac{Y}{P} \pi(2X) + \left(\frac{X}{P^2} + \frac{Y}{P} \right) \log\log X + \frac{X}{P^2} \pi(\sqrt{2Y}),$$

and we obtain

$$\sum_{N < p \leq 2X} C(X,Y,p^2) < \frac{3XY}{4P^3}$$

when N, X, Y are large enough.

But the total number of pairs $\langle x, y \rangle$ such that (6.2) holds and

$$X < x \le 2X, \quad Y < y \le 2Y$$

is asymptotic to XY/P^3. Hence there exists such a pair $\langle x, y \rangle$ with the property that $p^2 \nmid F(x, y)$ for every $p \le 2X$. Now by letting X vary we obtain infinitely many pairs $\langle x, y \rangle$ for which there exists an X such that

$$X < x \le 2X, \quad X^\alpha < y \le 2X^\alpha, \quad p^2 \mid F(x, y) \Rightarrow p > 2X, \qquad (6.5)$$

so that in particular x and y are squarefree.

From this we can deduce, using the abc-conjecture, that for infinitely many of these pairs $\langle x, y \rangle$ at least one of the numbers

$$xy(x^n + y^n), \quad xy(x^n - y^n)$$

is squarefree. For if

$$p_1^2 \mid xy(x^n + y^n), \quad p_2^2 \mid xy(x^n - y^n)$$

we would obtain $p_1 > 2X$, $p_2 > 2X$, so that actually $p_1 \nmid xy$, $p_2 \nmid xy$. Take $a = y^{2n}$, $b = x^{2n} - y^{2n}$. Then

$$Q(ab(a + b)) = Q(xy(x^{2n} - y^{2n})) \le \frac{xy(x^{2n} - y^{2n})}{p_1 p_2} < 2^{2n} Y^{(2n-1)\alpha + 1}.$$

Consequently

$$L_{a,b} > \frac{\log x^{2n}}{\log(2^{2n} Y^{(2n-1)\alpha+1})} > \frac{2n\alpha \log Y}{2n \log 2 + ((2n - 1)\alpha + 1) \log Y}$$

$$\to \frac{2n\alpha}{(2n - 1)\alpha + 1} > 1 \quad \text{as} \quad Y \to \infty,$$

since $\alpha > 1$. This contradicts the abc-conjecture.

We can now complete the proof of Theorem 4. If $xy(x^n - y^n)$ is squarefree for infinitely many pairs $\langle x, y \rangle$ appearing in (6.5), then put

$$a = y^n, \quad b = x^n - y^n.$$

Then

$$Q(ab(a + b)) = xy(x^n - y^n).$$

Otherwise $xy(x^n + y^n)$ is squarefree infinitely often, and we take $a = y^n$, $b = x^n$, so that $Q(ab(a + b)) = xy(x^n + y^n)$. In either case, we obtain

$$L_{a,b} = \frac{n\alpha}{(n + 1)\alpha + 1} + O_n\left(\frac{1}{\log Y}\right).$$

With α fixed, let $Y \to \infty$. Then let α vary over $(1, \infty)$. Since the set of limit points is closed, we get that all points of the interval $[n/(n+2), n/(n+1)]$ are limit points, and Theorem 4 now follows using

$$\bigcup_{n=1}^{\infty}\left[\frac{n}{n + 2}, \frac{n}{n + 1}\right] = [\tfrac{1}{3}, 1),$$

together with the observation of the introduction that the *abc*-conjecture implies that there are no limit points of $\mathcal{L}$ outside $[\tfrac{1}{3}, 1]$.

References

1 G.A. Baker and P. Graves-Morris: *Padé Approximants, Part I: Basic Theory* (Encyclopedia of Mathematics, Volume 13). Addison-Wesley (1981).

2 J. Browkin and J. Brzeziński: Some remarks on the *abc*-conjecture. *Math. Comp.* **62** (1994), 931–939.

3 J.W.S. Cassels: *An Introduction to the Geometry of Numbers.* Springer (1959).

4 P. Erdős: Arithmetical properties of polynomials. *J. London Math. Soc.* **28** (1953), 416–425.

5 M. Filaseta: Short interval results for *k*-free values of irreducible polynomials. *Acta Arith.* **64** (1993), 249–270.

6 P.X. Gallagher: The large sieve and probabilistic Galois theory. In *Proc. Sympos. Pure Math.* **24** , 91–101. American Math. Soc. (1973).

7 F. Gouvêa and B. Mazur: The square-free sieve and the rank of elliptic curves. *J. Amer. Math. Soc.* **4** (1991), 1–23.

8 G. Greaves: Power-free values of binary forms. *Quart. J. Math. Oxford (2)* **43** (1992), 45–65.

9 C. Hooley: On the power free values of polynomials. *Mathematika* **14** (1967), 21–26.

10 M.N. Huxley and M. Nair: Power free values of polynomials III. *Proc. London Math. Soc.* **41** (1980), 66–82.

11 S. Lang: Old and new conjectured diophantine inequalities. *Bull. Amer. Math. Soc.* **23** (1990), 37–75.

12 T. Nagell: Zur Arithmetik der Polynome. *Abh. Math. Sem. Hamburg Univ.* **1** (1922), 179–184.

13 C.L. Stewart and R. Tijdeman: On the Oesterlé-Masser conjecture. *Monatsh. Math.* **102** (1986), 251–257.

14 B.M.M. de Weger: Algorithms for Diophantine Equations. *C.W.I. Tract* **65**, 126. Amsterdam (1989).

4. The Values of Binary Linear Forms at Prime Arguments

J. Brüdern, R. J. Cook and A. Perelli

1. Introduction

We shall be concerned with the values of the linear form $\lambda_1 p_1 + \lambda_2 p_2$ where λ_1, λ_2 are non-zero real numbers and p_1, p_2 are prime variables. If λ_1/λ_2 is rational one can write $\lambda_1 p_1 + \lambda_2 p_2 = \lambda(ap_1 + bp_2)$ with some real λ and coprime integers a, b. It remains to determine the values of $ap_1 + bp_2$. There is an obviously necessary congruence condition modulo $2|ab|$, and a sign condition if $ab > 0$, and it is expected that if these conditions are met for an integer n then $n = ap_1 + bp_2$ is indeed soluble, perhaps with the exception of some small values of n. A standard application of the Hardy-Littlewood method shows the conjecture to be true for at least almost all n, in the sense usually adopted in analytic number theory. We illustrate this in the most classical but typical situation $a = b = 1$. Goldbach's conjecture asserts that the values taken by $p_1 + p_2$ are precisely all even numbers $n \geq 4$, and the numbers of the form $p + 2$ with prime p. In a seminal paper, Montgomery and Vaughan [10] have shown that

$$E(X) = \#\{n \leq X : 2n \neq p_1 + p_2\}$$

satisfies

$$E(X) \ll X^{1-\delta} \tag{1}$$

for some $\delta > 0$. The result may be extended to the case when integer coefficients are present without any major new difficulty. Attempts have been made to give an explicit value for δ, the latest published result of Chen [2] to the effect that $\delta = \frac{1}{25}$ is permissible has, however, been criticised. If the Riemann Hypothesis is true for all Dirichlet L-functions (referred to as GRH hereafter) then (1) holds for any $\delta < \frac{1}{2}$. This is a classical result due to Hardy and Littlewood [6].

The case when λ_1/λ_2 is irrational has received less attention so far. Qualitatively, one may expect that for $\lambda_1/\lambda_2 < 0$ the values of $\lambda_1 p_1 + \lambda_2 p_2$ are dense on the real line, and that for $\lambda_1 > 0$, $\lambda_2 > 0$ the gaps between these values tend to zero near infinity. This can be put in quantitative form. In the case of two positive coefficients an optimistic conjecture would be that $|\lambda_1 p_1 + \lambda_2 p_2 - v| \ll v^{\epsilon-1}$ is always soluble. In the case $\lambda_1 \lambda_2 < 0$

Sieve Methods, Exponential Sums, and their Applications in Number Theory
Greaves, G.R.H., Harman, G., Huxley, M.N., Eds. ©Cambridge University Press, 1996

small values of $\lambda_1 p_1 + \lambda_2 p_2$ are of particular interest, and in quantitative form this can be phrased as a problem in Diophantine approximation. As an example, we take $\lambda_1 = \lambda > 0$, $\lambda_2 = -1$. A difficult problem is to determine those exponents θ for which the inequality

$$\left| \lambda - \frac{p_2}{p_1} \right| \leq \frac{1}{p_1^{\theta}}$$

has infinitely many solutions in pairs of primes (p_1, p_2). Any $\theta < 2$ should be permissible here, and Harman [8] proved slightly more than this for almost all λ, in the sense of Lebesgue measure.

Our goal is a result comparable to the exceptional set estimate (1) of Montgomery and Vaughan, in the irrational case. It is no longer plausible to count exceptional integers, and we therefore average over suitably spaced sets of real numbers to mimic the role played by even integers in the work of Montgomery and Vaughan. Let $C > c > 0$ be real numbers. A set $\mathcal{V} \subset [0, \infty)$ is called (C, c)-spaced if every closed interval in $[0, \infty)$ of length C contains at least one element of $\mathcal{V}$, and if for any distinct $v_1, v_2 \in \mathcal{V}$ one has $|v_1 - v_2| > c$. The natural numbers are $(1, \frac{1}{2})$-spaced.

Theorem 1. *Let λ_1, λ_2 be positive real numbers. Suppose that λ_1/λ_2 is irrational and algebraic. Let $\mathcal{V}$ be (C, c)-spaced, for suitable C, c. Let $\delta > 0$. Then the number of $v \in \mathcal{V}$ with $v \leq X$ for which*

$$\left| \lambda_1 p_1 + \lambda_2 p_2 - v \right| < v^{-\delta}$$

has no solution in primes p_1, p_2 does not exceed $O\left(X^{\frac{2}{3} + 2\delta + \epsilon} \right)$, for any $\epsilon > 0$.

Any (C, c)-spaced set satisfies

$$X \ll \#\{v \in \mathcal{V} : v \leq X\} \ll X$$

whence the bound in the theorem is non-trivial for $\delta < \frac{1}{6}$. For small values of δ our exceptional set estimate is much stronger than (1). One reason for this is that our method depends implicitly on the distribution of the zeros of the Riemann zeta function, whereas in the more arithmetical situation of Goldbach's problem zeros of L-functions have to be considered with some uniformity with respect to the modulus.

Most of the conditions in the theorem can be relaxed. We write

$$\lambda = \lambda_1/\lambda_2$$

and denote the infinite sequence of the denominators of the convergents of λ by q_m, arranged in increasing order. We only require that

$$q_{m+1} \ll q_m^{1+\epsilon}. \tag{2}$$

For algebraic λ this holds by Roth's theorem, but is also true for almost all λ in the sense of Lebesgue measure. Even if (2) were known only with some finite value of ϵ it would be possible to obtain a weaker but still non-trivial version of the theorem. Unfortunately this does not cover all irrational values of λ. Perhaps it is possible to extend the results to all irrational λ by a method allied to that of Pitman [11].

Similarly, the spacing condition on $\mathcal{V}$ can be considerably weakened in various ways. At the end of the paper, we shall indeed formulate a version of the theorem which is more precise in this respect.

Subject to GRH the exponent $\frac{2}{3}$ in Theorem 1 can be replaced by $\frac{1}{2}$. This corresponds to the result of Hardy and Littlewood on Goldbach's problem mentioned earlier. Again we postpone a discussion of the consequences of GRH to the end of the paper.

Before we embark on the proof of Theorem 1 we mention in passing a different type of approximation to Goldbach's conjecture. One might ask for representations of all large even n and weaken the condition that both variables be prime instead. Chen [1] has shown that indeed all large even natural numbers are the sum of a prime and a number with at most two prime factors. Results of this type are also known for the related problem of Diophantine approximation. Improving earlier work of Vaughan [13] in this direction, Harman [7] proved that for λ_1/λ_2 irrational and negative, and any real v, the inequality

$$\left|\lambda_1 p + \lambda_2 P_3 - v\right| < p^{-1/300}$$

has infinitely many solutions in primes p and numbers P_3 with at most three prime factors.

2. Preliminaries

In this section we collect various known results, sometimes modified to suit the needs of the present communication. We are mainly concerned with the exponential sum

$$S(\alpha) = S(\alpha, X) = \sum_{p \leq X} e(\alpha p) \log p.$$

We begin by examining this sum in a neighbourhood of the origin. The routine approach is to approximate $S(\alpha)$ with

$$I(\alpha) = I(\alpha, X) = \sum_{n \leq X} e(\alpha n),$$

and we proceed to estimate the error $S(\alpha) - I(\alpha)$ in mean square. Let

$$\vartheta(X) = \sum_{p \leq X} \log p$$

and

$$J(X,h) = \int_1^X \left|\vartheta(x+h) - \vartheta(x) - h\right|^2 dx. \tag{3}$$

Then we have

Lemma 1. *Let* $1/X \leq Y \leq \frac{1}{2}$. *Then*

$$\int_{-Y}^Y |S(\alpha) - I(\alpha)|^2 \, d\alpha \ll \frac{\log^2 X}{Y} + Y^2 X + Y^2 J(X, 1/Y).$$

Proof. By Gallagher's Lemma ([4], Lemma 1) we have

$$\int_{-Y}^Y |S(\alpha) - I(\alpha)|^2 \, d\alpha \ll Y^2 \int_{-\infty}^\infty \left| \sum_{\substack{|n-x| \leq 1/2Y \\ 1 \leq n \leq X}} (\kappa(n) - 1) \right|^2 dx,$$

where $\kappa(n) = \log n$ if n is prime and $\kappa(n) = 0$ otherwise. The integrand on the right hand side vanishes unless $-1/2Y \leq x \leq X + 1/2Y$. We split this interval into the regions

$$E_1 = \left[-\frac{1}{2Y}, \frac{1}{2Y}\right], \quad E_2 = \left[\frac{1}{2Y}, X - \frac{1}{2Y}\right], \quad E_3 = \left[X - \frac{1}{2Y}, X + \frac{1}{2Y}\right].$$

We then rewrite the previous bound as

$$\int_{-Y}^Y |S(\alpha) - I(\alpha)|^2 \, d\alpha \ll Y^2(\mathcal{J}_1 + \mathcal{J}_2 + \mathcal{J}_3), \tag{4}$$

where

$$\mathcal{J}_k = \int_{E_k} \left| \sum_{\substack{|n-x| \leq 1/2Y \\ 1 \leq n \leq X}} (\kappa(n) - 1) \right|^2 dx.$$

The trivial estimate gives

$$\mathcal{J}_1 + \mathcal{J}_3 \ll \frac{\log^2 X}{Y^3},$$

and we have

$$\mathcal{J}_2 = \int_{1/2Y}^{X-1/2Y} \left| \vartheta\left(x + \frac{1}{2Y}\right) - \vartheta\left(x - \frac{1}{2Y}\right) - \frac{1}{Y} + O(1) \right|^2 dx$$

$$\leq J(X, 1/Y) + O(X).$$

The lemma now follows from (4).

The integral $J(X, h)$ can be bounded in terms of zero densities of the Riemann zeta function. Based on Huxley's density estimate [9] the work of section 6 in Saffari and Vaughan [12] shows that

$$J(X, h) \ll X h^2 / \log^A X \tag{5}$$

for any $A > 0$, uniformly in $h \geq X^{1/6} \log^B X$ for some $B = B(A)$. We take $A = 2$ and then conclude as follows.

Lemma 2. *Let* $\frac{1}{6} < \theta < 1$ *and* $Y = X^{-\theta}$. *Then*

$$\int_{-Y}^{Y} |S(\alpha) - I(\alpha)|^2 \ll \frac{X}{\log^2 X}.$$

With later applications in mind we also record here the simple bounds

$$\int_{-\frac{1}{2}}^{\frac{1}{2}} |I(\alpha)|^2 \, d\alpha \ll X, \qquad \int_{-\frac{1}{2}}^{\frac{1}{2}} |S(\alpha)|^2 \, d\alpha \ll X \log X. \tag{6}$$

Lemma 3. *Let* $1 \leq A \leq X^{1/5}/\log^5 X$ *and suppose that* $|S(\alpha)| \geq X/A$. *Then there are coprime integers* a, q *with*

$$1 \leq q \ll A^2 \log^8 X, \qquad |q\alpha - a| \ll A^2 X^{-1} \log^8 X.$$

Proof. This follows from Theorem 3.1 of Vaughan [14] and Dirichlet's theorem on Diophantine approximation, via a standard argument.

Our final lemma bounds the measure of the intersection of certain "major arcs". The technique is due to Watson [15] and has been used since in related situations by various writers.

Lemma 4. *For* $1 \leq Q \leq \frac{1}{2}\sqrt{X}$ *let* $\mathcal{E}(Q)$ *denote the union of all intervals* $|q\alpha - a| \leq Q/X$ *with* $1 \leq q \leq Q$ *and* $(a, q) = 1$. *Let* λ_1, λ_2 *be non-zero real numbers such that* $\lambda = \lambda_1/\lambda_2$ *is irrational. Let* b/r *be a convergent to the continued fraction expansion of* λ. *Let* $1 \leq Q_i \leq \frac{1}{2}\sqrt{X}$ *and suppose that* $y > 0$ *satisfies* $|\lambda_i|y \geq 2Q_i/X$ *for* $i = 1, 2$. *Then the measure of the set*

$$\mathcal{R}_y(Q_1, Q_2) = \left\{ \alpha \, : \, y < |\alpha| \leq 2y, \quad \lambda_i \alpha \in \mathcal{E}(Q_i) \text{ for } i = 1, 2 \right\}$$

does not exceed $O\left(yX^{-1}Q_1Q_2 r^{-1}\right)$ *providing* r *satisfies the inequality*

$$r \leq \frac{\epsilon_0 X}{Q_1 Q_2},$$

in which ϵ_0 *denotes a certain constant depending only on* λ_1, λ_2.

Proof. Let $\alpha \in \mathfrak{K}_y(Q_1, Q_2)$. Then there are coprime numbers a_i, q_i with $1 \leq q_i \leq Q_i$ and

$$\left| \lambda_i \alpha - \frac{a_i}{q_i} \right| \leq \frac{Q_i}{q_i X} \tag{7}$$

for $i = 1, 2$. Note that $a_i = a_i(\alpha)$, $q_i = q_i(\alpha)$ are uniquely determined by α. In the opposite direction, for a given quadruple a_1, q_1, a_2, q_2 the inequalities (7) define an interval $I(a_1, q_1, a_2, q_2)$ of length

$$\ll \min\left(\frac{Q_1}{q_1 X}, \frac{Q_2}{q_2 X} \right) \ll \frac{1}{X} \left(\frac{Q_1 Q_2}{q_1 q_2} \right)^{\frac{1}{2}}. \tag{8}$$

By (7) and the condition $|\lambda_i| y \geq 2Q_i / X$ we see that $y \ll |a_i / q_i| \ll y$, and in particular that $a_1 a_2 \neq 0$. Next, we observe that

$$a_2 q_1 \lambda - a_1 q_2 = \frac{a_2 / q_2}{\lambda_2 \alpha} q_1 q_2 \left(\lambda_1 \alpha - \frac{a_1}{q_1} \right) - \frac{a_1 / q_1}{\lambda_2 \alpha} q_1 q_2 \left(\lambda_2 \alpha - \frac{a_2}{q_2} \right).$$

It follows that for some constant K depending only on λ_1, λ_2 one has

$$\left| a_2 q_1 \lambda - a_1 q_2 \right| < \frac{K Q_1 Q_2}{X} < \frac{1}{4r} \tag{9}$$

(take $\epsilon_0 < 1/4K$). By Legendre's law of best approximation we deduce that $|a_2 q_1| \geq r$, and that

$$\left| a_2(\alpha) q_1(\alpha) - a_2(\alpha') q_1(\alpha') \right| \geq r$$

for any $\alpha, \alpha' \in \mathfrak{K}_y(Q_1, Q_2)$ when the products $a_2(\alpha) q_1(\alpha)$ and $a_2(\alpha') q_1(\alpha')$ are distinct. Let $s \in \mathbb{Z}$. We consider the consequences of the assumption that $(s - \frac{1}{2})r \leq a_2 q_1 < (s + \frac{1}{2})r$. We have seen that $s = 0$ is impossible, and for $s \neq 0$ the value of $a_2 q_1$ is unique (if it exists at all). For a given value of $a_2 q_1$ the value of $a_1 q_2$ is uniquely determined by (9). A divisor argument now shows that the number of distinct quadruples a_1, q_1, a_2, q_2 is $O(X^\epsilon)$. Since $q_2 \gg a_2 / y$ and $|a_2| q_1 \gg |s| r$ we have

$$q_1 q_2 \gg \frac{q_1 |a_2|}{y} \gg \frac{|s| r}{y}$$

for the quadruples under consideration. By (8) the total length of all intervals $I(a_1, q_1, a_2, q_2)$ with $a_2 q_1 \in \left[(s - \frac{1}{2})r, (s + \frac{1}{2})r \right)$ does not exceed

$$\ll X^{\epsilon - 1} \left(\frac{Q_1 Q_2 y}{|s| r} \right)^{\frac{1}{2}}.$$

Moreover, from $|a_2/q_2| \ll y$ we see that $|a_2|q_1 \ll q_1 q_2 y \ll Q_1 Q_2 y$, so that the maximal value of $|s|$ for which a pair a_2, q_1 can exist satisfies $|s| \leq K Q_1 Q_2 y / r$ with some constant K. We may suppose that

$$\frac{K Q_1 Q_2 y}{r} \geq 1,$$

for otherwise the set $\mathfrak{R}_y(Q_1, Q_2)$ is empty and there is nothing to show. We finally infer that

$$\int_{\mathfrak{R}_y(Q_1, Q_2)} d\alpha \ll X^{\epsilon-1} \sum_{1 \leq s \leq K Q_1 Q_2 y / r} \left(\frac{Q_1 Q_2 y}{sr} \right)^{\frac{1}{2}} \ll \frac{X^{\epsilon-1} Q_1 Q_2 y}{r}$$

as required.

3. The Fourier transform method initiated

In this section we shall approach the theorem via a Fourier transform method originating in the work of Davenport and Heilbronn [3]. We shall actually prove more than the theorem. Let $0 < \tau < 1$. For any $v \in \mathbb{R}$ it is our intention to estimate the sum

$$\mathcal{N}_v = \sum_{p_1, p_2 \leq X} \Upsilon(\lambda_1 p_1 + \lambda_2 p_2 - v) \log p_1 \log p_2 \qquad (10)$$

where

$$\Upsilon(\xi) = \Upsilon_\tau(\xi) = \max(0, \tau - |\xi|).$$

Hence $\mathcal{N}_v$ counts the solutions of $|\lambda_1 p_1 + \lambda_2 p_2 - v| \leq \tau$ with a non-negative weight. The key idea of Davenport and Heilbronn was to bring in the function

$$K(\alpha) = \left(\frac{\sin \pi \tau \alpha}{\pi \alpha} \right)^2.$$

Using the identity

$$\int_{-\infty}^{\infty} K(\alpha) e(\alpha \xi) \, d\alpha = \Upsilon(\xi) \qquad (11)$$

we have

$$\mathcal{N}_v = \int_{-\infty}^{\infty} S(\lambda_1 \alpha) S(\lambda_2 \alpha) e(-v\alpha) K(\alpha) \, d\alpha. \qquad (12)$$

The main contribution to this integral should arise from a neighbourhood of the origin. The next lemma makes this precise.

Lemma 5. *Let $\lambda_1 \geq \lambda_2 \geq 1$. Let $Y = X^{-1/3}$, and write*

$$\mathcal{I}S_Y(v) = \int_{-Y}^{Y} S(\lambda_1\alpha)S(\lambda_2\alpha)e(-v\alpha)K(\alpha)\,d\alpha.$$

Then, uniformly for all $0 < \tau < 1$ and all $v \in \left[\frac{1}{2}X, X\right]$ one has

$$\mathcal{I}S_Y(v) \gg \tau^2 X.$$

Proof. We begin with replacing $S(\lambda_1\alpha)$ with $I(\lambda_1\alpha)$. This yields

$$\mathcal{I}S_Y(v) = \int_{-Y}^{Y} I(\lambda_1\alpha)S(\lambda_2\alpha)e(-v\alpha)K(\alpha)\,d\alpha + O(E)$$

with

$$E = \int_{-Y}^{Y} \left|\left(S(\lambda_1\alpha) - I(\lambda_1\alpha)\right)S(\lambda_2\alpha)\right|K(\alpha)\,d\alpha.$$

By Schwarz's inequality, (6), Lemma 2 and the obvious bound

$$K(\alpha) \ll \min\left(\tau^2, |\alpha|^{-2}\right) \tag{13}$$

it is immediate that $E \ll \tau^2 X/\log X$. A similar argument applied to $S(\lambda_2\alpha)$ now reveals that

$$\mathcal{I}S_Y(v) = \int_{-Y}^{Y} I(\lambda_1\alpha)I(\lambda_2\alpha)e(-v\alpha)K(\alpha)\,d\alpha + O\left(\frac{\tau^2 X}{\log X}\right).$$

To avoid unneccessary complication in the sequel we replace the sums $I(\alpha)$ with the related integrals

$$\tilde{I}(\alpha) = \int_{1}^{X} e(\alpha\beta)\,d\beta.$$

Write

$$\mathcal{I}I_Y(v) = \int_{-Y}^{Y} I(\lambda_1\alpha)I(\lambda_2\alpha)e(-v\alpha)K(\alpha)\,d\alpha.$$

Euler's summation formula gives $I(\alpha) - \tilde{I}(\alpha) \ll 1 + X|\alpha|$, and repeating the procedure that was used to replace $S(\lambda_j\alpha)$ with $I(\lambda_j\alpha)$ yields

$$\mathcal{I}I_Y(v) = \int_{-Y}^{Y} \tilde{I}(\lambda_1\alpha)\tilde{I}(\lambda_2\alpha)e(-v\alpha)K(\alpha)\,d\alpha + O\left(\frac{\tau^2 X}{\log X}\right).$$

We now compare the integral on the right hand side with the complete integral

$$\mathcal{N}_v^* = \int_{-\infty}^{\infty} \tilde{I}(\lambda_1\alpha)\tilde{I}(\lambda_2\alpha)e(-v\alpha)K(\alpha)\,d\alpha.$$

By partial integration one has $\tilde{I}(\alpha) \ll 1/|\alpha|$, so that by (13) the resulting error is at most

$$\ll \tau^2 \int_Y^{\infty} \frac{d\alpha}{\alpha^2} \ll \frac{\tau^2}{Y}.$$

To evaluate $\mathcal{N}_v^*$ we use (11) to see that

$$\mathcal{N}_v^* = \int_1^X \int_1^X \Upsilon(\lambda_1\xi_1 + \lambda_2\xi_2 - v)\,d\xi_1\,d\xi_2.$$

Now $\Upsilon(\alpha) \geq \frac{1}{2}\tau$ for $|\alpha| \leq \frac{1}{2}\tau$. Let $3 \leq v \leq X$. For $2 \leq \xi_1 \leq v/\lambda_1$ the inequality $|\lambda_1\xi_1 + \lambda_2\xi_2 - v| \leq \frac{1}{2}\tau$ defines an interval for ξ_2 of length $\gg \tau$ contained in $[1, X]$. It follows that $\mathcal{N}_v^* \gg \tau^2 v$. The lemma is now immediate.

We wish to conclude from Lemma 5, (10) and (12) that the inequality $|\lambda_1 p_1 + \lambda_2 p_2 - v| \leq \tau$ is soluble by showing that the contribution to the integral (12) arising from $|\alpha| \geq Y$ is negligible. This is easy for rather large values of $|\alpha|$. We write

$$Y_0 = \frac{X \log X}{\tau^2} \tag{14}$$

and then deduce from (13) and the simple bound $|S(\alpha)| \ll X$ that

$$\int_{|\alpha| \geq Y_0} |S(\lambda_1\alpha)S(\lambda_2\alpha)|K(\alpha)\,d\alpha \ll X^2 \int_{Y_0}^{\infty} \frac{d\alpha}{\alpha^2} \ll \frac{X\tau^2}{\log X}, \tag{15}$$

which is acceptable.

4. An average estimate

By the results of the previous section it remains to discuss the contribution of the pair of intervals

$$\mathfrak{m} = \{\alpha : Y < |\alpha| \leq Y_0\}$$

to the integral (12); here Y_0 is still given by (14), and $Y = X^{-1/3}$ as in Lemma 5. Reasonable control is possible only on average over v.

Lemma 6. *Let $0 < \tau < 1$. Suppose that $\mathcal{V}$ is a set of real numbers such that $|v_1 - v_2| > 2\tau$ whenever $v_1, v_2 \in \mathcal{V}$ are distinct. Let λ_1, λ_2 be non-zero real numbers with λ_1/λ_2 algebraic and irrational. Then, in the notation introduced above,*

$$\sum_{v \in \mathcal{V}} \left| \int_{\mathfrak{m}} S(\lambda_1 \alpha) S(\lambda_2 \alpha) e(-v\alpha) K(\alpha) \, d\alpha \right|^2 \ll \tau^2 X^{\frac{8}{3}+\epsilon}.$$

Proof. We work in $L_2(\mathbb{R})$ and denote the standard inner product by

$$(f, g) = \int f \bar{g} \, d\alpha.$$

For $v \in \mathbb{R}$ define

$$k_v(\alpha) = \frac{\sin \pi \tau \alpha}{\pi \alpha} e(v\alpha).$$

Then $k_v \in L_2(\mathbb{R})$, and by (11) we have $(k_v, k_{v'}) = \Upsilon(v - v')$. Hence the set of functions $\tau^{-1/2} k_v$ with $v \in \mathcal{V}$ is an orthonormal family. Bessel's inequality yields

$$\sum_{v \in \mathcal{V}} |(\phi, k_v)|^2 \leq \tau(\phi, \phi)$$

for any function $\phi \in L_2(\mathbb{R})$. We take

$$\phi(\alpha) = \begin{cases} S(\lambda_1 \alpha) S(\lambda_2 \alpha) k_0(\alpha) & \text{if } \alpha \in \mathfrak{m} \\ 0 & \text{otherwise} \end{cases}$$

to deduce that the left hand side of the inequality proposed in the lemma does not exceed

$$\tau \int_{\mathfrak{m}} |S(\lambda_1 \alpha) S(\lambda_2 \alpha)|^2 K(\alpha) \, d\alpha. \tag{16}$$

We estimate this with the aid of Lemmata 3 and 4. Let

$$\mathfrak{m}_i = \left\{ \alpha \in \mathfrak{m} : |S(\lambda_i \alpha)| \leq X^{\frac{5}{6}} \right\}.$$

Then

$$\int_{\mathfrak{m}_1} |S(\lambda_1 \alpha) S(\lambda_2 \alpha)|^2 K(\alpha) \, d\alpha \leq X^{\frac{5}{3}} \int_{-\infty}^{\infty} |S(\lambda_2 \alpha)|^2 K(\alpha) \, d\alpha,$$

and the easy bound

$$\int_{-\infty}^{\infty} |S(\lambda \alpha)|^2 K(\alpha) \, d\alpha = \sum_{p, p' \leq X} \Upsilon(\lambda(p - p')) \log p \log p' \ll \tau \sum_{p \leq X} \log^2 p$$

yields the estimate

$$\int_{\mathfrak{m}_1} \left| S(\lambda_1 \alpha) S(\lambda_2 \alpha) \right|^2 K(\alpha) \, d\alpha \ll \tau X^{\frac{8}{3}} \log X. \tag{17}$$

By symmetry the same bound holds for $\mathfrak{m}_2$ in place of $\mathfrak{m}_1$. It therefore remains to discuss the set $\mathcal{L}$ of all $\alpha \in \mathfrak{m}$ where

$$\left| S(\lambda_1 \alpha) \right| \geq X^{\frac{5}{6}}, \qquad \left| S(\lambda_2 \alpha) \right| \geq X^{\frac{5}{6}}$$

simultaneously. For $y > 0$, $A_1, A_2 \geq 1$ let

$$\mathcal{L}(A_1, A_2, y) = \left\{ y < |\alpha| \leq 2y \ : \ X/A_i < \left| S(\lambda_i \alpha) \right| \leq 2X/A_i \ \text{ if } \ i = 1, 2 \right\}.$$

By a familiar dyadic dissection argument there are numbers

$$Y < y \leq Y_0, \qquad 1 \leq A_1, A_2 \leq X^{\frac{1}{6}}$$

such that

$$\int_{\mathcal{L}} \left| S(\lambda_1 \alpha) S(\lambda_2 \alpha) \right|^2 K(\alpha) \, d\alpha \ll \log^3 X \int_{\mathcal{L}(A_1, A_2, y)} \left| S(\lambda_1 \alpha) S(\lambda_2 \alpha) \right|^2 K(\alpha) \, d\alpha$$

By Lemma 3, $\mathcal{L}(A_1, A_2, y) \subset \mathfrak{K}_y(Q_1, Q_2)$ for some $Q_i \ll A_i^2 \log^8 X$. Let $\eta > 0$ be sufficiently small. By (2) we can pick a convergent b/r to λ_1/λ_2 with

$$\frac{X^{1-2\eta}}{A_1^2 A_2^2} < r < \frac{X^{1-\eta}}{A_1^2 A_2^2}$$

(note that the left hand side of this inequality tends to infinity with X if η is sufficiently small). Lemma 4 is applicable and shows that

$$\int_{\mathcal{L}(A_1, A_2, y)} d\alpha \ll A_1^4 A_2^4 X^{3\eta - 2} y.$$

Recalling (13) we deduce that

$$\int_{\mathcal{L}(A_1, A_2, y)} \left| S(\lambda_1 \alpha) S(\lambda_2 \alpha) \right|^2 K(\alpha) d\alpha \ll A_1^4 A_2^4 X^{2+3\eta} y \min(\tau^2, y^{-2}).$$

The right hand side is maximal for $y = \tau^{-1}$. We insert the upper bounds for A_i and infer that

$$\int_{\mathcal{L}} \left| S(\lambda_1 \alpha) S(\lambda_2 \alpha) \right|^2 K(\alpha) d\alpha \ll \tau X^{\frac{8}{3} + 3\eta}.$$

Since $\eta > 0$ was arbitrary, the lemma now follows from (16) and (17).

It will be observed that the choice $Y = X^{-1/3}$ is fairly arbitrary, any $Y = X^\nu$ with $-\frac{2}{3} < \nu < -\frac{1}{6}$ would serve equally well here.

5. Conclusion

It is now easy to distil the following result which contains Theorem 1.

Theorem 2. *Let $0 < \tau < 1$. Suppose that $\mathcal{V}(X)$ is a set of real numbers contained in $[\frac{1}{2}X, X]$ such that $|v_1 - v_2| > 2\tau$ whenever $v_1, v_2 \in \mathcal{V}$ are distinct. Let $\lambda_1, \lambda_2 \geq 1$ be real numbers and λ_1/λ_2 be algebraic and irrational. Then the number of $v \in \mathcal{V}$ for which the inequality*

$$\left|\lambda_1 p_1 + \lambda_2 p_2 - v\right| \leq \tau$$

has no solution in primes p_1, p_2 does not exceed

$$\ll \frac{X^{\frac{2}{3}+\epsilon}}{\tau^2}.$$

Proof. We use the notation from the earlier sections. Suppose that $\mathcal{N}_v = 0$ for some $v \in \mathcal{V}$. Then, by Lemma 5 and (15) we must have

$$\left|\int_{\mathfrak{m}} S(\lambda_1\alpha)S(\lambda_2)e(-v\alpha)K(\alpha)\,d\alpha\right| \gg \tau^2 X. \qquad (18)$$

Hence

$$\#\{v \in \mathcal{V} : \mathcal{N}_v = 0\} \ll \frac{1}{\tau^4 X^2} \sum_{v \in \mathcal{V}} \left|\int_{\mathfrak{m}} S(\lambda_1\alpha)S(\lambda_2\alpha)e(-v\alpha)K(\alpha)\,d\alpha\right|^2,$$

and the theorem is immediate from Lemma 6.

We now deduce Theorem 1 subject to the additional condition

$$\lambda_1 > \lambda_2 \geq 1.$$

We may suppose that $\delta < \frac{1}{6}$, for otherwise the estimate in Theorem 1 is trivial. Take $\tau = X^{-\delta}$ in Theorem 2. Then

$$\#\left\{v \in \mathcal{V} : \tfrac{1}{2}X < v \leq X, \ |\lambda_1 p_1 + \lambda_2 p_2 - v| < X^{-\delta} \text{ not soluble}\right\} \ll X^{\frac{2}{3}+2\delta+\epsilon}.$$

To obtain Theorem 1, replace X by $2^{-l}X$ and sum over $1 \leq l \ll \log X$. The condition $\lambda_i \geq 1$ is easily removed. We may assume $\lambda_1 \geq \lambda_2$ by symmetry. If $\lambda_2 < 1$ consider the inequality

$$\left|\frac{\lambda_1}{\lambda_2}p_1 + p_2 - \frac{v}{\lambda_2}\right| \leq \frac{1}{\lambda_2 v^\delta}$$

and note that the numbers v/λ_2 are still suitably spaced.

6. Implications of GRH

We briefly indicate possible improvements based on GRH. The main difference is a stronger form of Lemma 3 which is now available.

Lemma 7. *Assume GRH. Let $\epsilon > 0$. Let $1 \leq A \leq X^{1/4}/(\log X)^{3/2+\epsilon}$. Suppose that $|S(\alpha)| \geq X/A$. Then there are coprime integers q, a with*

$$1 \leq q \leq A \log^\epsilon X, \qquad |q\alpha - a| \leq \frac{A \log^\epsilon X}{X}.$$

Proof. This follows immediately from Lemma 5 of Goldston [5].

We can now follow the argument leading to Theorem 2. The unconditional Lemma 5 is still enough for our present needs. In the work of §4 we redefine

$$\mathfrak{m}_i = \left\{ \alpha \in \mathfrak{m} : |S(\lambda_i \alpha)| \leq X^{\frac{3}{4}} (\log X)^{\frac{3}{2}+\epsilon} \right\}.$$

Then the analogue of (17) is

$$\int_{\mathfrak{m}_i} \left| S(\lambda_1 \alpha) S(\lambda_2 \alpha) \right|^2 K(\alpha) \, d\alpha \ll \tau X^{\frac{5}{2}} \log^{4+2\epsilon} X.$$

Let $1 \leq A_i \leq X^{1/4}/\log^4 X$. With $\mathfrak{L}(A_1, A_2, y)$ as above we have

$$\mathfrak{L}(A_1, A_2, y) \subset \mathfrak{K}_y \left(A_1 \log^\epsilon X, A_2 \log^\epsilon X \right),$$

and the argument of the previous section yields

$$\int_{\mathfrak{L}} \left| S(\lambda_1 \alpha) S(\lambda_2 \alpha) \right|^2 K(\alpha) \, d\alpha \ll \tau X^{2+\epsilon}$$

where this time $\mathfrak{L}$ is the set of all $\alpha \in \mathfrak{m}$ with $|S(\lambda_i \alpha)| \geq X^{3/4}(\log X)^{\frac{3}{2}+\epsilon}$ for $i = 1, 2$. This establishes the following result.

Theorem 3. *Assume GRH. Let $0 < \tau < 1$. Suppose that $\mathcal{V}(X)$ is a set of real numbers contained in $\left[\frac{1}{2}X, X \right]$ such that $|v_1 - v_2| > 2\tau$ whenever $v_1, v_2 \in \mathcal{V}$ are distinct. Let $\lambda_1, \lambda_2 \geq 1$ be real numbers and λ_1/λ_2 be algebraic and irrational. Then the number of $v \in \mathcal{V}$ for which $|\lambda_1 p_1 + \lambda_2 p_2 - v| \leq \tau$ has no solution in primes p_1, p_2 does not exceed*

$$\ll \frac{X^{\frac{1}{2}} \log^{4+\epsilon} X}{\tau^2}.$$

References

1 J.R. Chen: On the representation of larger even integers as the sum of a prime and the product of at most two primes. *Sci. Sinica* **16** (1973), 157–176.
2 J.R. Chen: The exceptional set of Goldbach numbers II. *Sci. Sinica* **26** (1983), 714–731.
3 H. Davenport and H. Heilbronn: On indefinite quadratic forms in five variables. *J. London Math. Soc.* **21** (1946), 185–193.
4 P.X. Gallagher: A large sieve density estimate near $\sigma = 1$. *Invent. Math.* **11** (1970), 329–339.
5 D.A. Goldston: On Hardy and Littlewood's contribution to the Goldbach conjecture. In *Proc. Amalfi Conf. on Analytic Number Theory*, 115–155 (E. Bombieri, A. Perelli, S. Salerno and U. Zannier, eds.). Univ. di Salerno (1992).
6 G.H. Hardy and J.E. Littlewood: Some problems of "Partitio Numerorum". *Proc. London Math. Soc.* (2) **22** (1923), 46–56.
7 G. Harman: Diophantine approximation with a prime and an almost prime. *J. London Math. Soc.* (2) **29** (1984), 13–22.
8 G. Harman: Metric Diophantine approximation with two restricted variables III: Two prime numbers. *J. Number Theory* **29** (1988), 364–375.
9 M.N. Huxley: On the difference between consecutive primes. *Invent. Math.* **15** (1972), 164–170.
10 H.L. Montgomery and R.C. Vaughan: The exceptional set in Goldbach's problem. *Acta Arith.* **27** (1975), 353–370.
11 J. Pitman: Pairs of diagonal inequalities. *Recent Progress in Analytic Number Theory* Vol. 2, 183–215 (H. Halberstam and C. Hooley, eds.). Academic Press (1981).
12 B. Saffari and R.C. Vaughan: On the fractional part of x/n and related sequences II. *Ann. Inst. Fourier* **27** (1977), 1–30.
13 R.C. Vaughan: Diophantine approximation by prime numbers III. *Proc. London Math. Soc.* (3) **33** (1976), 177–192.
14 R.C. Vaughan: *The Hardy-Littlewood Method.* Cambridge University Press (1981).
15 G.L. Watson: On indefinite quadratic forms in five variables. *Proc. London Math. Soc.* (3) **3** (1953) 170–181.

5. Some Applications of Sieves of Dimension exceeding 1

H. Diamond and H. Halberstam

We combine the sieves of [1], [2] with the weighted sieve procedure of [3], Chapter 10, to formulate a general theorem about the incidence of almost-primes in integer sequences; and we illustrate the quality of this machinery with many concrete results about almost-primes representable by polynomials with integer or prime arguments.

Chapter 10.1 of [3] describes a weighted sieve procedure which, in combination with upper and lower sieve estimates of dimension $\kappa > 1$, leads to a general result of the following kind: given a finite integer sequence $\mathcal{A}$ and a set $\mathcal{P}$ of primes, then subject only to some rather weak conditions on the pair $\mathcal{A}$, $\mathcal{P}$, one can assert that $\mathcal{A}$ contains a large number of almost-primes P_r — numbers having at most r prime factors counted according to multiplicity — with r relatively small, made up of primes from $\mathcal{P}$. Since [3] is currently out of print and higher dimensional sieves superior to those described in Chapter 7 of [3] are now available (see [1], [2]), we state here an improved version of such a general result and describe various applications to integer sequences generated by reducible polynomials that are superior to those presented in Chapter 10.3 of [3].

Let $\mathcal{A}$ be a finite integer sequence whose members are not necessarily all positive or distinct. Let $\mathcal{P}$ be a set of primes, $\mathcal{P}^c$ its complement with respect to the set of all primes, and suppose that no member of $\mathcal{A}$ has a prime factor from $\mathcal{P}^c$ — symbolically, that

$$(a, \mathcal{P}^c) = 1 \quad \text{for all} \quad a \in \mathcal{A}.$$

We make some basic assumptions about the pair $\mathcal{A}$, $\mathcal{P}$: loosely speaking, we require the 'probability' of a member a from $\mathcal{A}$ being divisible by a prime p from $\mathcal{P}$ to be no larger than κ/p on average, where $\kappa > 1$ is a constant (see (III) below); and we require $\mathcal{A}$ to be well distributed among the arithmetic progressions $0 \bmod d$ as a runs over an extensive range of squarefree numbers coprime with $\mathcal{P}^c$ (see (IV) below). We proceed to make these assumptions precise.

Sieve Methods, Exponential Sums, and their Applications in Number Theory
Greaves, G.R.H., Harman, G., Huxley, M.N., Eds. ©Cambridge University Press, 1996

Suppose there exists an approximation X to the cardinality $|\mathcal{A}|$ of $\mathcal{A}$ and a non-negative multiplicative arithmetic function $\omega(\cdot)$ satisfying

$$\omega(1) = 1, \quad \omega(p) = 0 \quad \text{if} \quad p \in \mathcal{P}^c, \qquad \text{(I)}$$

$$0 \leq \omega(p) < p \quad \text{if} \quad p \in \mathcal{P}, \qquad \text{(II)}$$

and for some constants $\kappa > 1$, $A \geq 2$,

$$\prod_{z_1 \leq p < z} \left(1 - \frac{\omega(p)}{p}\right)^{-1} \leq \left(\frac{\log z}{\log z_1}\right)^{\kappa} \left(1 + \frac{A}{\log z_1}\right) \quad \text{if} \quad 2 \leq z_1 < z, \quad \text{(III)}$$

such that the 'remainders'

$$R_d := \left|\{a \in \mathcal{A} : a \equiv 0 \bmod d\}\right| - \frac{\omega(d)}{d} X$$

are small on average in the sense that for some constants τ, $0 < \tau \leq 1$, $A_1 \geq 1$ and $A_2 \geq 2$,

$$\sum_{\substack{d < X^\tau (\log X)^{-A_1} \\ (d, \mathcal{P}^c) = 1}} \mu^2(d) 4^{\nu(d)} |R_d| \leq A_2 \frac{X}{(\log X)^{\kappa+1}} \qquad \text{(IV)}$$

(here $\nu(d)$ denotes the number of prime factors of d).

Before we can state our results we introduce a constant μ such that

$$\max_{a \in \mathcal{A}} |a| \leq X^{\tau\mu}, \qquad \text{(V)}$$

where τ is the constant from (IV) above; and we quote from [2]

Theorem 0. *Let $\kappa > 1$. Then there exist numbers $\alpha = \alpha_\kappa > 2$ and $\beta = \beta_\kappa > 2$ such that the simultaneous differential-difference system*

$$F(u) = 1/\sigma(u) \qquad \text{if} \quad 0 < u \leq \alpha \qquad \text{(i)}$$

$$f(u) = 0 \qquad \text{if} \quad 0 < u \leq \beta \qquad \text{(ii)}$$

$$\left(u^\kappa F(u)\right)' = \kappa u^{\kappa-1} f(u-1) \qquad \text{if} \quad \alpha < u \qquad \text{(iii)}$$

$$\left(u^\kappa f(u)\right)' = \kappa u^{\kappa-1} F(u-1) \qquad \text{if} \quad \beta < u \qquad \text{(iv)}$$

has continuous solutions $F = F_\kappa$, $f = f_\kappa$ with the properties

$$F(u) = 1 + O(e^{-u}), \quad f(u) = 1 + O(e^{-u}),$$

$F(u)$ *decreases monotonically toward* 1 *as* $u \to \infty$,

$f(u)$ *increases monotonically toward* 1 *as* $u \to \infty$.

Then

Theorem 1. *Let $\mathcal{A}$ and $\mathcal{P}$ be as described above. For any two real numbers u and v satisfying*

$$\tau^{-1} < u < v, \quad \beta_\kappa < \tau v,$$

we have

$$|\{P_r : P_r \in \mathcal{A}\}| \gg X \prod_{p < X^{1/v}} \left(1 - \frac{\omega(p)}{p}\right)$$

provided only that

$$r > \tau\mu u - 1 + \frac{\kappa}{f_\kappa(\tau v)} \int_1^{v/u} F_\kappa(\tau v - s)\left(1 - \frac{u}{v}s\right)\frac{ds}{s}. \tag{1}$$

For values of κ that are of modest size, values of the functions f_κ and F_κ can be tabulated (cf. Theorem 0 above) — the authors' computers house such tables, compiled by Drs. D. Bradley and F. Wheeler — and the integral on the right of (1) evaluated numerically with great precision. In practice, u is chosen so that τu is near 1 and v so that τv is larger even than α_κ.

Theorem 1 combines Theorem 10.1 of [3] with the main result of [1]. It should be noted that, by virtue of the appendix in the latter, Theorem 1 holds under weaker conditions than Theorem 10.1.

If now we apply the analysis in section 10.3 of [3], we obtain results about almost-primes representable by polynomials. We state only a handful of these that appear to us to have intrinsic interest.

Let $H_1(n), \ldots, H_g(n)$ $(g > 1)$ be distinct irreducible polynomials with integer coefficients and write $H(n) = H_1(n) \ldots H_g(n)$. Let h be the degree of H, let $\rho(p)$ denote the number of solutions of

$$H(n) \equiv 0 \bmod p \quad (n = 0, 1, \ldots, p - 1)$$

and suppose that

$$\rho(p) < p \quad \text{for all primes} \quad p.$$

In these circumstances take

$$\mathcal{A} = \{H(n) : 1 \leq n \leq x\}$$

and $\mathcal{P}$ to be the set of all primes. Then $X = x$, $\omega(d) = \rho(d)$, $\kappa = g$, $\tau = 1$, and $H(n) \ll x^h$ implies that any constant $\mu > h$ is admissible once x is taken large enough. Hence, if x is sufficiently large, we deduce from Theorem 1 that, if (cf. Theorem 10.4 of [3]) u and v are real numbers satisfying

$$1 < u < v \quad \text{and} \quad \beta_g < v,$$

then

$$|\{n : 1 \leq n \leq x, H(n) = P_r\}| \gg \frac{x}{(\log x)^g} \tag{2}$$

provided

$$r > hu - 1 + \frac{g}{f_g(v)} \int_1^{v/u} F_g(v - s) \left(1 - \frac{u}{v}s\right) \frac{ds}{s}. \tag{3}$$

We illustrate this result with some special cases.

1. Take $H_i(n) = a_i n + b_i (i = 1, \ldots, g)$ (cf. Theorem 10.5 and Table 3 of [3]), and suppose that

$$\prod_{i=1}^{g} a_i \prod_{1 \le r < t \le g} (a_r b_t - a_t b_r) \ne 0,$$

also that $H(n)$ has no fixed prime divisor. Here $h = g$, and for $2 \le g \le 10$ we obtained the following admissible values of r:

g	2	3	4	5	6	7	8	9	10
r	5	8	12	16	20	25	29	34	39

Table 1

In the most interesting case $g = 2$, (3) required $r > 4.065\ldots$, so that we just missed $r = 4$. A more delicate weighting procedure, or a better 2-dimensional sieve, should give $r = 4$.

2. Take $H_1, \ldots, H_g$ to be of the same degree, call it k, so that $h = gk$ (cf. Table 4, [3] p. 286). We have carried out the numerical calculations for $2 \le g \le 7$, $2 \le k \le 7$, and arrived at the following table of admissible choices of r:

g \ k	2	3	4	5	6	7
2	7	9	11	14	16	18
3	12	16	19	22	25	28
4	17	22	27	31	35	40
5	23	29	35	40	46	51
6	29	36	43	50	57	63
7	35	44	52	60	68	75

Table 2

We just missed better results in the cases $g = 2$, $k = 5$ and 6, and $g = 3$, $k = 3$.

3. Consider the cases $g = 2$, $h \geq 3$, when $H(n) = H_1(n)H_2(n)$ and h is the degree of H (cf. Table 5, [3] p. 287). Here $\beta_2 = 4.2664\ldots$, and the following table gives the least admissible values of r when $3 \leq h \leq 10$:

h	3	4	5	6	7	8	9	10
r	6	7	8	9	10	11	12	14

Table 3

We just missed 13 when $h = 10$.

Let us turn now to the existence of almost-primes in sequences generated by polynomials with *prime* arguments. This time $H_i(n)$ $(\neq \pm n)$, $i = 1, \ldots, g$ $(g > 1)$, are distinct irreducible polynomials with integer coefficients, $H(n) = H_1(n) \ldots H_g(n)$ as before and h is again the degree of H. This time we require not only that

$$\rho(p) < p \quad \text{for all primes} \quad p$$

but also that

$$\rho(p) < p - 1 \quad \text{if} \quad p \nmid H(0).$$

(cf. [3], Theorem 10.6). Now

$$\mathcal{A} = \big\{ H(p) : p \leq x \big\},$$

$\mathcal{P}$ is again the set of all primes,

$$X = \operatorname{li} x, \qquad \omega(p) = \frac{p}{p-1}\,\rho_1(p)$$

where

$$\rho_1(p) = \begin{cases} \rho(p) & \text{if} \quad p \nmid H(0) \\ \rho(p) - 1 & \text{if} \quad p \mid H(0), \end{cases}$$

$\kappa = g$ and (IV) holds by Bombieri's theorem with $\tau = \frac{1}{2}$. Consequently (V) is valid with any $\mu > 2h$, supposing x to be large. Apply Theorem 1, with x sufficiently large: writing $U = \tau u = \frac{1}{2}u$ and $V = \tau v = \frac{1}{2}v$, we obtain

$$\big| \{ p : p \leq x, H(p) = P_r \} \big| \gg \frac{x}{(\log x)^{g+1}}$$

for any natural number r satisfying

$$r > 2hU - 1 + \frac{g}{f_g(V)} \int_1^{V/U} F_g(V-s) \left(1 - \frac{U}{V}s \right) \frac{ds}{s},$$

where

$$1 < U < V \quad \text{and} \quad \beta_g < V.$$

We illustrate this result with two special cases.

4. Let $H_1, \ldots, H_g$ all have the same degree k, so that $h = gk$ (cf. Table 6 of [3], p. 291). For $g = 2$, 3, 4 and $2 \le k \le 7$ the following table gives admissible values of r:

g \ k	2	3	4	5	6	7
2	11	16	20	24	28	32
3	19	25	32	38	44	50
4	27	35	44	52	61	69

Table 4

In the case $g = 2$, $k = 3$ we just failed to reach $r = 15$.

5. Consider the cases $g = 2$, $3 \le h \le 8$, when $H(p) = H_1(p)H_2(p)$ and h is the degree of H. We obtained the following admissible values of r:

h	3	4	5	6	7	8
r	9	11	14	16	18	20

Table 5

When $h = 5$ and 6 we just missed $r = 13$ and 15 respectively.

We conclude with comments about a problem posed at the 1993 West Coast Number Theory conference and reported to us by Kevin Ford.

"For which r are there only finitely many primitive Pythagorean triples (x, y, z) such that the number of distinct prime factors of xyz is r?"

Ford checked that there are only 30 such triples with $r \le 5$, and he observed that a higher dimensional sieve ought to apply to the problem, as indeed it does.

Consider the typical triple $x = 2mn$, $y = m^2 - n^2$, $z = m^2 + n^2$ with m, n coprime natural numbers of opposite parity, and secure coprimality of m, n by taking $n = 1$. Then Theorem 1 applies to

$$H(m) = m(m + 1)(m - 1)(m^2 + 1);$$

or, rather, can be made to apply since 2, 3, 5 are fixed prime factors of $H(m)$. We excluded 2, 3, 5 from $\mathcal{P}$; with $g = 4$, $h = 5$, Theorem 1 shows that there exist infinitely many triples with $r \leq 17$. (Theorem 10.4 of [3] gives 19.) One should be able to do better by not setting $n = 1$.

Ford argues heuristically that there are probably infinitely many triples with $r = 6$.

Let $m = 2^a$, $n = 3^b 5^c < m/2$. The probability that $m + n$, $m - n$ and $m^2 + n^2$ are simultaneously primes is $\gg 1/a^3$. Summing over b, c, a yields a divergent series. A computer search using PARI yielded 49 triples with $m = 2^a$, $n = 3^b 5^c$ (without the condition $n < m/2$) and $m, n < 10^{18}$. The one with largest z corresponded to $(a, b, c) = (40, 24, 9)$. The triple is

$$x = \qquad 606\,512\,811\,305\,166\,962\,688\,000\,000\,000$$
$$y = 304\,284\,832\,292\,768\,849\,232\,879\,419\,897\,559\,449$$
$$z = 304\,284\,832\,295\,186\,700\,872\,108\,678\,246\,971\,801\,.$$

References

1 H. Diamond, H. Halberstam, and H.-E. Richert: Combinatorial sieves of dimension exceeding one. *J. Number Theory* **28** (1988), 346–306.

2 H. Diamond, H. Halberstam, and H.-E. Richert: Combinatorial sieves of dimension exceeding one II. In *Analytic Number Theory, Proceedings of a Conference in Honor of Heini Halberstam*, 265–308 (B.C. Berndt, H.G. Diamond and A.J. Hildebrand, eds.). Birkhäuser (1996).

3 H. Halberstam and H.-E. Richert: *Sieve Methods*. Academic Press (1974).

6. Representations by the Determinant and Mean Values of L-Functions

W. Duke, J. Friedlander and H. Iwaniec

Introduction and Statement of Results

The determinant equation

$$\det \begin{pmatrix} m_1 & m_2 \\ n_1 & n_2 \end{pmatrix} = \Delta \tag{1}$$

appears in the context of various problems in analytic number theory. One would like to have, for a fixed integer $\Delta \neq 0$, a good asymptotic formula for the number of solutions of (1) as the entries vary over general sequences of integers.

To get a hold on the problem we specify the sizes:

$$\begin{aligned} M_1 < m_1 \leq 2M_1, \quad N_1 < n_1 \leq 2N_1, \\ M_2 < m_2 \leq 2M_2, \quad N_2 < n_2 \leq 2N_2. \end{aligned} \tag{2}$$

If we make no further restrictions on the entries, this is a problem from the spectral theory of $\mathrm{GL}_2(\mathbb{Z})$ automorphic forms. We should like to be able to treat the case where all four entries are from arbitrary sequences or, what amounts to the same thing, to evaluate the weighted sum

$$S_\Delta(M, N) = \sum_{m_1 n_2 - m_2 n_1 = \Delta} \sum \sum \sum f(m_1) g(m_2) a_{n_1} b_{n_2} \tag{3}$$

with f, g, a, b general functions supported in the box (2). In this generality the problem seems quisquose and would, just for example, have inopinate implications to the twin prime problem.

In this paper we are able to treat the case where the lower row has general weights a_{n_1}, b_{n_2}. For the upper row we require f, g to be smooth functions supported on $[M_1, 2M_1]$, $[M_2, 2M_2]$ with derivatives satisfying

$$f^{(j)} \ll \eta^j M_1^{-j}, \qquad g^{(j)} \ll \eta^j M_2^{-j}, \tag{4}$$

for all $j \geq 0$, some $\eta \geq 1$, and with the implied constant depending on j.

W. Duke and H. Iwaniec are supported in part by NSF grant DMS-9500797.
J. Friedlander is supported in part by NSERC grant A5123.

Sieve Methods, Exponential Sums, and their Applications in Number Theory
Greaves, G.R.H., Harman, G., Huxley, M.N., Eds. ©Cambridge University Press, 1996

The most interesting case is when the entries in the lower row are larger than the smoothed ones in the upper row. In the opposite case an analysis via known Fourier techniques directly gives the asymptotic formula. Here we prove:

Theorem 1. *Let* $\Delta \neq 0$, a_{n_1}, b_{n_2} *be complex numbers for*

$$N_1 < n_1 \leq 2N_1, \quad N_2 < n_2 \leq 2N_2$$

and f, g *smooth functions supported on* $[M_1, 2M_1]$, $[M_2, 2M_2]$ *and with derivatives satisfying* (4). *Then*

$$S_\Delta(M, N) = \sum\sum_{(n_1,n_2)|\Delta} \frac{(n_1, n_2)}{n_1 n_2} a_{n_1} b_{n_2} \int f\left(\frac{x+\Delta}{n_2}\right) g\left(\frac{x}{n_1}\right) dx$$

$$+ O\left(\eta^{\frac{19}{8}} \|a\| \|b\| \left(\frac{M_1 N_2}{M_2 N_1} + \frac{M_2 N_1}{M_1 N_2} \right)^{\frac{19}{8}} \right.$$

$$\left. \times (N_1 N_2)^{\frac{3}{8}} (N_1 + N_2)^{\frac{11}{48}} (M_1 M_2 N_1 N_2)^\varepsilon \right), \quad (5)$$

where the implied constant depends only on ε, *and* $\| \ \|$ *denotes the* L_2 *norm.*

This result should be compared with the trivial bound

$$S_\Delta(M, N) \ll \|a\| \|b\| (M_1 M_2)^{\frac{1}{2}} (M_1 M_2 N_1 N_2)^\varepsilon$$

that follows from Cauchy's inequality.

In the special case $\eta = 1$, $M_1 = M_2 = M$, $N_1 = N_2 = N$, the theorem simplifies to

$$S_\Delta(M, N) = \sum\sum_{(n_1,n_2)|\Delta} \frac{(n_1, n_2)}{n_1 n_2} a_{n_1} b_{n_2} \int f\left(\frac{x+\Delta}{n_2}\right) g\left(\frac{x}{n_1}\right) dx$$

$$+ O\left(\|a\| \|b\| N^{\frac{47}{48}} (MN)^\varepsilon \right).$$

Here the error term is smaller than the above trivial bound provided that $M > N^{47/48}$.

The main tool in the proof of Theorem 1 is a new estimate [1] for bilinear forms of Kloosterman fractions. The particular variant needed, which follows quickly from Theorem 2 of that paper, we record here as

Proposition. *Let α_m for $M < m \le 2M$ and β_n for $N < n \le 2N$ be arbitrary complex numbers, and $k \ne 0$ an integer. Then, for any $\varepsilon > 0$, we have*

$$\sum\sum_{(m,n)=1} \alpha_m \beta_n \, e\left(k\frac{\overline{m}}{n} + \frac{X}{mn}\right)$$

$$\ll \|\alpha\| \, \|\beta\| \left(1 + \frac{|X|}{MN}\right)(|k| + MN)^{\frac{3}{8}}(M+N)^{\frac{11}{48}+\varepsilon}$$

where $e(t) = e^{2\pi i t}$, $\overline{m}m \equiv 1 \pmod{n}$, and the implied constant depends only on ε.

We apply Theorem 1 to obtain a mean-value theorem for character sums and L-functions. The classical mean-value theorem for Dirichlet polynomials asserts that

$$\sum_{\chi \,(\mathrm{mod}\, q)} \left|\sum_{n \le N} \lambda_n \chi(n)\right|^2 = (\varphi(q) + O(N)) \sum_{\substack{n \le N \\ (n,q)=1}} |\lambda_n|^2,$$

where the implied constant is absolute; cf. Theorem 6.2 in [3]. The result is best possible when $N < q$, in which case the error term is superfluous. In the case $N > q$ the result was improved [2] for sequences of triple convolution type $\lambda = \alpha * \beta * f$ with smooth f and with α, β quite general but having support specially located. Theorem 1 allows us to remove this last restriction and treat the convolution $\lambda = a * f$ for a general sequence a.

More precisely, we consider an arbitrary sequence a_n of complex numbers defined for $N < n \le 2N$, $(n,q) = 1$, and a smooth function $f(m)$ supported on $M < m \le 2M$ and having derivatives satisfying (4).

Theorem 2. *We have*

$$\sum_{\chi \ne \chi_0 \,(\mathrm{mod}\, q)} \left|\sum_m f(m)\chi(m)\right|^2 \left|\sum_n a_n \chi(n)\right|^2 \ll \eta^{\frac{19}{8}} q^\varepsilon \left(q + N^{\frac{95}{48}}\right) M \|a\|^2.$$

Here the point is that we obtain a non-trivial bound with N as large as $q^{48/95}$, and $\frac{48}{95} > \frac{1}{2}$.

Since Dirichlet L-functions may be well approximated by sums of the type occurring in Theorem 2, we may deduce a mean-value theorem for the former.

Theorem 3. *For $\mathrm{Re}\, s = \frac{1}{2}$ we have*

$$\sum_{\chi \,(\mathrm{mod}\, q)} |L(s,\chi)|^2 \left|\sum_n a_n \chi(n)\right|^2 \ll |s|^{\frac{19}{8}} q^\varepsilon \left(q + N^{\frac{95}{48}}\right) \|a\|^2.$$

Note that by choosing $a_n = \overline{\psi}(n)$ (amplification technique) for a given non-principal ψ mod q, we deduce the bound

$$L(s, \psi) \ll |s|^{\frac{19}{16}} q^{\frac{47}{190}+\varepsilon},$$

which is a little stronger than the convexity bound in the q-aspect.

Proof of Proposition

Theorem 2 of [1] is precisely the case $X = 0$. To derive the general case we separate the variables m, n in $e(X/mn)$ by Fourier or Mellin inversion.

Proof of Theorem 1

We split the summation in (3) in accordance with the greatest common divisor d of n_1, n_2 getting

$$S_\Delta(M, N) = \sum_{d|\Delta} \sum \sum_{(n_1, n_2)=1} a_{dn_1} b_{dn_2} \sum \sum_{m_1 n_2 - m_2 n_1 = \Delta/d} f(m_1) g(m_2).$$

To the inner double sum we apply Poisson's formula as follows:

$$\sum_{m_1} \sum_{m_2} = \sum_{m_2 \equiv -\Delta \overline{n}_1/d \,(\mathrm{mod}\, n_2)} f\left(\frac{m_2 n_1}{n_2} + \frac{\Delta}{dn_2}\right) g(m_2)$$

$$= \sum_h e\left(h\frac{\Delta}{d}\frac{\overline{n}_1}{n_2}\right) I_h(n_1, n_2)$$

where

$$I_h(n_1, n_2) = \frac{1}{dn_1 n_2} \int f\left(\frac{x+\Delta}{dn_2}\right) g\left(\frac{x}{dn_1}\right) e\left(\frac{hx}{dn_1 n_2}\right) dx.$$

The main term in (5) arises as the contribution from the term $h = 0$, as is easily seen after a change of variables.

By partial integration $j+1$ times we have (remember n_i is about N_i/d)

$$I_h(n_1, n_2) \ll \frac{\eta}{|h|} \left(\frac{\eta}{d|h|}\left(\frac{N_1}{M_1} + \frac{N_2}{M_2}\right)\right)^j.$$

Put

$$H = \frac{\eta}{d}\left(\frac{N_1}{M_1} + \frac{N_2}{M_2}\right), \qquad D = M_1 N_2 + M_2 N_1.$$

Hence, the contribution to $S_\Delta(M, N)$ from terms with $|h| > HD^\varepsilon$ is

$$\ll \eta D^{-j\varepsilon} \sum_{d|\Delta} \sum_{(n_1,n_2)=1} \sum |a_{dn_1} b_{dn_2}| \leq \eta D^{-j\varepsilon} \|a\|_1 \|b\|_1,$$

and by Cauchy's inequality, and since j is arbitrary, this is admissible for Theorem 1.

The remaining contribution to $S_\Delta(M, N)$ is

$$\leq \sum_{d|\Delta} \sum_{1 \leq |h| \leq HD^\varepsilon} \int_0^{4D} \left| \sum_{(n_1,n_2)=1} \sum \frac{a_{dn_1} b_{dn_2}}{dn_1 n_2} \right.$$

$$\left. \times f\left(\frac{x+\Delta}{dn_2}\right) g\left(\frac{x}{dn_1}\right) e\left(\frac{hx}{dn_1 n_2} + \frac{h\Delta}{d}\frac{\overline{n}_1}{n_2}\right) \right| dx.$$

By the Proposition, this is bounded by

$$\ll \sum_{d|\Delta} \sum_{1 \leq h \leq H} \frac{dD}{N_1 N_2} \left(1 + \frac{dhD}{N_1 N_2}\right) \|a\| \|b\| \left(\frac{hD}{d} + \frac{N_1 N_2}{d^2}\right)^{\frac{3}{8}} \left(\frac{N_1 + N_2}{d}\right)^{\frac{11}{48}} D^\varepsilon$$

$$\ll \eta^{\frac{19}{8}} \left(\frac{M_1 N_2}{M_2 N_1} + \frac{M_2 N_1}{M_1 N_2}\right)^{\frac{19}{8}} \|a\| \|b\| (N_1 N_2)^{\frac{3}{8}} (N_1 + N_2)^{\frac{11}{48}} D^\varepsilon.$$

Here we have used the bound $|\Delta| \leq 4D$; in the opposite case the sum and the main term in (5) both vanish. This completes the proof of Theorem 1.

Proof of Theorem 2

Now we derive Theorem 2 from Theorem 1. We may assume that

$$M < q^{10}, \quad N < q^{10}$$

since otherwise the result is trivial. We define

$$S(\chi) = \left| \sum_m f(m)\chi(m) \right|^2 \left| \sum_n a_n \chi(n) \right|^2$$

and consider

$$S = \sum_{\chi \,(\mathrm{mod}\,q)} S(\chi) = \varphi(q) \sum \sum_{\substack{m_1 n_2 \equiv m_2 n_1 \,(\mathrm{mod}\,q) \\ (m_1 m_2, q)=1}} \sum \sum f(m_1)\overline{f}(m_2)\overline{a}_{n_1} a_{n_2}.$$

The diagonal terms where $m_1 n_2 = m_2 n_1$ contribute

$$S_0 \ll \|a\|^2 q M (MN)^\varepsilon.$$

The remaining terms give a contribution

$$S^* = \varphi(q) \sum_{l|q} \mu(l) \sum_{r \neq 0} \sum \sum \sum_{m_1 n_2 - m_2 n_1 = qr/l} f(lm_1) \overline{f}(lm_2) \overline{a}_{n_1} a_{n_2}$$

$$= \varphi(q) \sum_{l|q} \mu(l) \sum_{1 \leq |r| \leq R} \left\{ \sum \sum_{(n_1,n_2)|r} \frac{(n_1,n_2)}{n_1 n_2} \overline{a}_{n_1} a_{n_2} \int f\left(\frac{lx + qr}{n_2}\right) \overline{f}\left(\frac{lx}{n_1}\right) dx \right.$$

$$\left. + O\left(\eta^{\frac{19}{8}} \|a\|^2 N^{\frac{47}{48}} (MN)^\varepsilon\right) \right\} \quad (6)$$

by Theorem 1, where we have put $R = 3MN/lq$. Here the contribution to S^* from the error term is

$$\ll \eta^{\frac{19}{8}} \|a\|^2 MN^{\frac{95}{48}+\varepsilon}.$$

For the main term we write $d = (n_1, n_2)$, $r = ds$ and sum first over s getting by Poisson summation

$$\sum_{s \neq 0} f\left(\frac{lx + dqs}{n_2}\right) = \int f\left(\frac{lx + dqy}{n_2}\right) dy + O(\eta).$$

Hence the contribution to S^* from the main term in (6) is

$$\varphi(q) \sum_{l|q} \mu(l) \sum_d \sum \sum_{(n_1,n_2)=1} \frac{\overline{a}_{dn_1} a_{dn_2}}{n_1 n_2} \iint f\left(\frac{lx + qy}{n_2}\right) \overline{f}\left(\frac{lx}{n_1}\right) dx\, dy$$

$$+ O(\eta \|a\|^2 qMN^\varepsilon)$$

$$= \left(\frac{\varphi(q)}{q}\right)^2 |\hat{f}(0)|^2 \left(\sum_n a_n\right)^2 + O(\eta \|a\|^2 qMN^\varepsilon).$$

Combining the above results, we conclude that

$$S = \left(\frac{\varphi(q)}{q}\right)^2 |\hat{f}(0)|^2 \left(\sum_n a_n\right)^2 + O\left(\eta^{\frac{19}{8}} \|a\|^2 M(q + N^{\frac{95}{48}})(MN)^\varepsilon\right).$$

The contribution of the principal character is

$$S(\chi_0) = \left| \sum_{(m,q)=1} f(m) \right|^2 \left(\sum_n a_n\right)^2$$

and

$$\sum_{(m,q)=1} f(m) = \frac{\varphi(q)}{q} \hat{f}(0) + O(\eta\tau(q)).$$

Subtracting this contribution, we conclude the proof of Theorem 2.

Proof of Theorem 3

To prove Theorem 3 we first note that the contribution from χ_0 is bounded by

$$\ll \left(|s|\tau(q)\right)^2 N\|a\|^2,$$

which is admissible. For the non-principal characters we wish to apply Theorem 2. We approximate

$$L(s,\chi) = \sum_{m=1}^{\infty} \frac{\chi(m)}{m^s}$$

with the aid of a smooth partition of unity getting sums of the type

$$M^{-\frac{1}{2}} \sum_m f(m)\chi(m),$$

where f satisfies (4) with $\eta = |s|$ and where $M < q^{10}$. For $M \geq q^{10}$ a trivial bound suffices. The number of partial sums is thus $\ll \log q$ and an application of Theorem 2 to each yields Theorem 3.

We thank the Institute for Advanced Study for providing excellent working conditions during the preparation of this paper.

References

1 W. Duke, J. Friedlander, and H. Iwaniec: Bilinear forms with Kloosterman fractions (preprint). (1995).

2 J. Friedlander and H. Iwaniec: A mean-value theorem for character sums. *Michigan Math. J.* **39** (1992), 153–159.

3 H. L. Montgomery: *Topics in Multiplicative Number Theory* (Lecture Notes in Math. **227**). Springer (1971).

7. On the Montgomery-Hooley Asymptotic Formula

D. A. Goldston and R. C. Vaughan

1. Introduction

Let Λ denote von Mangoldt's function. That is, when n is of the form p^k where p is a prime number and k is a natural number then $\Lambda(n) = \log p$ and otherwise $\Lambda(n) = 0$. Further let

$$\psi(x, q, a) = \sum_{\substack{n \leq x \\ n \equiv a \,(\mathrm{mod}\, q)}} \Lambda(n) \tag{1.1}$$

and let $V(x, Q)$ denote the variance

$$V(x, Q) = \sum_{q \leq Q} \sum_{\substack{a=1 \\ (a,q)=1}}^{q} \left| \psi(x, q, a) - \frac{x}{\phi(q)} \right|^2. \tag{1.2}$$

Then, as a refinement of work of Barban [1], Davenport and Halberstam [3] and Gallagher [5], Montgomery [13] obtained the asymptotic formula for $V(x, Q)$,

$$V(x, Q) = Qx \log Q + O(Qx + x^2/\log^A x) \quad \text{if} \quad Q \leq x. \tag{1.3}$$

Montgomery's method depended upon a result of Lavrik [12] in additive number theory and was, therefore, dependent upon Vinogradov's theory of exponential sums over primes. In [8], [9], [10] Hooley introduced and developed a simpler yet more sophisticated method which led *inter alia* to the asymptotic formula

$$V(x, Q) = Qx \log Q - cQx + O(Q^{\frac{5}{4}} x^{\frac{3}{4}} + x^2/\log^A x) \quad \text{if} \quad Q \leq x \tag{1.4}$$

and to

$$V(x, Q) = Qx \log Q - cQx + O(Q^{\frac{5}{4}} x^{\frac{3}{4}} + x^{\frac{3}{2}+\epsilon}) \quad \text{if} \quad Q \leq x \tag{1.5}$$

on the assumption of the generalised Riemann hypothesis. Here c is an absolute constant. Very recently Friedlander and Goldston [4] have obtained

R.C. Vaughan is supported by an EPSRC Senior Fellowship.

Sieve Methods, Exponential Sums, and their Applications in Number Theory
Greaves, G.R.H., Harman, G., Huxley, M.N., Eds. ©Cambridge University Press, 1996

some very interesting results by a process at least partially based on considerations concerning the additive theory of primes. In particular they are able to improve (1.5) when Q is not close to x and obtain

$$V(x,Q) = Qx \log Q - cQx + O\left(\min(Q^{\frac{3}{2}} x^{\frac{1}{2}} (\log x)^{\frac{3}{2}}, Qx)\right)$$
$$+ O\left(x^{\frac{3}{2}} \log^6 x\right) \quad \text{if} \quad Q \le x. \tag{1.6}$$

On receipt of their preprint it occurred to one of us (Vaughan) to check the contents of a letter written to Hugh Montgomery dated 19th June 1973. Consideration of the ideas in it was shelved at that time, partly because other things were being pursued and partly because Professor Hooley was then developing his exciting new ideas. It was thus rather amusing to see that with the ideas sketched in that letter, when one assumes the generalised Riemann hypothesis, it is possible to obtain superior results to those obtained for $V(x,Q)$ either by Hooley or Friedlander and Goldston. The idea is to work via the additive theory of primes, but instead of appealing to, or at least developing separately, the appropriate additive theory, to view the "difficult part" of $V(x,Q)$, namely S_1 below, as an additive problem and treat it directly. This has the advantage that there is no need for an application of the theory of exponential sums over primes à la Vinogradov, as there is inbuilt a "type I" sum, $F(\alpha)$ below, which can be estimated directly and more effectively.

Theorem 1.1. *Suppose that the generalised Riemann hypothesis holds and let*

$$U(x,Q) = V(x,Q) - Qx \log Q - cxQ \tag{1.7}$$

where

$$c = \gamma + \log 2\pi + 1 + \sum_p \frac{\log p}{p(p-1)}. \tag{1.8}$$

Then
(i) when $1 \le Q \le x$ one has

$$U(x,Q) \ll Q^2 (x/Q)^{\frac{1}{4}+\epsilon} + x^{\frac{3}{2}} (\log 2x)^{\frac{5}{2}} (\log \log 3x)^2, \tag{1.9}$$

(ii) there is an absolute constant C such that when $x/Q \to \infty$ with

$$C x^{\frac{5}{7}} (\log 2x)^{\frac{10}{7}} (\log \log 3x)^{\frac{8}{7}} < Q \le x$$

one has

$$U(x,Q)/Q^2 = \Omega_\pm \left((x/Q)^{\frac{1}{4}}\right). \tag{1.10}$$

If one is prepared to make the further assumptions that the non-trivial zeros ρ of the Riemann zeta function are simple and $\zeta'(\rho)$ does not get too small, then the ϵ in (1.9) can be replaced by 0. We remark further upon this at the end of §8.

We further observe that without any unproved assumption the method readily gives (1.4) with the error term $x^{3/4}Q^{5/4}$ replaced by $x^{1/2}Q^{3/2}$ and the argument is then as short and simple as any used hitherto. With only a modest amount of extra effort it is possible to obtain

$$U(x,Q) \ll \frac{x^2}{\log^A x} + x^{\frac{1}{2}}Q^{\frac{3}{2}} \exp\left(-\frac{c'(\log 2x/Q)^{\frac{3}{5}}}{(\log\log 3x/Q)^{\frac{1}{5}}}\right).$$

The following theorem gives more precise results when Q is very close to x.

Theorem 1.2. *Suppose that the generalised Riemann hypothesis holds and that*

$$C_1 = \frac{\zeta(2)\zeta(3)}{\zeta(6)} \tag{1.12}$$

and

$$C_2 = \left(\gamma - \tfrac{3}{2} - \sum_p \frac{\log p}{p^2 - p + 1}\right)C_1. \tag{1.13}$$

Then
(i) *when $\frac{1}{3}x < Q \leq \frac{1}{2}x$ one has*

$$V(x,Q) = x^2\left(C_1 \log \frac{x}{Q} + C_2 - 2\right) + Qx(\log x + 5) - 5Q^2 + O(x^{\frac{3}{2}}\log^3 2x),$$

(ii) *when $\frac{1}{2}x < Q \leq x$ one has*

$$V(x,Q) = x^2\left(C_1 \log \frac{x}{Q} + C_2 - 1\right) + Qx(\log x + 1) - Q^2 + O(x^{\frac{3}{2}}\log^3 2x).$$

A number of recent papers in additive prime number theory have stated or claimed results on the generalised Riemann hypothesis which are based on an oversight. For example, this oversight occurs in [7], where a neglected term requires that the term $\log^2 qN$ in equation (2.40) be replaced by $\log^4 qN$. Therefore, the paper has been made as self contained as possible in this regard. In particular all the basic material following from that hypothesis is contained in §4. None of the lemmas listed there are new, although some of the proofs are.

2. Preliminaries

We have

$$V(x, Q) = 2S_1 - S_2 + S_3 + 2E(x, Q) - E_1 + O\left(x \log 2x\right)$$

where

$$S_1 = \sum_{q \leq Q} \sum_{\substack{m < n \leq x \\ q|n-m}} \Lambda(m)\Lambda(n), \tag{2.1}$$

$$S_2 = \sum_{q \leq Q} \frac{x^2}{\phi(q)}, \qquad S_3 = Q \sum_{m \leq x} \Lambda(m)^2, \tag{2.2}$$

$$E_1 = \sum_{q \leq Q} \left(\sum_{\substack{m < n \leq x \\ q|n-m;(mn,q)>1}} 2\Lambda(m)\Lambda(n) + \sum_{\substack{n \leq x \\ (n,q)>1}} \left(\Lambda(n)^2 - \frac{2x}{\phi(q)}\Lambda(n)\right) \right)$$

and

$$E(x, Q) = \sum_{q \leq Q} \frac{x}{\phi(q)} \left(x - \psi(x)\right). \tag{2.3}$$

A simple estimation gives

$$E_1 \ll \sum_{q \leq Q} \sum_{\substack{p,s \\ p|q;p^s \leq x}} \left(\left(\sum_{t < s} \log^2 p\right) + \log^2 p + \frac{x}{\phi(q)} \log p \right)$$

$$\ll \sum_{q \leq Q} \sum_{p|q} \left(\log^2 2x + \frac{x \log 2x}{\phi(q)} \right).$$

Thus

$$V(x, Q) = 2S_1 - S_2 + S_3 + 2E(x, Q) + O\left((Q + x)\log^3 x\right). \tag{2.4}$$

For future reference observe that it may be supposed that x is sufficiently large, and that $Q \geq \sqrt{x}$, for the contrary case is immediate from the case $Q = \sqrt{x}$.

The bulk of this memoir is concerned with a detailed analysis of S_1.

3. The Farey dissection and an exponential sum

Let

$$F(\alpha) = \sum_{q \leq Q} \sum_{r \leq x/q} e(\alpha q r), \tag{3.1}$$

$$G(\alpha) = \sum_{n \leq x} \Lambda(n)e(\alpha n). \tag{3.2}$$

Then

$$S_1 = \int_0^1 F(\alpha)|G(\alpha)|^2 d\alpha. \tag{3.3}$$

For any arbitrary function $F : \mathbb{N} \to \mathbb{C}$ one has

$$\sum_{q \leq Q} \sum_{r \leq x/q} f(qr) = \sum_{q \leq \sqrt{x}} \sum_{r \leq x/q} f(qr) + \sum_{r \leq \sqrt{x}} \sum_{\sqrt{x} < q \leq \min(Q, x/r)} f(qr).$$

For convenience an expression of the kind on the right is written in the form

$$\sum_{l \leq \sqrt{x}} \left(\sum_{m \leq x/l} + \sum_{\sqrt{x} < m \leq \min(Q, x/l)} \right) f(lm).$$

Thus

$$F(\alpha) = F_q(\alpha) + H_q(\alpha) \tag{3.4}$$

where

$$F_q(\alpha) = \sum_{\substack{l \leq \sqrt{x} \\ q \mid l}} \left(\sum_{m \leq x/l} + \sum_{\sqrt{x} < m \leq \min(Q, x/l)} \right) e(\alpha l m) \tag{3.5}$$

and $H_q(\alpha)$ is the corresponding multiple sum with $q \nmid l$. On performing the inner summation in $H_q(\alpha)$ one has

$$H_q(\alpha) \ll \sum_{\substack{l \leq \sqrt{x} \\ q \nmid l}} \min \left(\frac{x}{l}, \frac{1}{\|\alpha l\|} \right).$$

For a given a and q with $(a, q) = 1$ put $\beta = \alpha - a/q$. Then

$$\|\alpha l\| \geq \|a l / q\| - |\beta| l$$

and so when

$$l \leq \sqrt{x}, \quad q \nmid l, \quad |\beta| \leq \frac{1}{2q\sqrt{x}}$$

one has

$$H_q(\alpha) \ll \sum_{\substack{l \leq \sqrt{x} \\ q \nmid l}} \frac{1}{\|a l / q\|} \ll \left(\frac{\sqrt{x}}{q} + 1 \right) q \log 2q$$

and so

$$H_q(\alpha) \ll (\sqrt{x} + q) \log 2q \quad \text{when} \quad |\beta| \leq \frac{1}{2q\sqrt{x}}. \tag{3.6}$$

Suppose that R satisfies

$$2\sqrt{x} \leq R \leq \tfrac{1}{2}x \tag{3.7}$$

and consider a typical interval $\mathfrak{M}(q, a)$ associated with the element a/q of the Farey dissection of order R, namely, when $1 \leq a \leq q \leq R$ and $(a, q) = 1$,

$$\mathfrak{M}(q, a) = \left(\frac{a + a_-}{q + q_-}, \frac{a + a_+}{q + q_+} \right]$$

where $q_\pm$ is defined by $aq_\pm \equiv \mp 1 \pmod q$, $R - q < q_\pm \le R$ and $a_\pm$ is defined by $a_\pm = (aq_\pm \pm 1)/q$. We observe that

$$\left| \frac{a + a_\pm}{q + q_\pm} - \frac{a}{q} \right| = \frac{1}{q(q + q_\pm)}$$

lies in $\left[1/(2qR), 1/(qR) \right)$.

We have

$$S_1 = \sum_{q \le R} \sum_{\substack{a=1 \\ (a,q)=1}}^{q} \int_{\mathfrak{M}(q,a)} F(\alpha) |G(\alpha)|^2 d\alpha$$

and

$$\sum_{q \le R} \sum_{\substack{a=1 \\ (a,q)=1}}^{q} \int_{\mathfrak{M}(q,a)} H_q(\alpha) |G(\alpha)|^2 d\alpha \ll R \log x \int_0^1 |G(\alpha)|^2 d\alpha.$$

We note also that $F_q(\alpha) = 0$ when $q > \sqrt{x}$. Hence

$$S_1 = S_4 + O(Rx \log^2 x) \qquad (3.8)$$

where

$$S_4 = \sum_{q \le \sqrt{x}} \sum_{\substack{a=1 \\ (a,q)=1}}^{q} \int_{\mathfrak{M}(q,a)} F_q(\alpha) |G(\alpha)|^2 d\alpha. \qquad (3.9)$$

Let $\beta = \alpha - a/q$. Then, by (3.5),

$$F_q(\alpha) \ll \sum_{m \le \sqrt{x}/q} \min \left(\frac{x}{qm}, \frac{1}{\|\beta qm\|} \right). \qquad (3.10)$$

Hence

$$F_q(\alpha) \ll \frac{x \log(2\sqrt{x}/q)}{q + qx|\beta|} \quad \text{if} \quad q \le \sqrt{x}, \quad |\beta| \le \frac{1}{2q\sqrt{x}}. \qquad (3.11)$$

Suppose that $q \le \sqrt{x}$ and define

$$\mathfrak{N}(q,a) = \left[\frac{a}{q} - \frac{1}{2qR}, \frac{a}{q} + \frac{1}{2qR} \right]. \qquad (3.12)$$

Then $\mathfrak{N}(q,a) \subset \mathfrak{M}(q,a)$ and for $\alpha \in \mathfrak{M}(q,a) \backslash \mathfrak{N}(q,a)$ one has

$$F_q(\alpha) \ll R \log x.$$

Moreover, the same conclusion holds when $\alpha \in \mathfrak{N}(q, a)$ and $q > x/R$. Thus, by (3.8) and (3.9),

$$S_1 = S_5 + O\left(Rx\log^2 x\right) \tag{3.13}$$

where

$$S_5 = \sum_{\substack{q \le x/R}} \sum_{\substack{a=1 \\ (a,q)=1}}^{q} \int_{\mathfrak{N}(q,a)} F_q(\alpha)|G(\alpha)|^2 d\alpha. \tag{3.14}$$

4. Auxiliary lemmas in multiplicative number theory

The following lemmas are required in the proof of Theorem 1.1. The notation $\sum_{\chi(\bmod\, q)}^{*}$ is used to denote summation over the primitive characters modulo q, and given any character χ the function $E(\chi)$ is taken to be 1 when χ is principal and to be 0 otherwise.

Lemma 1. *Suppose that* $0 \le P \le Q$ *and* $f : [P, Q] \to \mathbb{R}_{\ge 0}$ *is such that* f' *exists and is continuous on* $[P, Q]$ *and* $f(u)/u$ *is decreasing. Suppose further that* $X \ge 1$. *Then*

$$\sum_{P < q \le Q} \frac{f(q)}{\phi(q)} \sum_{\chi(\bmod\, q)}^{*} \left| \sum_{n \le X}(\Lambda(n)\chi(n) - E(\chi)) \right|^2$$

$$\ll \left(\frac{Xf(P)}{P} + Pf(P) + \int_P^Q f(u)\,du \right) X\log 2X.$$

This is a variant of a result already contained in Gallagher [5].

Proof. First observe that it may be supposed that $P \ge 1$, for otherwise the term $q = 1$ is $\ll X^2$ and an appeal can be made to the case $P = 1$.

Let

$$T(u) = \sum_{q \le u} \frac{q}{\phi(q)} \sum_{\chi(\bmod\, q)}^{*} \left| \sum_{n \le X} \Lambda(n)\chi(n) \right|^2.$$

Then, by the large sieve inequality, in the form given, for example, by Theorem 4 of §27 of [2],

$$T(u) \ll \left(X + u^2\right)X\log 2X.$$

Hence the sum in question is

$$\frac{f(Q)T(Q)}{Q} - \frac{f(P)T(P)}{P} + \int_P^Q \left(-\frac{f(u)}{u}\right)' T(u)\,du$$

$$\ll \frac{f(Q)(X + Q^2)X\log 2X}{Q} + \int_P^Q \left(-\frac{f(u)}{u}\right)' (X + u^2)X\log 2X\,du$$

$$= \frac{f(P)(X + P^2)X\log 2X}{P} + \int_P^Q f(u)\,2X\log 2X\,du$$

as required.

Let

$$D(X,\chi) = \sum_{n \leq X} (\Lambda(n)\chi(n) - E(\chi)). \tag{4.1}$$

The following lemma summarises all the consequences of the generalised Riemann hypothesis which are required for the proof of Theorem 1.1.

Lemma 2. *Suppose that χ is a primitive character modulo q and that the generalised Riemann hypothesis holds for the Dirichlet L-function defined for $\operatorname{Re} s > 1$ by*

$$L(s,\chi) = \sum_{n=1}^{\infty} \frac{\chi(n)}{n^s}.$$

Suppose further that $X \geq 1$. Then

$$D(X,\chi) \ll X^{\frac{1}{2}} \log^2 2Xq \tag{4.2}$$

and

$$\int_0^X |D(y,\chi)|^2 dy \ll X^2 \log^2 2q. \tag{4.3}$$

Moreover, suppose that $0 \leq H \leq X$. Then

$$\int_0^X |D(y+H,\chi) - D(y,\chi)|^2 dy \ll X(H+1)\log^2 \frac{3qX}{H+1}. \tag{4.4}$$

The main idea of the proof of (4.4) is taken from the proof of Lemma 6 of [14].

Proof. The conclusion (4.2) is immediate from §§18 and 20 of [2].

To prove (4.3) and (4.4) it can certainly be supposed that $X \geq 2$. By (9) and (10) of §17 and by (4) and (5) of §19 of [2] when $2 \leq z \leq X$ and $T = X^2 q^2$,

$$\sum_{n \leq z} (\Lambda(n)\chi(n) - E(\chi)) = -f(z) - b(\chi) + O(\log X), \tag{4.5}$$

where $b(\chi)$ depends only on χ and

$$f(z) = {\sum_{\rho}}' \frac{z^{\rho}}{\rho} \tag{4.6}$$

with $\rho = \frac{1}{2} + i\gamma$ a typical non-trivial zero of $L(s,\chi)$ and $\sum'$ indicating a sum over all such ρ with $|\gamma| \leq T$.

The integral on the left hand side of (4.3) is

$$\ll \int_2^X \left| \sum_{\sqrt{y} < n \leq y} (\Lambda(n)\chi(n) - E(\chi)) \right|^2 dy + X^2$$

$$\ll \int_0^X |f(y) - f(\sqrt{y})|^2 dy + X^2.$$

By squaring our the integrand and integrating term by term this integral becomes $\Sigma_0 - 2\operatorname{Re}\Sigma_1 + \Sigma_2$ where Σ_0, Σ_1 and Σ_2 are sums over ρ_1, ρ_2 with $|\gamma_j| \leq T$ and general terms

$$\frac{X^{\frac{1}{2}\rho_1 + \frac{1}{2}\bar{\rho}_2 + 1}}{\rho_1\bar{\rho}_2(\frac{1}{2}\rho_1 + \frac{1}{2}\bar{\rho}_2 + 1)}, \quad \frac{X^{\rho_1 + \frac{1}{2}\bar{\rho}_2 + 1}}{\rho_1\bar{\rho}_2(\rho_1 + \frac{1}{2}\bar{\rho}_2 + 1)}, \quad \frac{X^{\rho_1 + \bar{\rho}_2 + 1}}{\rho_1\bar{\rho}_2(\rho_1 + \bar{\rho}_2 + 1)},$$

respectively. Thus

$$\Sigma_0 - 2\operatorname{Re}\Sigma_1 + \Sigma_2 \ll \Sigma_3 + \Sigma_4$$

where Σ_3 is over ρ_1, ρ_2 as before but with the general term now

$$\frac{X^{\frac{7}{4}}}{(1 + |\gamma_1|)(1 + |\gamma_2|)(1 + |2\gamma_1 - \gamma_2|)}$$

and Σ_4 over likewise but with the additional constraint $|\gamma_1| \leq |\gamma_2|$ and with general term

$$\frac{X^2}{(1 + |\gamma_1|)(1 + |\gamma_2|)(1 + |\gamma_1 - \gamma_2|)}.$$

Clearly $\Sigma_4 \ll \Sigma_5 + \Sigma_6$ where Σ_5 has the additional constraint $|\gamma_2| \geq 2|\gamma_1|$ and general term

$$\frac{X^2}{(1 + |\gamma_1|)(1 + |\gamma_2|)^2}$$

and Σ_6 has the additional constraint $|\gamma_1| \leq |\gamma_2| \leq 2|\gamma_1|$ and general term

$$\frac{X^2}{(1 + |\gamma_1|)^2(1 + |\gamma_1 - \gamma_2|)}.$$

By (1) and (2) of §15 and (1) of §16 of [2] the number $M(n, \chi)$ of non-trivial zeros of $L(s, \chi)$ with $n - 1 \leq |\gamma| \leq n + 1$ satisfies

$$M(n, \chi) \ll \log 2qn. \qquad (4.7)$$

We have

$$\Sigma_3 \ll \sum_{n \geq 1} \sum_{m \geq 1} \frac{X^{\frac{7}{4}} \log 2qm \log 2qn}{mn(1 + |2n - m|)}.$$

For a given n the terms in the inner sum with $m \leq n$ or $m > 3n$ contribute

$$\ll \frac{X^{\frac{7}{4}} \log^2 2qn \log 2n}{n^2},$$

and likewise those with $n < m \leq 3n$. Thus

$$\Sigma_3 \ll X^{\frac{7}{4}} \log^2 2q.$$

Similarly

$$\Sigma_5 + \Sigma_6 \ll X^2 \log^2 2q.$$

This completes the proof of (4.3).

The conclusion (4.4) is trivial when $H \leq 2$ and follows from (4.3) when $X \leq 4H$. Thus it may be supposed that $H \geq 2$ and $X > 4H$. Moreover it then suffices to show that

$$\int_{4H}^{X} \left| \sum_{y < n \leq y+H} (\Lambda(n)\chi(n) - E(\chi)) \right|^2 dy \ll XH \log^2 \frac{3qX}{H}$$

and this will follow if it is shown first that for $X > 4H$

$$\int_{\frac{1}{2}X}^{X} \left| \sum_{y < n \leq y+H} (\Lambda(n)\chi(n) - E(\chi)) \right|^2 dy \ll XH \log^2 \frac{3qX}{H}.$$

By (4.5) the left hand side is

$$\ll \int_{\frac{1}{2}X}^{X} |f(y+H) - f(y)|^2 dy + X \log^2 X.$$

The second term is acceptable and it remains to deal with the first. For h with $0 \leq h \leq H$, by a change of variable, the first term is

$$\int_{\frac{1}{2}X-h}^{X-h} |f(z+h+H) - f(z+h)|^2 dz$$

$$= \frac{1}{H} \int_0^H dh \int_{\frac{1}{2}X-h}^{X-h} |f(z+h+H) - f(z+h)|^2 dz.$$

The integrand here is

$$\ll \left| f(z + h + H) - f(z) \right|^2 + \left| f(z + h) - f(z) \right|^2$$

and the interval of integration in the inner integral lies in $[\frac{1}{4}X, X]$. Thus the above repeated integral is

$$\ll \frac{1}{H} \int_0^{2H} dh \int_{\frac{1}{4}X}^{X} \left| f(z + h) - f(z) \right|^2 dz.$$

By interchanging the order of integration and substituting $h = zu$ the above becomes

$$\frac{1}{H} \int_{\frac{1}{4}X}^{X} dz \int_0^{2H/z} \left| f(z + zu) - f(z) \right|^2 z \, du$$

$$\ll \frac{X}{H} \int_{\frac{1}{4}X}^{X} dz \int_0^{8H/X} \left| f(z + zu) - f(z) \right|^2 du.$$

Once more the order of integration is inverted. Observe that for $z \in [\frac{1}{4}X, X]$ one has $2X - z \geq X$. Thus the above is

$$\ll \frac{1}{H} \int_0^{8H/X} du \int_0^{2X} (2X - z) \left| f(z + zu) - f(z) \right|^2 dz.$$

By (4.6),

$$f(z + zu) - f(z) = {\sum_\rho}' z^\rho \frac{(1 + u)^\rho - 1}{\rho}.$$

Hence the inner integral is

$${\sum_{\rho_1'}}' {\sum_{\rho_2}}' \frac{(2X)^{\rho_1 + \bar{\rho}_2 + 2}}{(\rho_1 + \bar{\rho}_2 + 1)(\rho_1 + \bar{\rho}_2 + 2)} \cdot \frac{(1 + u)^{\rho_1} - 1}{\rho_1} \cdot \frac{(1 + u)^{\bar{\rho}_2} - 1}{\bar{\rho}_2}.$$

Therefore

$$\int_{\frac{1}{2}X}^{X} \left| f(y + H) - f(y) \right|^2 dy \ll \Sigma_7$$

where Σ_7 is over ρ_1, ρ_2 with $|\gamma_1| \leq |\gamma_2|$ and has the general term

$$\frac{X^2}{(1 + |\gamma_1 - \gamma_2|)^2} \min\left(\frac{H}{X}, \frac{1}{1 + |\gamma_1|} \right) \min\left(\frac{H}{X}, \frac{1}{1 + |\gamma_2|} \right).$$

By (4.7) the terms with $|\gamma_1| \le \frac{1}{2}|\gamma_2|$ contribute

$$\ll {\sum_{\rho_1}}' \; {\sum_{\substack{\rho_2 \\ |\gamma_2| \ge 2|\gamma_1|}}}' \frac{XH}{(1+|\gamma_2|)^3} \ll {\sum_{\rho_1}}' XH \frac{\log 2q(1+|\gamma_1|)}{(1+|\gamma_1|)^2} \ll XH \log^2 2q.$$

Similarly the terms with $|\gamma_1| > \frac{1}{2}|\gamma_2|$ contribute

$$\ll {\sum_{\rho_1}}' \; {\sum_{\substack{\rho_2 \\ |\gamma_1| \le |\gamma_2| < 2|\gamma_1|}}}' \frac{X^2}{(1+|\gamma_1 - \gamma_2|)^2} \min\left(\frac{H^2}{X^2}, \frac{1}{(1+|\gamma_1|)^2}\right).$$

By considering separately the two cases $|\gamma_1| \le X/H$ and $|\gamma_1| > X/H$ and arguing in the same vein as before one finds that this is

$$\ll XH \log^2 \frac{2qX}{H}.$$

This completes the proof of (4.4), whence Lemma 2.

Lemma 3. *Suppose that the Riemann hypothesis holds. Then*

$$\sum_{m \le x} |\psi(x-m) - x + m|^2 \ll x^2, \qquad \sum_{m \le x} |\psi(m) - m|^2 \ll x^2$$

and

$$\sum_{m \le x} \Lambda^2(m) = x \log x - x + O\left(x^{\frac{1}{2}} \log^3 x\right).$$

Proof. The function $\psi(y)$ satisfies $\psi(y) = \psi([x-m]) = \psi(x-m)$ when $[x-m] \le y < [x-m]+1$. Hence

$$\int_{[x-m]}^{[x-m]+1} |\psi(y) - y|^2 \, dy$$
$$= |\psi(x-m) - x + m|^2 + (2\{x-m\} - 1)(\psi(x-m) - x + m)$$
$$+ \{x-m\}^2 - \{x-m\} + \tfrac{1}{3}.$$

Similarly

$$\int_m^{m+1} |\psi(y) - y|^2 \, dy = |\psi(m) - m|^2 - \psi(m) + m + \tfrac{1}{3}.$$

The first two estimates now follow easily from (4.3).

To prove the third observe that

$$\sum_{m \le x} \Lambda^2(m) = \psi(x) \log x - \int_1^x \frac{\psi(u)}{u} \, du + O\left(x^{\frac{1}{2}} \log x\right).$$

The conclusion then follows from (4.2).

Lemma 4. *Suppose that* $X \geq 1$. *Then*

$$\sum_{q \leq X} \frac{1}{\phi(q)} = (\log X + \gamma)C_1 - \sum_{r=1}^{\infty} \frac{\mu(r)^2 \log r}{r\phi(r)} + O\left(\frac{\log 2X}{X}\right)$$

where $C_1 = \zeta(2)\zeta(3)/\zeta(6)$ *as in* (1.12).

Proof. We have

$$\frac{1}{\phi(q)} = \frac{1}{q} \sum_{r|q} \frac{\mu(r)^2}{\phi(r)}.$$

Thus

$$\sum_{q \leq X} \frac{1}{\phi(q)} = \sum_{r \leq X} \frac{\mu(r)^2}{r\phi(r)} \left(\log \frac{X}{r} + \gamma + O\left(\frac{r}{X}\right)\right)$$

$$= (\log X + \gamma) \sum_{r=1}^{\infty} \frac{\mu(r)^2}{r\phi(r)} - \sum_{r=1}^{\infty} \frac{\mu(r)^2 \log r}{r\phi(r)} + O\left(\frac{\log 2X}{X}\right).$$

The lemma follows at once.

5. The exponential sum G

Let

$$J(\beta) = \sum_{n \leq x} e(\beta n), \tag{5.1}$$

$$K(\beta, q, a) = G\left(\beta + \frac{a}{q}\right) - \frac{\mu(q)}{\phi(q)} J(\beta) \tag{5.2}$$

and

$$I(q) = \left[-\frac{1}{2qR}, \frac{1}{2qR}\right]. \tag{5.3}$$

Then, by (3.5), (3.12) and (3.14),

$$S_5 = S_6 + 2S_7 + S_8 \tag{5.4}$$

where

$$S_6 = \sum_{q \leq x/R} \frac{\mu(q)^2}{\phi(q)} \int_{I(q)} F_q(\beta) |J(\beta)|^2 d\beta, \tag{5.5}$$

$$S_7 = \sum_{q \leq x/R} \frac{\mu(q)}{\phi(q)} \int_{I(q)} F_q(\beta) \operatorname{Re}\left(\overline{J(\beta)} \sum_{\substack{a=1 \\ (a,q)=1}}^{q} K(\beta, q, a)\right) d\beta, \tag{5.6}$$

$$S_8 = \sum_{\substack{q \leq x/R}} \sum_{\substack{a=1 \\ (a,q)=1}}^{q} \int_{I(q)} F_q(\beta) |K(\beta, q, a)|^2 d\beta. \qquad (5.7)$$

Let

$$G_q(\alpha) = \sum_{\substack{n \leq x \\ (n,q)=1}} \Lambda(n) e(\alpha n).$$

Then, by (3.2),

$$G(\alpha) - G_q(\alpha) \ll \sum_{\substack{p,t \\ p|q; p^t \leq x}} \log p \ll \log 2q \log x. \qquad (5.8)$$

Moreover

$$G_q\left(\beta + \frac{a}{q}\right) = \sum_{\substack{r=1 \\ (r,q)=1}}^{q} e\left(\frac{ar}{q}\right) \sum_{n \leq x} \frac{\Lambda(n) e(\beta n)}{\phi(q)} \sum_{\chi \pmod q} \chi(n) \bar{\chi}(r)$$

$$= \frac{1}{\phi(q)} \sum_{\chi \pmod q} \chi(a) \tau(\bar{\chi}) \sum_{n \leq x} \Lambda(n) \chi(n) e(\beta n) \qquad (5.9)$$

where τ denotes the Gauss sum

$$\tau(\chi) = \sum_{r=1}^{q} \chi(r) e(r/q) \qquad (5.10)$$

and satisfies $\tau(\chi_0) = \mu(q)$. Let

$$K(\chi, \beta) = \sum_{n \leq x} \left(\Lambda(n)\chi(n) - E(\chi)\right) e(\beta n). \qquad (5.11)$$

Then, by (5.1), (5.2), (5.8) and (5.9),

$$K(\beta, q, a) = \sum_{\chi \pmod q} \frac{\tau(\bar{\chi})}{\phi(q)} \chi(a) K(\chi, \beta) + O\left(\log 2q \log x\right).$$

Hence, by the orthogonality of Dirichlet characters

$$\sum_{\substack{a=1 \\ (a,q)=1}}^{q} |K(\beta, q, a)|^2 \ll \sum_{\chi \pmod q} \frac{|\tau(\bar{\chi})|^2}{\phi(q)} |K(\chi, \beta)|^2 + q \log^2 2q \log^2 x.$$

Therefore, by (3.11) and (5.7),

$$S_8 \ll I_1 + \frac{x \log^6 x}{R}$$

where

$$I_1 = \sum_{q \leq x/R} \sum_{\chi \pmod q} \frac{|\tau(\bar{\chi})|^2}{\phi(q)} \int_{I(q)} |F_q(\beta) K(\chi, \beta)^2| \, d\beta.$$

Let χ^* denote the primitive character which induces χ. Then

$$K(\chi, \beta) \ll |K(\chi^*, \beta)| + \log 2q \log x.$$

Thus, by (3.11)

$$S_8 \ll I_2 + \frac{x \log^6 x}{R}$$

where

$$I_2 = \sum_{q \leq x/R} \sum_{\chi \pmod q} \frac{|\tau(\bar{\chi})|^2}{\phi(q)} \int_{I(q)} \frac{x \log(2\sqrt{x}/q)}{q(1 + x|\beta|)} |K(\chi^*, \beta)|^2 d\beta.$$

Write $q = mr$ where r is the conductor of χ. Then by pages 66-67 of §9 of [2] $|\tau(\bar{\chi})| \leq \sqrt{r}$. Thus

$$I_2 \leq \sum_{r \leq x/R} \sum_{m \leq x/rR} \sum_{\chi^* \pmod r}^* \frac{r}{\phi(mr)} \int_{I(mr)} \frac{x \log(2\sqrt{x}/mr)}{mr(1 + x|\beta|)} |K(\chi^*, \beta)|^2 d\beta.$$

Hence $S_8 \ll I_3 + (x/R) \log^6 x$ where

$$I_3 = \sum_{r \leq x/R} \sum_{\chi \pmod r}^* \frac{\log(2\sqrt{x}/r)}{\phi(r)} \int_{I(r)} \frac{x}{1 + x|\beta|} |K(\chi, \beta)|^2 d\beta.$$

Let $\delta_j = 2^j/x$. Then the integral above is

$$\ll \sum_j \frac{1}{\delta_j} \int_{-\delta_j}^{\delta_j} |K(\chi, \beta)|^2 d\beta$$

where the sum is over $j \geq 0$ with $\delta_j \leq 1/rR$.
     By Lemma 1 of [6]

$$\int_{-\delta_j}^{\delta_j} |K(\chi, \beta)|^2 d\beta \ll \delta_j^2 \int_{-\infty}^{\infty} \left| \sum_{\substack{y < n \leq y+1/(2\delta_j) \\ n \leq x}} (\Lambda(n)\chi(n) - E(\chi)) \right|^2 dy.$$

For brevity δ is used to mean δ_j. Then by (4.1) the right hand side here is

$$\delta^2 \int_{-\frac{1}{2\delta}}^{0} \left| D\left(y + \frac{1}{2\delta}, \chi\right) \right|^2 dy + \delta^2 \int_0^{x - \frac{1}{2\delta}} \left| D\left(y + \frac{1}{2\delta}, \chi\right) - D(y, \chi) \right|^2 dy$$

$$+ \delta^2 \int_{x - \frac{1}{2\delta}}^{x} |D(x, \chi) - D(y, \chi)|^2 dy.$$

Hence, by Lemma 2 and a change of variable, it is

$$\ll \delta x \log^2 2x\delta + \delta |D(x,\chi)|^2 + \delta^2 \int_0^{\frac{1}{2\delta}} |D(x-z,\chi)|^2 dz.$$

Therefore

$$\int_{I(r)} \frac{x}{1+x|\beta|} |K(\chi,\beta)|^2 d\beta \ll x \log \frac{2x}{rR} \log^2 x + \left(\log \frac{2x}{rR}\right) |D(x,\chi)|^2$$

$$+ \int_0^{\frac{x}{2}} \min\left(\frac{1}{z}, \frac{1}{rR}\right) |D(x-z,\chi)|^2 dz$$

and so

$$S_8 \ll I_4 + I_5 + \frac{x^2}{R} \log^2 x \log\left(\frac{R}{\sqrt{x}}\right)$$

where

$$I_4 = \sum_{r \leq x/R} \sideset{}{^*}\sum_{\chi \,(\text{mod } r)} \frac{\log(2\sqrt{x}/r)}{\phi(r)} \left(\log \frac{2x}{rR}\right) |D(x,\chi)|^2,$$

$$I_5 = \sum_{r \leq x/R} \sideset{}{^*}\sum_{\chi \,(\text{mod } r)} \frac{\log(2\sqrt{x}/r)}{\phi(r)} \int_0^{\frac{1}{2}x} \min\left(\frac{1}{z}, \frac{1}{rR}\right) |D(x-z,\chi)|^2 dz.$$

Let $P = x^{\frac{1}{2}}/(\log x)^{\frac{3}{2}}$ and suppose henceforward that $R \leq x/P$. Then, by Lemma 2, the terms in I_4 and I_5 with $r \leq P$ each contribute

$$\ll P x \log^4 x \log \frac{2\sqrt{x}}{P} \log \frac{2x}{PR} \ll x^{\frac{3}{2}} (\log x)^{\frac{5}{2}} (\log\log x)^2.$$

By Lemma 1 the contribution of the remaining terms to I_4 is

$$\ll x^{\frac{3}{2}} (\log x)^{\frac{5}{2}} (\log\log x)^2.$$

In I_5 the remaining terms are $\ll I_6 + I_7 + I_8$ where

$$I_6 = \int_{PR}^{x} \frac{1}{z} \sum_{P < r \leq z/R} \sideset{}{^*}\sum_{\chi \,(\text{mod } r)} \frac{\log(2\sqrt{x}/r)}{\phi(r)} |D(x-z,\chi)|^2 dz,$$

$$I_7 = \int_{PR}^{x} \sum_{z/R < r \leq x/R} \sideset{}{^*}\sum_{\chi \,(\text{mod } r)} \frac{\log(2\sqrt{x}/r)}{rR\phi(r)} |D(x-z,\chi)|^2 dz,$$

$$I_8 = \int_0^{PR} \sum_{P < r \leq x/R} \sideset{}{^*}\sum_{\chi \,(\text{mod } r)} \frac{\log(2\sqrt{x}/r)}{rR\phi(r)} |D(x-z,\chi)|^2 dz.$$

By Lemma 1,

$$I_6 \ll \int_{PR}^x \frac{1}{z} \left(\frac{x}{P} \log\log x + P \log\log x + \frac{z}{R} \log \frac{2R\sqrt{x}}{z} \right) x \log x \, dz,$$

$$I_7 \ll \int_{PR}^x \left(\frac{xR}{z^2} \log \frac{2R\sqrt{x}}{z} + \frac{\log^2(2\sqrt{x}/P)}{R} \right) x \log x \, dz,$$

$$I_8 \ll \int_0^{PR} \left(\frac{x \log 2\sqrt{x}/P}{P^2 R} + \frac{\log^2(2\sqrt{x}/P)}{R} \right) x \log x \, dz.$$

Thus

$$I_6 + I_7 + I_8 \ll x^{\frac{3}{2}} (\log x)^{\frac{5}{2}} (\log\log x)^2.$$

Therefore

$$S_8 \ll x^{\frac{3}{2}} (\log x)^{\frac{5}{2}} (\log\log x)^2. \tag{5.12}$$

Now consider S_7. If one thinks of writing K in terms of the Dirichlet characters modulo q then one sees that the principal character is essentially missing, and so one expects to get essentially a zero contribution when averaging over a. More precisely the contribution here should only depend on the primes which divide q and the error term in the prime number theorem. To see this is so, observe that if $(n, q) = 1$ then

$$\sum_{\substack{a=1 \\ (a,q)=1}}^q e\left(\frac{an}{q} \right) = \mu(q).$$

Thus, by (3.2) and (5.2)

$$\sum_{\substack{a=1 \\ (a,q)=1}}^q K(\beta, q, a) = \mu(q) K(\beta, 1, 0) + E$$

where $E \ll q \log q \log x$. Thus, by (3.11) and (5.1), the term E contributes to S_7 an amount

$$\ll \sum_{q \le x/R} \frac{\log^2 x}{\phi(q)} \int_0^1 \frac{x^2 \log(2\sqrt{x}/q)}{(1 + x\beta)^2} \, d\beta.$$

Therefore

$$S_7 = S_9 + O\left(x^{\frac{3}{2}}\right) \tag{5.13}$$

where

$$S_9 = \sum_{q \le x/R} \frac{\mu(q)^2}{\phi(q)} \int_{I(q)} F_q(\beta) \operatorname{Re}\left(\overline{J(\beta)} K(\beta, 1, 0) \right) d\beta.$$

It is convenient first to add in the terms with $x/R < q \leq \sqrt{x}$. By (3.11) the error introduced is

$$\ll \sum_{q > x/R} \frac{\mu(q)^2}{q\phi(q)} \int_0^1 \frac{x^3 \log x}{(1 + x\beta)^2} \, d\beta \ll xR \log x.$$

Now the interval of integration is completed to a unit interval. By (3.10) the error introduced here is

$$\ll \sum_{q \leq \sqrt{x}} \frac{\mu(q)^2}{\phi(q)} \int_{1/2qR}^{\frac{1}{2}} \sum_{m \leq \sqrt{x}/q} \min\left(\frac{x}{mq}, \frac{1}{\|\beta qm\|}\right) \frac{x}{\beta} \, d\beta.$$

By an inversion of the order of integration and summation and a change of variable the integral here is

$$\sum_{m \leq \sqrt{x}/q} \int_{m/2R}^{mq/2} \min\left(\frac{x}{mq}, \frac{1}{\|\gamma\|}\right) \frac{x}{\gamma} \, d\gamma.$$

In the integral the part from $[m/2R, 1/2]$ contributes $\ll xR/m$ and the part from $\left[n - \frac{1}{2}, n + \frac{1}{2}\right]$ contributes $(x/n)\log(2x/mq)$. Hence the error is

$$\ll \sum_{q \leq \sqrt{x}} \frac{\mu(q)^2}{\phi(q)} \left(xR \log \frac{2\sqrt{x}}{q} + \sum_{m \leq \sqrt{x}/q} x \log 2mq \log \frac{2x}{mq}\right)$$

$$\ll xR \log^2 x + x^{\frac{3}{2}} \log^2 x.$$

Therefore, by (3.7),
$$S_9 = I_9 + xR \log^2 x \tag{5.14}$$

where

$$I_9 = \sum_{q \leq \sqrt{x}} \frac{\mu(q)^2}{\phi(q)} \int_{-\frac{1}{2}}^{\frac{1}{2}} F_q(\beta) \operatorname{Re}\left(\overline{J(\beta)} K(\beta, 1, 0)\right) d\beta.$$

Let

$$c_n = \sum_{\substack{l \mid n \\ l \leq \sqrt{x}}} \frac{l}{\phi(l)} + \sum_{\substack{l \leq \sqrt{x}, \sqrt{x} < m \leq Q \\ lm = n}} \frac{l}{\phi(l)}.$$

Then, by (3.5),

$$\sum_{q \leq \sqrt{x}} \frac{\mu(q)^2}{\phi(q)} F_q(\beta) = \sum_{n \leq x} c_n e(n\beta).$$

Thus, by (3.1), (5.1) and (5.2),

$$2I_9 = \int_{-\frac{1}{2}}^{\frac{1}{2}} \sum_{m \leq x} c_m e(m\beta)\big(J(-\beta)K(\beta,1,0) + J(\beta)K(-\beta,1,1)\big)d\beta$$

$$= \sum_{\substack{m,n \\ m+n \leq x}} c_m\big(\Lambda(n) - 1\big) + \sum_{n \leq x}\big(\Lambda(n) - 1\big)\sum_{m < n} c_m.$$

The first sum here is

$$\sum_{m \leq x} c_m\big(\psi(x-m) - [x-m]\big),$$

and c_m satisfies $c_m \ll f(m)$ where $f(m) = \sum_{n|m} n/\phi(n)$. Moreover f satisfies $\sum_{m \leq x} f(m)^2 \ll x\log^3 x$. Hence, by Cauchy's inequality and Lemma 3

$$\sum_{m \leq x} c_m\big(\psi(x-m) - [x-m]\big) \ll x^{\frac{3}{2}}\log^{\frac{3}{2}} x.$$

The second sum is

$$\sum_{m \leq x} c_m\big(\psi(x) - [x] - \psi(m) + m\big)$$

$$= \big(\psi(x) - x\big)\sum_{m \leq x} c_m - \sum_{m \leq x} c_m\big(\psi(m) - m - \{x\}\big)$$

and the second sum here can be treated as above. Therefore

$$2I_9 = \big(\psi(x) - x\big)\sum_{m \leq x} c_m + O\big(x^{\frac{3}{2}}\log^{\frac{3}{2}} x\big).$$

The sum here is

$$\sum_{l \leq \sqrt{x}} \frac{l}{\phi(l)}\left[\frac{x}{l}\right] + \sum_{l \leq x/Q} \frac{l}{\phi(l)}\big(Q - [\sqrt{x}]\big) + \sum_{x/Q < l \leq \sqrt{x}} \frac{l}{\phi(l)}\left(\left[\frac{x}{l}\right] - [\sqrt{x}]\right)$$

$$= x\sum_{l \leq \sqrt{x}} \frac{1}{\phi(l)} + x\sum_{x/Q < l \leq \sqrt{x}} \frac{1}{\phi(l)} + O(x).$$

Therefore, by Lemma 4,

$$\sum_{m \leq x} c_m = C_1 x \log Q + O(x) = x\sum_{q \leq Q} \frac{1}{\phi(q)} + O(x).$$

Hence, by (2.3), $2I_9 = -E(x,Q) + O\big(x^{\frac{3}{2}}\log^2 x\big)$. Therefore, by (3.13), (5.4), (5.12), (5.13) and (5.14)

$$S_1 = S_6 - E(x,Q) + O\big(xR\log^2 x + x^{\frac{3}{2}}(\log x)^{\frac{5}{2}}(\log\log x)^2\big). \qquad (5.15)$$

6. The transition to the main term

By (3.5), (5.1) and (5.3), the interval of integration in (5.5) can be
extended to $[-\frac{1}{2}, \frac{1}{2}]$ with an error

$$\ll \frac{x \log x}{q} \int_{1/2qR}^{\frac{1}{2}} \frac{d\beta}{\beta^2} \ll xR \log x.$$

Hence, by (5.5) and (5.15),

$$S_1 = S_{10} - E(x, Q) + O\left(xR \log^2 x + x^{\frac{3}{2}} (\log x)^{\frac{5}{2}} (\log \log x)^2\right) \qquad (6.1)$$

where

$$S_{10} = \sum_{q \leq x/R} \frac{\mu(q)^2}{\phi(q)} \int_{-\frac{1}{2}}^{\frac{1}{2}} F_q(\beta) |J(\beta)|^2 d\beta. \qquad (6.2)$$

We also have

$$\sum_{x/R < q \leq \sqrt{x}} \frac{\mu(q)^2}{\phi(q)} \int_{-\frac{1}{2}}^{\frac{1}{2}} F_q(\beta) |J(\beta)|^2 d\beta \ll \sum_{x/R < q \leq \sqrt{x}} \frac{\mu(q)^2}{q\phi(q)} x^2 \log x \ll xR \log x.$$

Thus on taking $R = Cx^{\frac{1}{2}}$ for a suitable constant C, consistent with (3.7),
one obtains

$$S_1 = S_{11} - E(x, Q) + O\left(x^{\frac{3}{2}} (\log x)^{\frac{5}{2}} (\log \log x)^2\right), \qquad (6.3)$$

where

$$S_{11} = \sum_{q \leq \sqrt{x}} \frac{\mu(q)^2}{\phi(q)} \int_{-\frac{1}{2}}^{\frac{1}{2}} F_q(\beta) |J(\beta)|^2 d\beta. \qquad (6.4)$$

7. The main term

Let

$$N = [x]. \qquad (7.1)$$

Then, by (5.1),

$$|J(\beta)|^2 = \sum_{|h| \leq x} (N - |h|) e(\beta h).$$

Thus, by (3.5) and (6.5),

$$S_{11} = \sum_{q \leq \sqrt{x}} \frac{\mu(q)^2}{\phi(q)} \sum_{\substack{l \leq \sqrt{x} \\ q|l}} \left(\sum_{m \leq x/l} + \sum_{\sqrt{x} < m \leq \min(Q, x/l)} \right) (N - lm)$$

$$= \sum_{l \leq \sqrt{x}} \frac{l}{\phi(l)} \left(\sum_{m \leq x/l} + \sum_{\sqrt{x} < m \leq \min(Q, x/l)} \right) (N - lm).$$

We can replace N by x with a total error

$$\ll \sum_{l \le \sqrt{x}} \frac{x}{\phi(l)}.$$

Thus

$$S_{11} = S_{12} + O(x \log x) \tag{7.2}$$

where

$$S_{12} = \sum_{l \le \sqrt{x}} \frac{l}{\phi(l)} \left(\sum_{m \le x/l} + \sum_{\sqrt{x} < m \le \min(Q, x/l)} \right)(x - lm).$$

The first sum over m is $\frac{1}{2}x^2/l + O(x)$ and the second is

$$\begin{cases} xQ - x^{\frac{3}{2}} - \frac{1}{2}l(Q^2 - x) + O(x) & \text{when} \quad l \le x/Q \\ \frac{1}{2}x^2/l - x^{\frac{3}{2}} + \frac{1}{2}lx + O(x) & \text{when} \quad x/Q < l \le \sqrt{x}. \end{cases}$$

Thus, by (7.2),

$$S_{11} = \sum_{l \le \sqrt{x}} \frac{1}{\phi(l)} \left(x^2 - lx^{\frac{3}{2}} + \frac{1}{2}l^2 x \right)$$

$$- \sum_{l \le x/Q} \frac{1}{\phi(l)} \left(\frac{1}{2}x^2 - xQl + \frac{1}{2}Q^2 l^2 \right) + O(x^{\frac{3}{2}}).$$

Therefore, by (6.4),

$$S_1 = \frac{1}{2}x^2 \sum_{l \le \sqrt{x}} \frac{1}{\phi(l)} + \frac{1}{2}xW(\sqrt{x}) - \frac{1}{2}Q^2 W(x/Q)$$

$$- E(x, Q) + O\left(x^{\frac{3}{2}}(\log x)^{\frac{5}{2}}(\log \log x)^2 \right), \tag{7.3}$$

where

$$W(X) = \sum_{l \le X} \frac{1}{\phi(l)}(X - l)^2. \tag{7.4}$$

By (2.2) and Lemma 3,

$$S_3 = Qx \log x - Qx + O\left(Qx^{\frac{1}{2}} \log^3 x \right).$$

Hence, by (2.4) and (2.2)

$$V(x, Q) = Qx \log x - Qx + xW(\sqrt{x}) - Q^2 W(x/Q) - \sum_{\sqrt{x} < q \le Q} \frac{x^2}{\phi(q)}$$

$$+ O\left(x^{\frac{3}{2}}(\log x)^{\frac{5}{2}}(\log \log x)^2 + Qx^{\frac{1}{2}}(\log x)^3 \right). \tag{7.5}$$

By Lemma 4,

$$\sum_{\sqrt{x} < q \le Q} \frac{1}{\phi(q)} = C_1 \log \frac{Q}{\sqrt{x}} + O(x^{-\frac{1}{2}} \log x)$$

where

$$C_1 = \frac{\zeta(2)\zeta(3)}{\zeta(6)} \tag{7.6}$$

as in (1.12). Therefore, by (7.5),

$$V(x, Q) = Qx \log x - Qx + xW(\sqrt{x}) - Q^2 W(x/Q) - C_1 x^2 \log(Q/\sqrt{x})$$
$$+ O(x^{\frac{3}{2}} (\log x)^{\frac{5}{2}} (\log \log x)^2 + Qx^{\frac{1}{2}} \log^3 x). \tag{7.7}$$

8. Completion of the proof of Theorem 1.1

In order to conclude the deliberations a good asymptotic formula for the function W is required. To this end another lemma is established.

Lemma 5. *Let*

$$C_2 = \left(\gamma - \tfrac{3}{2} - \sum_p \frac{\log p}{p^2 - p + 1}\right) C_1 \tag{8.1}$$

$$C_3 = \gamma + \log 2\pi + \sum_p \frac{\log p}{p(p-1)} \tag{8.2}$$

where C_1 (and therefore C_2) are as in (1.12), (1.13), and define

$$E(X) = W(X) - C_1 X^2 \log X - C_2 X^2 - X \log X - C_3 X.$$

Then (without any unproved hypothesis)

$$E(X) = \Omega_\pm(X^{\frac{1}{4}}), \tag{8.3}$$

and on the Riemann hypothesis

$$E(X) = O(X^{\frac{1}{4}+\epsilon}). \tag{8.4}$$

Proof. Let

$$D(s) = \sum_{n=1}^{\infty} \frac{1}{\phi(n)n^s} \quad \text{if} \quad \mathrm{Re}\, s > 0.$$

Then

$$D(s) = \zeta(s+1) \prod_p \left(1 + \frac{1}{p^{s+2} - p^{s+1}} \right) \tag{8.5}$$

$$= \zeta(s+1)\zeta(s+2) \prod_p \left(1 + \frac{p^{s+1} - 1}{p^{2s+4} - p^{2s+3}} \right) \tag{8.6}$$

$$= \frac{\zeta(s+1)\zeta(s+2)}{\zeta(2s+4)} T(s) \tag{8.7}$$

where

$$T(s) = \prod_p \left(1 + \frac{1}{(p-1)(p^{s+2}+1)} \right). \tag{8.8}$$

The product in (8.8) converges absolutely and locally uniformly for $\mathrm{Re}\, s > -2$ and so (8.7) affords an analytic continuation of $D(s)$ to that open half plane. We also observe that then (8.5) holds for $\mathrm{Re}\, s > -1$ and (8.6) for $\mathrm{Re}\, s > -\frac{3}{2}$. Thus $D(s)$ is meromorphic in $\mathrm{Re}\, s > -2$ with simple poles at $s = 0$ and $s = -1$ and poles at $\frac{1}{2}\rho - 2$ where ρ is a typical non-trivial zero of $\zeta(s)$.

The proof of (8.3) is an easy consequence of the ideas expounded by Ingham in Chapter V, §§3,4 of [11]. For completeness we give a brief outline. Let

$$F(s) = \int_1^\infty \frac{E(X)}{X^{s+3}} \, dX \quad \text{if} \quad \mathrm{Re}\, s > 0. \tag{8.9}$$

Then

$$F(s) = \frac{2D(s)}{s(s+1)(s+2)} - \frac{C_1}{s^2} - \frac{C_2}{s} - \frac{1}{(s+1)^2} - \frac{C_3}{s+1} \tag{8.10}$$

and this affords an analytic continuation for F to the half plane $\mathrm{Re}\, s > -2$. Moreover, by (8.5), (8.6) the principal parts of

$$\frac{2D(s)}{s(s+1)(s+2)}$$

at $s = 0$ and $s = -1$ are respectively

$$\frac{C_1}{s^2} + \frac{C_2}{s}, \quad \frac{B_1}{(s+1)^2} + \frac{B_2}{(s+1)}$$

where

$$B_1 = -2\zeta(0), \quad B_2 = -2\left(\zeta'(0) + \zeta(0) \sum_p \frac{\log p}{p(p-1)} + \gamma\zeta(0) \right).$$

As a simple deduction from formulae established in §12 of [2] one has

$$\zeta(0) = -\tfrac{1}{2}, \quad \zeta'(0) = -\tfrac{1}{2}\log 2\pi.$$

Therefore, by (7.6), (8.1), (8.2), (8.7) and (8.8), $F(s)$ is analytic in that region except possibly at the zeros of $\zeta(2s+4)$. In particular $F(s)$ is analytic at all real s with $s \geq -\tfrac{7}{4}$.

Let $\rho = \tfrac{1}{2} + i\gamma$ denote the zero of $\zeta(s)$ with $\gamma > 0$ and γ minimal. Then it is a simple zero and $\zeta(\tfrac{1}{4} + \tfrac{1}{2}i\gamma) \neq 0$ (see §15.1 of [15]), and so $D(s)$, and therefore $F(s)$, has a simple pole at $w = -\tfrac{7}{4} + \tfrac{1}{2}i\gamma$. Let z denote the residue of $F(s)$ at w and let $\eta = |z|$. Suppose that for some $X_0 \geq 1$ we have

$$E(X) \leq \tfrac{1}{4}\eta X^{\frac{1}{4}} \quad (X \geq X_0). \tag{8.11}$$

Let θ denote the abscissa of convergence of

$$G(s) = \int_1^\infty \frac{\tfrac{1}{4}\eta X^{\frac{1}{4}} - E(X)}{X^{s+3}}\, dX.$$

Then, by (8.10), for $\mathrm{Re}\, s > \max(-2, \theta)$ we have

$$G(s) = \frac{\eta}{4s + 7} - F(s)$$

and by (8.11) and Theorem H of Chapter V of [11] it follows that G has a singularity at θ. Hence $\theta \leq -\tfrac{7}{4}$. Therefore (8.9) holds for $\mathrm{Re}\, s > -\tfrac{7}{4}$. Let $\phi = -\arg z$. Then for $\sigma > -\tfrac{7}{4}$ we have

$$F(\sigma) + \mathrm{Re}\left(e^{i\phi} F(\sigma + \tfrac{1}{2}i\gamma)\right) = \int_1^\infty \frac{E(X)}{X^{\sigma+3}}\left(1 + \cos(\phi - \tfrac{1}{2}i\gamma \log X)\right) dX.$$

The contribution to the integral from the interval $[1, X_0]$ is a uniformly bounded function of σ for $\sigma \geq -2$ and when $\sigma > -\tfrac{7}{4}$ the contribution from $[X_0, \infty]$ does not exceed

$$\frac{\eta}{\left(2\sigma + \tfrac{7}{2}\right)X_0^{\sigma+\frac{7}{4}}} \leq \frac{\eta}{2\sigma + \tfrac{7}{2}}.$$

When $\sigma + \tfrac{7}{4}$ is positive and sufficiently small

$$-\mathrm{Re}\left(e^{i\phi} F(\sigma + \tfrac{1}{2}i\gamma)\right) \leq -\frac{3\eta}{4\sigma + 7}.$$

Hence

$$F(\sigma) \leq -\frac{\eta}{8\sigma + 14}.$$

and so F has a singularity at $-\frac{7}{4}$ which gives a contradiction, and so (8.11) must be false. Hence for every $X_0 \geq 1$ there is an $X > X_0$ such that $E(X) > \eta X^{1/4}/4$. Thus $E(X) = \Omega_+(X^{1/4})$. A concomitant argument shows that $E(X) = \Omega_-(X^{1/4})$.

We can now concentrate on (8.4), and henceforward assume the Riemann hypothesis. By Theorems 14.2, 14.14A of Chapter 14 of [15], the functional equation, and standard estimates for the Γ function one has

$$\frac{1}{\zeta(s)} \ll \tau^\epsilon \quad \text{if} \quad \sigma \geq \tfrac{1}{2} + \epsilon, \quad \zeta(s) - \frac{1}{s-1} \ll \tau^{\lambda(\sigma)+\epsilon} \quad \text{if} \quad \sigma \geq -2$$

where

$$\tau = 1 + |t|, \quad t = \operatorname{Im} s, \quad \lambda(\sigma) = \max\left(0, \tfrac{1}{2} - \sigma\right), \quad \sigma = \operatorname{Re} s.$$

Thus in the region $\sigma \geq \epsilon - \frac{7}{4}$, $|s| \geq \frac{1}{10}$, $|s+1| \geq \frac{1}{10}$ one has

$$D(s) \ll \tau^{\theta(\sigma)+\epsilon}$$

where

$$\theta(\sigma) = \begin{cases} 0 & \text{if} \quad -\tfrac{1}{2} \leq \sigma \\ -\sigma - \tfrac{1}{2} & \text{if} \quad -\tfrac{3}{2} \leq \sigma < -\tfrac{1}{2} \\ -2\sigma - 2 & \text{if} \quad \epsilon - \tfrac{7}{4} \leq \sigma < -\tfrac{3}{2}. \end{cases}$$

Now

$$E(X) = \frac{1}{2\pi i} \int_{2-i\infty}^{2+i\infty} F(s) X^{s+2} ds,$$

and in view of the above the line of integration may be moved to the line $\sigma = \epsilon - \frac{7}{4}$. Thus $E(X) \ll X^{\frac{1}{4}+\epsilon}$ and the lemma follows at once.

By (8.4)

$$xW(\sqrt{x}) - Q^2 W\left(\frac{x}{Q}\right)$$
$$= C_1 x^2 \log\sqrt{x} + x^{\frac{3}{2}} \log\sqrt{x} + C_3 x^{\frac{3}{2}} - C_1 x^2 \log(x/Q)$$
$$- Qx \log(x/Q) - C_3 Qx - Q^2 E(x/Q) + O(x^{\frac{9}{8}+\epsilon}).$$

Therefore, by (7.7),

$$V(x, Q) = Qx \log Q - (1 + C_3) Qx - Q^2 E(x/Q)$$
$$+ O(x^{\frac{3}{2}} (\log x)^{\frac{5}{2}} (\log\log x)^2 + Qx^{\frac{1}{2}} \log^3 x).$$

Then the first part of our theorem follows from (8.2) and (8.4), and the second part from (8.2) and (8.3).

Apropos the comment immediately after Theorem 1.1, we observe that if the non-trivial zeros ρ of ζ are simple and $\zeta'(\rho)$ does not get too small, then it follows readily from the above that there is a $\delta > 0$ such that

$$E(X) = \sum_{\rho} \frac{8\zeta(\tfrac{1}{2}\rho - 1)\zeta(\tfrac{1}{2}\rho)}{(\rho - 4)(\rho - 2)\rho\zeta'(\rho)} \, T(\tfrac{1}{2}\rho - 2)X^{\frac{1}{2}\rho} + O(X^{\frac{1}{4}-\delta}) \qquad (8.12)$$

and the series converges absolutely. In particular $E(x) \ll X^{\frac{1}{4}}$.

9. The proof of Theorem 1.2

To prove Theorem 1.2 we simply observe that in (7.7), instead of applying Lemma 4 to approximate $Q^2 W(x/Q)$ we evaluate this term exactly. When $\tfrac{1}{2}x < Q \leq x$ it is $x^2 - 2xQ + Q^2$ and when $\tfrac{1}{3}x < Q \leq \tfrac{1}{2}x$ it is $2x^2 - 6xQ + 5Q^2$. The other terms are treated as in §8.

References

1 M.B. Barban: The large sieve method and its applications in the theory of numbers. *Uspehi Mat. Nauk* **21** (1966), 51–102. English translation in *Russian Math. Surveys* **22** (1966), 49–103.

2 H. Davenport: *Multiplicative Number Theory* (2nd Ed., revised by H.L. Montgomery). Springer (1980).

3 H. Davenport and H. Halberstam: Primes in arithmetic progressions. *Michigan Math. J.* **13** (1966), 485–489.

4 J.B. Friedlander and D.A. Goldston: Variance of distribution of primes in residue classes (preprint). (1995).

5 P.X. Gallagher: The large sieve. *Mathematika* **14** (1967), 14–20 .

6 P.X. Gallagher: A large sieve density estimate near $\sigma = 1$. *Invent. Math.* **11** (1970), 329–339.

7 D.A. Goldston: On Hardy and Littlewood's contribution to the Goldbach Conjecture. In *Proc. Amalfi Conf. on Analytic Number Theory*, 115–155 (E. Bombieri, A. Perelli, S. Salerno and U. Zannier, eds.). Univ. di Salerno (1992).

8 C. Hooley: On the Barban-Davenport-Halberstam theorem I. *J. Reine Angew. Math.* **274/275** (1975), 206–223.

9 C. Hooley: On the Barban-Davenport-Halberstam theorem II. *J. London Math. Soc.* (2) **9** (1975), 625–636.

10 C. Hooley: On the Barban-Davenport-Halberstam theorem III. *J. London Math. Soc.* (2) **11** (1975), 399–407.

11 A.E. Ingham: *The Distribution of Prime Numbers.* Cambridge Tracts in Math. **30**. Reissued with an introduction by R.C. Vaughan, Cambridge Univ. Press (1990).

12 A.F. Lavrik: On the twin prime hypothesis of the theory of primes by the method of I.M. Vinogradov. *Soviet Math. Dokl.* **1** (1960), 700–702.

13 H.L. Montgomery: Primes in arithmetic progressions. *Michigan Math. J.* **17** (1970), 33–39.

14 B. Saffari and R.C. Vaughan: On the fractional parts of x/n and related sequences II. *Ann. de l'Inst. Fourier* **27** (1977), 1–30.

15 E.C. Titchmarsh: *The Theory of the Riemann Zeta-Function* (2nd Ed., revised by D.R. Heath-Brown). Oxford University Press (1986).

8. Franel Integrals

R. R. Hall

1. Introduction

Let

$$\overline{B}_1(x) = x - [x] - \tfrac{1}{2} \sim \frac{i}{2\pi} \sum_{m}^{*} \frac{e(mx)}{m}, \tag{1}$$

where $e(x) = \exp(2\pi i x)$. Here, and throughout this paper, $\sum^{*}$ denotes summation over $\mathbb{Z}$, or, where there are k variables, over $\mathbb{Z}^k$, undefined terms with zero denominators omitted. Also let $A_1, A_2, \ldots, A_n$ be non-zero integers. We shall study the multiple Franel integral

$$I(A_1, A_2, \ldots, A_n) = \int_0^1 \overline{B}_1(A_1 x)\overline{B}_1(A_2 x) \ldots \overline{B}_1(A_n x)\, dx \tag{2}$$

introduced by Greaves, Hall, Huxley and Wilson [2]. Originally the A_i were required to be positive; the extension to $\mathbb{Z}\backslash\{0\}$ will be convenient here and is easy because the Bernoulli function $\overline{B}_1$ is odd on $\mathbb{R}\backslash\mathbb{Z}$. We sometimes denote the integral (2) by $I(\mathbf{A})$ in which $\mathbf{A} = (A_1, \ldots, A_n)$ is an integer vector. Some generalizations are possible: we may introduce higher order Bernoulli functions (since these are polynomials in $\overline{B}_1$ the resulting integrals are related to the original ones), and we may consider higher dimensional integrals. For example a two dimensional integral is

$$I(\mathbf{A}_1, \mathbf{A}_2) = \int_0^1 \int_0^1 \overline{B}_1(A_{11}x_1 + A_{21}x_2) \ldots \overline{B}_1(A_{1n}x_1 + A_{2n}x_2)\, dx_1\, dx_2. \tag{3}$$

This integral occurs in an approximate reduction step suggested to the author by Andrew Granville, to appear in a later memoir. Broadly, a reduction step is a formula relating a Franel integral to another one in which the A_i are numerically smaller. Two such steps were given in [2].

We notice that all the Franel integrals discussed here take rational values: this may be seen by splitting the range of integration into boxes in which the integrand is a rational polynomial. Because $\overline{B}_1$ is (essentially) odd, the integrals are zero when n is odd, and we therefore restrict our attention to the case n even from now on.

Our primary purpose in this paper is to prove that the ordinary Franel integral (2) has an average order in the following sense.

Sieve Methods, Exponential Sums, and their Applications in Number Theory
Greaves, G.R.H., Harman, G., Huxley, M.N., Eds. ©Cambridge University Press, 1996

Theorem 1. *Let $n \geq 4$ be even. Then there exists a constant $H(n) > 0$ such that we have*

$$F_n(N) := \sum_{1 \leq A_1, A_2, \ldots, A_n \leq N} I(\mathbf{A})$$

$$= H(n)N^{n-1} + O(N^{n-1-\kappa} \log^{n-1} N), \qquad (4)$$

where $\kappa = 1/(\lambda(n) + 1)$, with $\lambda(n)$ as in (38). Thus $\kappa \sim (n \log n)^{-1}$. Furthermore,

$$\frac{1}{2^n} \geq H(n) \geq \frac{\zeta(n)}{(n+1)2^{n-1}\zeta(n+1)} + D(n), \qquad (5)$$

where

$$D(n) = 2^{1-n} \int_1^\infty \{y\}^n (1 - \{y\})^n \frac{dy}{y^n} \qquad (\{y\} = y - [y]). \qquad (6)$$

In the special case $n = 2$, (4) holds with $H(2) = \frac{1}{4}$ and error term $O(\log^2 n)$.

In the case $n = 2$ the right hand side of (5) is $0.244096\ldots$; in the general case we expect to have to have strict, but numerically slight, inequality on the right. The left-hand inequality is simpler and unlikely to be very precise. We remark that it would be possible to define alternative mean values by varying the norm applied to the vector $\mathbf{A}$. We have chosen the supremum norm because the sum in (4) is then given by the formula

$$F_n(N) = \int_0^1 F(x, N)^n \, dx \qquad (7)$$

in which

$$F(x, N) = \sum_{A=1}^N \overline{B}_1(Ax). \qquad (8)$$

We employ the integral (7) to obtain the lower bound for $H(n)$ when $n \geq 4$ by a Farey dissection method. The upper bound follows immediately from the bound $|F(x,N)| \leq \frac{1}{2}N$ together with $H(2) = \frac{1}{4}$. So far this analysis has not been successful in evaluating $H(n)$ or proving Theorem 1: the proof given here depends on different ideas altogether. Thus we do not know, for example, whether or not $H(n)$ is rational.

The scheme of the paper is as follows. In the next section we describe the connection between Franel integrals and a special class of zeta-values, which were introduced to the author by Zagier in e-mail conversations. We set out the basic theory of these zeta-values, which will be indispensable for the proof of Granville's reduction step, and employ them to give short proofs of reduction steps 1 and 2, described in [2]. In §3 we prove Theorem 1 when $n = 2$, using Franel's original evaluation of the second order integral (2). In §4 we establish the existence of the constant $H(n)$ such that (4) holds: the proof yields no information about the value of $H(n)$. In §5 we apply the Farey dissection (circle) method to the integral (7), to obtain the lower bound for $H(n)$. At this point we require information about the sum

$$\sum_{m \leq R} \overline{B}_1\left(\frac{ma}{q}\right) \qquad (R < q, \quad (a, q) = 1), \qquad (9)$$

and because this has ramifications which may be of independent interest, we develop the theory of this sum, for Bernoulli functions of arbitrary order, in the final section of the paper. We obtain the following result.

Theorem 2. *Let* $\overline{B}_n$ *denote the periodic extension to* $\mathbb{R}$ *of the nth degree Bernoulli polynomial on* $[0, 1]$, *and*

$$G_n\left(\frac{a}{q}, R\right) = \sum_{m=1}^{R} \overline{B}_n\left(\frac{ma}{q}\right) \qquad (10)$$

when $R < q$, $1 < a < q$, $(a, q) = 1$. *Then we have*

$$G_n\left(\frac{a}{q}, R\right) = H_n\left(\frac{a}{q}, R\right) + (-1)^n \left(\frac{a}{q}\right)^{n-1} G_n\left(\frac{q}{a}, S\right) \qquad (11)$$

where $S = [Ra/q] < a$ *and*

$$H_n\left(\frac{a}{q}, R\right)$$
$$= \frac{1}{n+1} \sum_{r=1}^{n+1} \binom{n+1}{r} (-1)^{n+1-r} B_{n+1-r}\left(\overline{B}_r\left(\frac{Ra}{q}\right) - B_r\right)\left(\frac{a}{q}\right)^{n-r}. \qquad (12)$$

Furthermore (11) *leads to a formula for* $G_n(a/q, R)$ *in terms of the Euclidean Algorithm. In the case* $n = 1$, *when* $G_1 = F$, *this is*

$$F(a/q, R) = \sum_{\mu=0}^{\nu} (-1)^\mu H_1\left(\frac{a_\mu}{a_{\mu-1}}, R_\mu\right) \qquad (13)$$

in which $R_0 = R$, $R_{\mu+1} = [R_\mu a_\mu / a_{\mu-1}]$, $a_0 = a$, $a_{-1} = q$ *and*

$$a_{\mu-2} = m_\mu a_{\mu-1} + a_\mu \quad (1 \leq \mu \leq \nu), \qquad a_\nu = 1. \qquad (14)$$

2. Zeta-values

We shall be concerned with sums of the form

$$Z = \sum_{n_1,\ldots,n_k}^{*} \prod_{j=0}^{k} \frac{1}{e_{1j}n_1 + e_{2j}n_2 + \cdots + e_{kj}n_k}, \tag{15}$$

where the e_j are integers. We follow Zagier in calling this sum a *generalized zeta-value*. We claim that the series is absolutely convergent if the $k \times (k+1)$ matrix $[e_{ij}]$ has the property that each of its $k \times k$ submatrices is non-singular; this point is quite important and we give a proof. For $1 \leq j \leq k$ put $e_{1j}n_1 + \cdots + e_{kj}n_k = m_j$. By hypothesis there exists a non-zero integer Δ, (the determinant of the submatrix on the left) such that, for each $i \leq k$, Δn_i is a linear combination over $\mathbb{Z}$ of the m_j. Hence we have

$$\Delta(e_{10}n_1 + e_{20}n_2 + \cdots + e_{k0}n_k) = c_1 m_1 + c_2 m_2 + \cdots + c_k m_k$$

where the c_j are integers, moreover they are non-zero. We argue this last point by contradiction. If we had say $c_h = 0$, then the equations

$$e_{1j}n_1 + e_{2j}N_2 + \cdots + e_{kj}n_k = m_j \qquad (1 \leq j \leq k, \; j \neq h)$$
$$\Delta(e_{10}n_1 + e_{20}n_2 + \cdots + e_{k0}n_k) = c_1 m_1 + c_2 m_2 + \cdots + c_k m_k$$

would, by our non-singularity hypothesis and the extra hypothesis $c_h = 0$, determine the n_i independently of m_h. This impossibility is the required contradiction, whence $c_h \neq 0$. The m_j are integers and so it will be sufficient for our claim to show that

$$\sum_{m_1,\ldots,m_k}^{*} \frac{1}{|m_1 \ldots m_k| \, |c_1 m_1 + \cdots + c_k m_k|} < \infty. \tag{16}$$

In fact the sum on the left is

$$\int_0^1 \prod_{j=1}^{k} \left\{ -2\log|1 - e(c_j\theta)| \right\} \left\{ -2\log|1 - e(\theta)| \right\} d\theta. \tag{17}$$

The zeta-value defined in (15) is zero whenever k is even, because there are $k+1$ factors in the denominator which change sign if we change the sign of all the n_i. We prove next that each ordinary Franel integral as in (2) is simply related to an associated zeta-value, and conversely.

We shall have $n = k+1$, so that k is odd. Let us expand each Bernoulli function in (2) in a Fourier series, and integrate formally term by term. We obtain

$$I(\mathbf{A}) = \left(\frac{i}{2\pi}\right)^n \sum_{\mathbf{m}\cdot\mathbf{A}=0}^{*} \frac{1}{m_1 m_2 \ldots m_n} \tag{18}$$

where $\mathbf{m} \cdot \mathbf{A}$ denotes the usual inner product in $\mathbb{R}^n$. The integer vectors $\mathbf{m}$ orthogonal to $\mathbf{A}$ lie in a lattice $\Lambda = \Lambda(\mathbf{A})$ say, with integer basis

$$\mathbf{e}_1, \ldots, \mathbf{e}_k \in \mathbb{Z}^n.$$

That is, the $\mathbf{m}$ are given by

$$\mathbf{m} = n_1 \mathbf{e}_1 + \cdots + n_k \mathbf{e}_k \qquad (n_i \in \mathbb{Z}, \quad 1 \le i \le k).$$

The series on the right of (18) is now formally equal to the zeta-value Z defined in (15). This analysis requires justification. Let us approximate the Bernoulli function in (1) by the trigonometric polynomial

$$\overline{B}_1^M(x) = \frac{i}{2\pi} \sum_{0 < |m| \le M} \frac{e(mx)}{m}, \tag{19}$$

denoting the integral corresponding to (2) by $I^M(\mathbf{A})$. We have

$$\overline{B}_1(x) - \overline{B}_1^M(x) \ll \min\left(1, M^{-1} |\csc \pi x|\right), \tag{20}$$

which implies that

$$I(\mathbf{A}) = I^M(\mathbf{A}) + O\left(\frac{\log M}{M}\right). \tag{21}$$

(The O-constant depends on n only). We apply the truncation $|m_j| \le M$ in (19) which involves, in turn, a restriction to a suitable partial sum (not a truncation) of the zeta series (15). If we assume temporarily that this series is absolutely convergent we may let $M \to \infty$ to obtain

$$I(\mathbf{A}) = \left(\frac{i}{2\pi}\right)^n Z \tag{22}$$

as required.

We have to show that the zeta-value obtained in this way is of the type defined, that is the matrix $[e_{ij}]$ has all $k \times k$ submatrices non-singular. Consider the $(k+1) \times (k+1)$ matrix E (say), obtained by adjoining the extra (top) row $(m_1, \ldots, m_n)$ to $[e_{ij}]$, that is, put $e_{0j} = m_j$ when $1 \le j \le n$, recalling that $n = k + 1$. Denote the determinant of this large matrix by $\det E$. If $\mathbf{m} \cdot \mathbf{A} = 0$ then $\mathbf{m}$ is a linear combination of the $\mathbf{e}_i$ and so $\det E = 0$. The converse is true because the $\mathbf{e}_i$ for $1 \le i \le k$ are independent. Now expand $\det E$ along the new row to obtain, say, $\det E = \mathbf{m} \cdot \mathbf{B}$. Then $\mathbf{A}$ and $\mathbf{B}$ are parallel integer vectors and we deduce that $\mathbf{A} = \lambda \mathbf{B}$, where $\lambda \ (\neq 0)$ is rational. The coordinates of $\mathbf{B}$ are therefore non-zero, and are cofactors of the large matrix, the corresponding minors being the matrices which we require to be non-singular. This completes the proof.

It is worthwhile to notice at this point that we have $\mathbf{B} = \pm\mathbf{A}$ when $\mathrm{hcf}(A_1, A_2, \ldots, A_n) = 1$. One way to see this is to show that $|\mathbf{B}|$ is independent of the choice of basis, and then construct a special basis for which $|\mathbf{B}|$ may be evaluated simply. Let $\mathbf{f}_j$ $(1 \leq j \leq k)$ be another basis, and $U = [u_{ij}]$ be the $k \times k$ matrix representing our change of basis, so that U is unimodular. Let F be the $(k+1) \times (k+1)$ matrix corresponding to E, (with $f_{0l} = m_l$). Then $E = U^*F$ where we add a top row, and left column, of noughts to U, with 1 in the top left corner, to obtain U^*. Clearly

$$\det E = \det U \det F = \pm \det F.$$

This achieves the first step. The special basis has $e_{ij} = 0$ if $j < i$, and $e_{ii} = h_{i+1}/h_i$ where

$$h_i = \mathrm{hcf}(A_i, A_{i+1}, \ldots, A_n) \qquad (1 \leq i \leq k).$$

(The reader may easily check that such a basis exists). In this case the nth coordinate of $\mathbf{B}$ is

$$(-1)^k e_{11} e_{22} \ldots e_{kk} = (-1)^k A_n,$$

whence $\mathbf{B} = (-1)^k \mathbf{A}$ as claimed. This argument shows in addition that the k-dimensional measure of a fundamental parallelepiped in $\Lambda(\mathbf{A})$ is $|\mathbf{A}|$, (provided $\mathrm{hcf}\{A_i\} = 1$). Replace the top row of the matrix E by the unit row vector $\mathbf{a} = \mathbf{A}/|\mathbf{A}|$. Then $\det E = \pm|\mathbf{A}|$. Since $\mathbf{a}$ is perpendicular to $\Lambda(\mathbf{A})$, we have $|\det E| = |\Lambda(\mathbf{A})|$, ($|\det E|$ equals the n-dimensional measure of a slab of unit thickness whose base, in $\Lambda(\mathbf{A})$, is a fundamental parallelepiped). This gives the result stated.

We remark in passing an observation of Zagier that a zeta-value (15) is a rational multiple of π^{k+1}. The above process, relating $I(\mathbf{A})$ to a zeta-value, is reversible, and in most cases provides the easiest evaluation of Z. For example, let

$$Z = \sum{}^* \frac{1}{n_1 n_2 n_3 (n_1 + n_2 + n_3)}. \tag{23}$$

We have $k = 3$, $n = 4$, $\mathbf{e}_1 = (1, 0, 0, 1)$, $\mathbf{e}_2 = (0, 1, 0, 1)$, $\mathbf{e}_3 = (0, 0, 1, 1)$. The vector $\mathbf{A} = (1, 1, 1, -1)$ is orthogonal to these, and in this case $I(\mathbf{A}) = -\frac{1}{80}$. We conclude that $Z = -\frac{1}{5}\pi^4$.

At this point we give proofs of Reduction Steps 1 and 2, as described in [2], employing zeta-values. This illustrates the theory, and the new proof of Reduction Step 2 is certainly more transparent.

Reduction Step 1 asserts that if $\mathrm{hcf}(A_1, \ldots, A_n) = d$ then

$$I(\mathbf{A}) = I\left(\frac{A_1}{d}, \ldots, \frac{A_n}{d}\right).$$

We have only to note that the lattice Λ is unaltered when we scale $\mathbf{A}$. (It is not quite enough to argue from the purely formal identity (18)).

Reduction Step 2 asserts that if $(A_1, d) = 1$, where $d \mid \mathrm{hcf}(A_2, \ldots, A_n)$, then

$$I(\mathbf{A}) = \frac{1}{d} I\left(A_1, \frac{A_2}{d}, \frac{A_3}{d}, \ldots, \frac{A_n}{d}\right).$$

To see this, notice that the condition $\mathbf{m} \cdot \mathbf{A} = 0$ involves $d \mid m_1 A_1$ and so $d \mid m_1$. Hence $\mathbf{m} = (dl_1, l_2, l_3, \ldots, l_n)$ where $(l_1, \ldots, l_n)$ is orthogonal to $(A_1, A_2/d, \ldots, A_n/d)$. The lattice alters: in the new basis e_{i1} becomes de_{i1} for every i, so that Z is divided by d. As before this is formally intuitive from (18). The two-dimensional integral $I(A_1, A_2)$ is connected with a zeta-value in a similar fashion. In this case the lattice is $\Lambda(\mathbf{A}_1, \mathbf{A}_2)$ comprising the integer vectors $\mathbf{m}$ orthogonal to both of $\mathbf{A}_1, \mathbf{A}_2$.

3. Proof of Theorem 1 (the case $n = 2$)

This is a special case because we have the Franel-Kluyver formula [1],[6]:

$$I(A_1, A_2) = \frac{1}{12} \frac{(A_1, A_2)^2}{A_1 A_2}. \tag{24}$$

Now let $\phi_2(d)$ be the arithmetical function defined implicitly by the relation

$$n^2 = \sum_{d \mid n} \phi_2(d), \tag{25}$$

so that actually $\phi_2(d) = d^2 \prod_{p \mid d}(1 - 1/p^2)$. We write down the sum $12F_2(N)$ and apply (25), inverting summations, to obtain

$$12F_2(N) = \sum_{d \leq N} \frac{\phi_2(d)}{d^2} \left(\sum_{m \leq N/d} \frac{1}{m}\right)^2. \tag{26}$$

Let us define $L_j = 1 + 1/2 + 1/3 + \cdots + 1/j$. Then the inner sum in (26) is equal to L_j if and only if $N/(j+1) < d \leq N/j$, whence

$$12F_2(N) = \sum_{j \leq N} L_j^2 \sum_{N/(j+1) < d \leq N/j} \frac{\phi_2(d)}{d^2}$$

$$= L_1^2 T_N + (L_2^2 - L_1^2) T_{N/2} + \cdots + (L_N^2 - L_{N-1}^2) T_1, \tag{27}$$

where

$$T_x := \sum_{d \leq x} \frac{\phi_2(d)}{d^2} = \sum_{h \leq x} \frac{\mu(h)}{h^2} \left[\frac{x}{h}\right] = \frac{x}{\zeta(3)} + O(1). \tag{28}$$

We insert (28) into (27) which yields (with $L_0 = 0$)

$$12F_2(N) = \frac{N}{\zeta(3)} \sum_{j \leq N} \left(\frac{2}{j} L_{j-1} + \frac{1}{j^2}\right) \frac{1}{j} + O(\log^2 N). \tag{29}$$

We employ the familiar identity

$$\zeta(3) = \frac{1}{2^2} + \frac{1}{3^2}\left(1 + \tfrac{1}{2}\right) + \frac{1}{4^2}\left(1 + \tfrac{1}{2} + \tfrac{1}{3}\right) + \cdots \tag{30}$$

to obtain

$$12F_2(N) = 3N + O(\log^2 N). \tag{31}$$

This completes the proof of this part of Theorem 1.

4. Proof of Theorem 1 (the existence of $H(n)$ when $n \geq 4$)

Next we consider the case $n \geq 4$ (even), and set out to prove that there exists a constant $H(n)$ such that (4) holds. Our argument does not determine $H(n)$, indeed it does not show that $H(n) > 0$. In this section $n \geq 4$, unless stated otherwise.

Lemma 1. *Let Λ be a lattice of dimension k in $\mathbb{R}^n$, where $n \geq k$, and let $\mathbf{f}_1, \mathbf{f}_2, \ldots, \mathbf{f}_k$ be any set of linearly independent elements of Λ. Then Λ has a basis $\mathbf{e}_1, \mathbf{e}_2, \ldots, \mathbf{e}_k$ such that*

$$|\mathbf{e}_1| \leq |\mathbf{f}_1|, \qquad |\mathbf{e}_2| \leq |\mathbf{f}_1| + |\mathbf{f}_2|,$$

and, in general,

$$|\mathbf{e}_k| \leq 2^{k-2}|\mathbf{f}_1| + 2^{k-3}|\mathbf{f}_2| + \cdots + |\mathbf{f}_{k-1}| + |\mathbf{f}_k|. \tag{32}$$

Proof. We refer to Gruber and Lekkerkerker [4] (Chapter 1, Theorem 3, Corollary) which asserts that Λ has a basis such that for $1 \leq i \leq k$

$$\mathbf{f}_i = u_{i1}\mathbf{e}_1 + u_{i2}\mathbf{e}_2 + \cdots + u_{ii}\mathbf{e}_i \tag{33}$$

where $0 \leq u_{ij} < u_{ii}$ for $1 \leq j < i$. We invert the equations recursively, and (32) follows.

Throughout the remainder of this section, $k = n - 1$.

Lemma 2. *Let $\mathbf{m}$ be an integer vector in $\mathbb{R}^n$ and $\Lambda(\mathbf{m})$ be the lattice of integer vectors $\mathbf{A}$ such that $\mathbf{m} \cdot \mathbf{A} = 0$. Then Λ has a basis $\mathbf{e}_1, \ldots, \mathbf{e}_k$ such that*

$$|\mathbf{e}_i| \ll |\mathbf{m}|^{1/(k+1-i)} \qquad (1 \leq i \leq k). \tag{34}$$

Proof. We construct a set of linearly independent vectors $\mathbf{f}_i \in \Lambda(\mathbf{m})$ satisfying similar inequalities to (34), and apply Lemma 1. We use the box principle. Consider the $(2X + 1)^n$ numbers $r = \mathbf{m} \cdot \mathbf{x}$ where $\mathbf{x} \in \mathbb{Z}^n$ and $\max |x_j| \leq X$. Let $M = \max |m_j|$. Then $|r| \leq nMX$ and so, if $(2X + 1)^n > 2nMX + 1$, a value of r is repeated. Then, subtracting the two corresponding x's, we obtain a vector $\mathbf{f}_1$ with $|\mathbf{f}_1| \ll |\mathbf{m}|^{1/k}$, and which is perpendicular to $\mathbf{m}$. Let f_{1j} be any non-zero component. We repeat the construction except that we fix $x_j = 0$. This yields $\mathbf{f}_2$ with $|\mathbf{f}_2| \ll |\mathbf{m}|^{1/(k-1)}$, and with $\mathbf{f}_1$ and $\mathbf{f}_2$ independent, because $f_{2j} = 0 \neq f_{1j}$. Let f_{2l} be any non-zero component of f_2. We repeat the construction again, this time fixing $x_j = x_l = 0$. We obtain $\mathbf{f}_3$ such that $|\mathbf{f}_3| \ll |\mathbf{m}|^{1/(k-2)}$, moreover $\mathbf{f}_1$, $\mathbf{f}_2$ and $\mathbf{f}_3$ are independent. After k applications this process gives the required set of k vectors $\mathbf{f}_j$.

Lemma 3. *Let* $\mathbf{m} = (m_1, \ldots, m_n)$ *be an integer vector with non-zero components such that* $\mathrm{hcf}(m_1, \ldots, m_n) = 1$. *Then the number of solutions* $S(\mathbf{m}, N)$ *of the Diophantine equation*

$$m_1 A_1 + m_2 A_2 + \cdots + m_n A_n = 0, \tag{35}$$

subject to the constraints $0 < A_i \leq N$ *for* $1 \leq i \leq n$, *satisfies*

$$S(\mathbf{m}, N) \ll \frac{N^{n-1}}{\max\{|m_i| : 1 \leq i \leq n\}} + N^{n-2}. \tag{36}$$

Proof. This is a straightforward generalization of §2.3, Lemma 3 of [3], in which there is an extra condition $|m_i| \leq N$ when $1 \leq i \leq n$ and the final term in (36) is not needed. The induction goes through in a similar fashion (we start at $n = 2$).

Lemma 4. *Let* $n \geq 2$ *and let* $S(\mathbf{m}, N)$ *be defined as in the previous lemma, except that we now allow* $\mathrm{hcf}(m_1, \ldots, m_n) > 1$. *Then we have*

$$S(\mathbf{m}, N) = V(\mathbf{m}) N^{n-1} + O\big(|\mathbf{m}|^{\lambda(n)} N^{n-2}\big) \tag{37}$$

where $\lambda(2) = 0$, *and for* $n > 2$

$$\lambda(n) = (n - 2)\left(1 + \frac{1}{2} + \frac{1}{3} + \cdots + \frac{1}{n-2}\right); \tag{38}$$

$V(\mathbf{m})$ *is a function of* $\mathbf{m}$ *satisfying*

$$V(\mathbf{m}) \ll \frac{\mathrm{hcf}(m_1, \ldots, m_n)}{\max\{|m_i| : 1 \leq i \leq n\}}. \tag{39}$$

Remark. Clearly $S(\mathbf{m}, N)$ and $V(\mathbf{m})$ are independent of the highest common factor of the components.

Proof. The result for $n = 2$ is immediate: we have included it for completeness. Let $n > 2$ and $\mathbf{e}_1, \ldots, \mathbf{e}_k$ be the basis for $\Lambda(\mathbf{m})$ constructed in Lemma 2. Put $\mathbf{A} = x_1\mathbf{e}_1 + \cdots + x_k\mathbf{e}_k$. We have to count the integer vectors $\mathbf{x}$ in $\mathbb{R}^k$ for which the (pairs of) inequalities

$$0 < A_j = x_1 e_{1j} + x_2 e_{2j} + \cdots + x_k e_{kj} \leq N \qquad (1 \leq j \leq n) \tag{40}$$

hold. If we omit any one value of j in (40), (say j_0) and allow $\mathbf{x}$ to be a real vector, the resulting inequalities define a parallelepiped in $\mathbb{R}^k$. The intersection of these n parallelepipeds is a convex region, $\mathbb{K}$, say. We need an upper bound for the maximum distance from the origin to a point of $\mathbb{K}$, which we denote by R.

We have $R = |OP|$, where P is a vertex of one of the parallelepipeds defined above. The vertex P has coordinates $(y_1, \ldots, y_k)$ where the y_i satisfy k equations

$$y_1 e_{1j} + y_2 e_{2j} + \cdots + y_k e_{kj} = 0 \text{ or } N \qquad (1 \le j \le n, \ j \ne j_0). \qquad (41)$$

From the discussion following (22), we know that the matrix involved here is non-singular, indeed its determinant is $\pm m_{j0}$. We apply Cramer's rule and a familiar result of Hadamard (see pp. 212–3 in [9]) to deduce that for $1 \le i \le k$

$$|y_i| \ll N \prod_{h \ne i} |e_h| \le |\mathbf{m}|^\mu N, \qquad (42)$$

where $\mu = 1 + \frac{1}{2} + \frac{1}{3} + \cdots + \frac{1}{k-1}$ by (34). Hence

$$R = |\mathbf{y}| \ll |\mathbf{m}|^\mu N. \qquad (43)$$

The constant implicit in Vinogradov's notation $\ll$ depends on k, and therefore n, in the obvious way.

Let $\mathbb{S}$ be the sphere of radius R centred at $\mathbf{0}$ in which $\mathbb{K}$ is inscribed. The surface area of $\mathbb{S}$ (the $(k-1)$-dimensional measure of $\partial \mathbb{S}$) satisfies

$$|\partial \mathbb{S}| \ll |\mathbf{m}|^{(k-1)\mu} N^{k-1}. \qquad (44)$$

Next, the volume (k-dimensional measure) of $\mathbb{K}$ is $V(\mathbf{m}) N^k$, where $V(\mathbf{m})$ is the measure of $\mathbb{K}_0$, the region defined by the inequalities (40) with $N = 1$. The number of integer vectors $\mathbf{x} \in \mathbb{K}$ is therefore

$$V(\mathbf{m}) N^k + O\big(|\mathbf{m}|^{(k-1)\mu} N^{k-1}\big), \qquad (45)$$

and we note that $(k-1)\mu = \lambda(n)$. This proves (37).

It remains to prove (39). As we remarked above, $V(\mathbf{m})$ does not depend on the highest common factor of the components of $\mathbf{m}$, and so it will be sufficient to show that in the case $\mathrm{hcf}(m_1, \ldots, m_n) = 1$ we have

$$V(\mathbf{m}) = O\left(\frac{1}{\max\{|m_i| : 1 \le i \le n\}}\right). \qquad (46)$$

We apply Lemma 3, inserting the bound (36) into (37) and letting $N \to \infty$. This completes the proof.

Next we define $H(n)$ for every even $n \geq 4$, and show that (4) holds. We put

$$
\begin{aligned}
H(n) &= \left(\frac{i}{2\pi}\right)^n \sum_{m_1,\ldots,m_n}^{*} \frac{V(\mathbf{m})}{m_1 \ldots m_n} \\
&= \left(\frac{i}{2\pi}\right)^n \zeta(n) \sum_{\substack{m_1,\ldots,m_n \\ \mathrm{hcf}(m_1,\ldots,m_n)=1}}^{*} \frac{V(\mathbf{m})}{m_1 \ldots m_n}
\end{aligned}
\tag{47}
$$

and we establish first that the series on the right is absolutely convergent. For this we simply insert the bound

$$
V(\mathbf{m}) = O\big(|m_1 \ldots m_n|^{-1/n}\big)
\tag{48}
$$

which follows from (46) in the coprime case, into (47). This is all we need.

We employ the approximation described in (20) and (21) above, with $M = N^3$. By (21), this approximates each Franel integral to within $O(N^{-2})$, and so $F_n(N)$ to within $O(N^{n-2})$. We therefore have

$$
F_n(N) = \left(\frac{i}{2\pi}\right)^n \sum_{0<|m_i|\leq M} \frac{S(\mathbf{m},N)}{m_1 \ldots m_n} + O(N^{n-2}).
\tag{49}
$$

We split the sum on the right according to whether or not $\max |m_i| \leq L$ (where L is at our disposal: it is small compared to N); and we apply Lemma 4 to the first part, which therefore contributes

$$
\left(\frac{i}{2\pi}\right)^n N^{n-1} \sum_{0<|m_i|\leq L} \frac{V(\mathbf{m})}{m_1 \ldots m_n} + O(N^{n-2}L^{\lambda(n)}\log^{n-1}L).
\tag{50}
$$

We have

$$
\begin{aligned}
\sum_{\max|m_i|>L}^{*} \frac{V(\mathbf{m})}{m_1 \ldots m_n} &\ll \sum_{\max|m_i|>L}^{*} \frac{1}{|m_1 \ldots m_n|\max|m_i|} \sum_{d|\mathrm{hcf}(m_1,\ldots,m_n)} \phi(d) \\
&\ll \sum_{d=1}^{\infty} \frac{\phi(d)}{d^{n+1}} \sum_{\max l_i>L/d} \frac{1}{l_1 \ldots l_n \max l_i} \\
&\ll \sum_{d\leq L} \frac{1}{d^n} \sum_{l>L/d} \frac{\log^{n-1}l}{l^2} + \sum_{d>l} \frac{1}{d^n} \sum_{l=1}^{\infty} \frac{\log^{n-1}l}{l^2} \\
&\ll \frac{1}{L} \sum_{d\leq L} \frac{1}{d^{n-1}} \log^{n-1}(L/d) + L^{1-n} \\
&\ll \frac{\log^{n-1}L}{L}.
\end{aligned}
\tag{51}
$$

We now fix $L = N^{\kappa}$, where $\kappa = 1/(\lambda(n) + 1)$, and assemble (50), (51) and the formula (47) for $H(n)$. The first part of the sum (49) is equal to

$$H(n)N^{n-1} + O(N^{n-1-\kappa}\log^{n-1}N).$$ (52)

We apply Lemma 3 to the second part of the sum (49). We have

$$\sum_{L<\max|m_i|\leq M} \frac{S(\mathbf{m},N)}{|m_1\ldots m_n|} \ll \sum_{L<\max|m_i|\leq M} \frac{N^{n-1}\operatorname{hcf}(m_1,\ldots,m_n)}{|m_1\ldots m_n|\max|m_i|}$$

$$+ O(N^{n-2}\log^n M).$$ (53)

The sum on the right is similar to that in (51), and the whole sum is absorbed by the error term in (52). This completes the proof of Theorem 1, except for the information about $H(n)$ for $n \geq 4$.

5. Farey dissection

The previous discussion gives us no information about the constant $H(n)$ in the case $n > 2$, and for this we employ the alternative representation of $F_n(N)$ in (7). We have to consider the function $F_n(x, N)$ in (8). Denote by $\mathcal{F}_N$ the Farey sequence of order N. We write $\mathcal{F}_N = \{x_1, x_2, \ldots, x_H\}$ in which $x_1 = 1/N$ and $x_H = 1$, so that

$$H = \phi(1) + \phi(2) + \cdots + \phi(N) \sim \frac{3N^2}{\pi^2},$$

and we put $x_0 = 0$. We notice that for $x \in [0,1]\backslash\mathcal{F}_N$ the function $F(x, N)$ is differentiable, indeed linear with gradient $G = N(N+1)/2$. Moreover it is right continuous on $[0,1]$. Hence

$$\int_{x_{\nu-1}}^{x_\nu} F(x,N)^n\,dx = \frac{1}{(n+1)G}\Big(F(x_\nu - 0, N)^{n+1} - F(x_{\nu-1}, N)^{n+1}\Big).$$ (54)

F is periodic, whence, and from (7),

$$(n+1)GF_n(N) = \sum_{\nu=1}^{H}\Big(F(x_\nu - 0, N)^{n+1} - F(x_\nu, N)^{n+1}\Big).$$ (55)

Let $x_\nu = a/q, (a,q) = 1$. Raabe's formula gives

$$\sum_{n\,(\mathrm{mod}\,q)} \overline{B}_1\Big(\frac{na}{q}\Big) = -\tfrac{1}{2},$$ (56)

whence

$$F(x_\nu, N) = -\tfrac{1}{2}[N/q] + F\big(x_\nu, R(q)\big),$$ (57)

where $R(q) = N - q[N/q]$. At the point $x_\nu = a/q$ the function $\overline{B}_1(nx)$ has a jump discontinuity -1 if q divides n, else it is continuous. Therefore $F(x, N)$ jumps by $-[N/q]$, and $F(x, R(q))$ is continuous. We see from this and (57) that

$$F(x_\nu - 0, N) = \tfrac{1}{2}[N/q] + F(x_\nu, R(q)). \tag{58}$$

We insert (57) and (58) into (55) and sum over ν, with the understanding that q is a function of ν. We obtain

$$(n+1)GF_n(N)$$

$$= \sum_{h \equiv 0 \,(\mathrm{mod}\,2)} 2^{h-n} \binom{n+1}{h} \sum_{q \leq N} [N/q]^{n+1-h} \sum_{(a,q)=1} F(a/q, R(q))^h. \tag{59}$$

Notice that all the terms on the right-hand side of (59) are positive. Since $G \gg N^2$, we seek a main term with order of magnitude N^{n+1}, and we find that the sum over the whole range $0 < h < n$ is of smaller order. We have $|F(a/q, R(q))| < q$, whence the innermost sum is $< q^{h+1}$. Therefore the sum over q in (59) does not exceed

$$N^{n+1-h} \sum_{q \leq N} q^{2h-n} \ll \begin{cases} N^{n+1-h} & \text{if } 2h < n \\ N^{h+2} & \text{if } 2h \geq n. \end{cases} \tag{60}$$

We remark that since n is even, $2h - n = -1$ is impossible and so there are no logarithms. Furthermore $h < n$ implies $h \leq n - 2$ because the sum in (59) is restricted to even h. From (60), the terms in which $0 < h < n$ contribute $\ll N^n$ on the right hand side of (59), and so $\ll N^{n-2}$ towards $F_n(N)$. The term in which $h = 0$ contributes

$$2^{-n} \sum_{q \leq N} [N/q]^{n+1} \phi(q) = 2^{-n} \frac{\zeta(n)}{\zeta(n+1)} N^{n+1} + O(N^n), \tag{61}$$

and we deduce that

$$F_n(N) = \frac{\zeta(n)}{\zeta(n+1)} \cdot \frac{N^{n-1}}{(n+1)2^{n-1}} + \frac{2}{N^2} W_n(N) + O(N^{n-2}), \tag{62}$$

where

$$W_n(N) = \sum_{q \leq N} [N/q] \sum_{(a,q)=1} F(a/q, R(q))^n. \tag{63}$$

This last sum comprises positive terms and so (62) already establishes that $H(n) > 0$, indeed it supplies a lower bound. We note that $\zeta(2)/6\zeta(3) = .228072\ldots$ is only a little less than $H(2)$: we expect that in the general case the second term on the right of (62) will have order of magnitude N^{n-1}, with a numerically small constant.

In the next section of the paper we give an explicit formula for the sum $F(a/q, R(q))$ in terms of the continued fraction expansion of a/q. We have not succeeded in deriving an asymptotic formula for $W_n(N)$ from this, indeed the project seems pretty awkward. We evaluate part of the sum (63) here, and this shows that the first term on the right of (62) is certainly not a main term, for any n.

From (80) and (12), recalling that $G_1 = F$, we have the approximate formula (using curly brackets to denote fractional part)

$$F(a/q, R(q)) = -\frac{q}{2a}\left\{\frac{Na}{q}\right\}\left(1-\left\{\frac{Na}{q}\right\}\right)+O(a) \qquad (a > 0, \ (a,q) = 1).$$
(64)

This is useful when $a < \sqrt{q}$. Since F is odd on $\mathbb{R}\backslash\mathbb{Z}$ we can apply it to the range $q - \sqrt{q} < a < q$ as well. This introduces a negligible error when $q = 2$ because we count $a = 1$ twice. It is important for the next step that the right-hand side of (64) does not involve $R(q)$ explicitly: this happens because $R(q) \equiv N \pmod q$. We denote the main term in (64) by $f(a/q)$, suppressing N temporarily. From (64),

$$\sum_{(a,q)=1} F(a/q, R(q))^n \geq 2 \sum_{\substack{a<\sqrt{q} \\ (a,q)=1}} f(a/q)^n + O(q^{n-1}).$$
(65)

Notice that we need $n \geq 4$ in the calculation of the error term. We deduce from (63) and (65) that

$$W_n(N) \geq 2 \sum_{q\leq N} [N/q] \sum_{\substack{a<\sqrt{q} \\ (a,q)=1}} f(a/q)^n + O(N^n).$$
(66)

Observe now that $f(x)$ is a well-defined function of the positive real variable x. Let us drop the condition $(a, q) = 1$ in the inner sum in (66): for each fraction a/q in its lowest terms this introduces $[N/q]$ copy fractions ra/rq, to which we apply the condition $ra < \sqrt{rq}$, which is stiffer than we demanded of the parent a/q. Hence the inequality

$$W_n(N) \geq 2 \sum_{q\leq N} \sum_{a<\sqrt{q}} f(a/q)^n + O(N^n)$$
(67)

is valid, involving a less awkward sum. We invert the order of summation to obtain

$$W_n(N) \geq \frac{1}{2^{n-1}} \sum_{a<\sqrt{N}} \sum_{a^2<q\leq N} \left(\frac{q}{a}\{Na/q\}\left(1 - \{Na/q\}\right)\right)^n + O(N^n),$$ (68)

and we note that the condition $q > a^2$ may be dropped without affecting the error term, since $n \geq 4$.

We consider the resulting inner sum. MacLeod [7] studied the sum

$$\sum_{q \leq N} q^n \left\{ \frac{N}{q} \right\}^t \qquad (n \in \mathbb{Z}, \quad t \in \mathbb{N}) \tag{69}$$

in connection with divisor problems, and his results have been extended and sharpened by Mercier [8] and by Ishibashi and Kanemitsu [5]. We need an extension that introduces the extra variable $a \in \mathbb{N}$, and this may be achieved by writing

$$\sum_{q \leq N} q^n \left\{ \frac{Na}{q} \right\}^t = \sum_{q \leq Na} q^n \left\{ \frac{Na}{q} \right\}^t - \sum_{N < q \leq Na} q^n \left\{ \frac{Na}{q} \right\}^t, \tag{70}$$

since the first sum on the right is MacLeod's and in the second we note that $[Na/q]$ takes only the values $a - 1, a - 2, \ldots, 1$. Other methods are available. We shall be content here with a pure asymptotic formula and we state without proof that

$$\sum_{q \leq N} q^n \left\{ \frac{Na}{q} \right\}^t = (C_{n,t}(a) + o(1)) N^{n+1} \qquad (n, t \in \mathbb{N}), \tag{71}$$

where

$$
\begin{aligned}
C_{n,t}(a) &= a^{n+1} \int_a^\infty \frac{\{y\}^t}{y^{n+2}} \, dy \\
&= \frac{1}{(n+1)(t+1)} + a^{n+1} \int_1^\infty \left(\{y\}^t - \frac{1}{t+1} \right) \frac{dy}{y^{n+2}}.
\end{aligned} \tag{72}
$$

The right-hand term is $\ll 1/a$ by the second mean-value theorem. As far as I am aware this formula is new. We remark *en passant* the consequence that as $a \to \infty$, the fractional part $\{Na/q\}$ behaves as a random variable, that is the weighted mean value of $\{Na/q\}^t$ is asymptotic to $1/(t+1)$.

We do not require the uniformity in a in (71) because the sum over this variable in (68) is the beginning of a convergent series. We substitute (71) and (72) into (68) and we have

$$W_n(N) \geq (C(n) + o(1)) N^{n+1}, \tag{73}$$

where

$$
\begin{aligned}
C(n) &= \frac{1}{2^{n-1}} \sum_{a=1}^\infty a \int_a^\infty \frac{\{y\}^n (1 - \{y\})^n}{y^{n+2}} \, dy \\
&= \frac{1}{2^n} \int_1^\infty \frac{[y]([y] + 1)\{y\}^n (1 - \{y\})^n}{y^{n+2}} \, dy.
\end{aligned} \tag{74}
$$

We recall that in (62) the factor $2/N^2$ multiplies $W_n(N)$ and so to complete the proof of Theorem 1 it remains to show that $C(n) = \frac{1}{2}D(n)$. We leave this an exercise.

6. Bernoulli Sums; proof of Theorem 2

We write $\exp(t) = E(t)$ for convenience. We have

$$\frac{tE(\{x\}t)}{e^t - 1} = \sum_{n=0}^{\infty} \overline{B}_n(x)\frac{t^n}{n!} \tag{75}$$

and

$$\sum_{m=1}^{R} E\left(\left\{\frac{ma}{q}\right\}t\right) = \sum_{m<q/a} E\left(\frac{mat}{q}\right) + e^{-t} \sum_{q/a<m<2q/a} E\left(\frac{mat}{q}\right) + \cdots$$

$$+ e^{-St} \sum_{Sq/a<m\leq R} E\left(\frac{mat}{q}\right), \tag{76}$$

where $S = [Ra/q]$. Put $\theta = at/q$. None of hq/a for $1 \leq h \leq S$ is an integer because $R < q$. After some algebra (76) yields

$$\frac{t}{e^t - 1} \sum_{m=1}^{R} E\left(\left\{\frac{ma}{q}t\right\}\right) = \frac{te^\theta}{e^\theta - 1}\left(\frac{e^{R\theta-St}-1}{e^t-1} + \sum_{h=1}^{S} E\left(\left[\frac{hq}{a}\right]\theta - ht\right)\right) \tag{77}$$

and we notice that $[hq/a]\theta - ht = -\{hq/a\}\theta$.

Denote the left-hand side of (77) by $Z_R(a/q, t)$. Then (77) reads

$$Z_R(a/q, t) = \left(\frac{te^\theta}{e^\theta - 1}\right)\left(\frac{e^{R\theta-St} - 1}{e^t - 1}\right) + \frac{q}{a}Z_S\left(\frac{q}{a}, -\theta\right). \tag{78}$$

We equate coefficients of $t^n/n!$ in (78). This gives

$$G_n\left(\frac{a}{q}, R\right) = H_n\left(\frac{a}{q}, R\right) + (-1)^n\left(\frac{a}{q}\right)^{n-1} G_n\left(\frac{q}{a}, S\right), \tag{79}$$

with H_n as in the statement of the theorem. Since $S = [Ra/q] < a$ and the Bernoulli functions are bounded, an immediate corollary of (79) is

$$G_n\left(\frac{a}{q}, R\right) = H_n\left(\frac{a}{q}, R\right) + O\left(\frac{a^n}{q^{n-1}}\right), \tag{80}$$

which, in the case $n = 1$, $R \equiv N \pmod{q}$, is (64). This completes the proof of Theorem 2.

Postscript. It is very likely, but not certain, that $H(n)$ is irrational for $n \geq 4$. We have

$$H(4) = \frac{\pi^2 \zeta(3)}{36\zeta(5)} - \frac{7}{24} = .0261478806\ldots,$$

$$H(6) = \frac{\pi^4 \zeta(3)}{480\zeta(7)} - \frac{\pi^2 \zeta(5)}{8\zeta(7)} + \frac{33}{32} = .0045043884\ldots,$$

and in general

$$H(n) = \frac{\zeta(n)}{(n+1)2^{n-1}\zeta(n+1)} + \frac{\zeta(n)}{\zeta(n+1)}D(n), \qquad (81)$$

with $D(n)$ as in (6). The evaluation of $H(n)$ involves a new idea, and my present derivation of (81) is somewhat indirect, (as the case $n = 2$ might suggest). This matter will be clarified in a later memoir.

I should like to thank the referee for his careful reading of the paper, and for his helpful suggestions concerning the Geometry of Numbers.

References

1 J.Franel: Les suites de Farey et le problème des nombres premiers. *Göttinger Nachr.* (1924), 198–201.

2 G.R.H. Greaves, R.R. Hall, M.N. Huxley, and J.C. Wilson: Multiple Franel integrals. *Mathematika* 40 (1993), 50–69.

3 A.O. Gelfond and Yu.V. Linnik: *Elementary Methods in Analytic Number Theory.* Allen and Unwin (1966).

4 P.M. Gruber and C.G. Lekkerkerker: *Geometry of Numbers.* North-Holland (1987).

5 M. Ishibashi and S. Kanemitsu: Fractional part sums and divisor functions I. In *Number Theory and Combinatorics Japan 1984*, 119–183. World Sci. Publishing (1985).

6 J.C. Kluyver: An analytical expression for the greatest common divisor of two integers. *Proc. Royal Acad. Amsterdam* 5 (1903), 658–662.

7 R.A. MacLeod: Fractional part sums and divisor functions. *J. Number Theory* 14 (1982), 185–227.

8 A. Mercier: Sums containing the fractional parts of numbers. *Rocky Mountain J.* 15 (1985), 513–520.

9 E.T. Whittaker and G.N. Watson: *Modern Analysis*, 4th. edn. Cambridge University Press (1927).

9. Eratosthenes, Legendre, Vinogradov and beyond

The hidden power of the simplest sieve

Glyn Harman

In this note we consider how the simplest sieve, discovered in antiquity by Eratosthenes, has been developed into a powerful tool for modern multiplicative number theory. As is well known, the sieve of Eratosthenes (in the developed form known since the 13th century A.D.) provides a procedure for finding the primes between z and z^2 given knowledge of the primes up to z: the method is simply to list the numbers between z and z^2 and cross out all multiples of the primes less than z. The survivors of this sifting process must then be primes. If we write $\pi(x)$ for the number of primes not exceeding x, then the sieve of Eratosthenes was given an important re-interpretation as a formula, rather than an algorithm, by Legendre, namely

$$\pi(x) - \pi(\sqrt{x}) + 1 = \sum_{d \mid P(\sqrt{x})} \mu(d) \left[\frac{x}{d}\right], \tag{1}$$

where μ denotes the Möbius function, $[\,]$ represents integer part, and

$$P(z) = \prod_{p < z} p.$$

Now suppose we wanted to use (1) to count the number of primes up to x. We might then perform the typical number-theoretic trick of writing

$$[x/d] = x/d - \{x/d\},$$

where $\{\cdot\}$ denotes fractional part. We then have

$$\pi(x) = \sum_{d \mid P(\sqrt{x})} \mu(d) \frac{x}{d} - \sum_{d \mid P(\sqrt{x})} \mu(d) \left\{\frac{x}{d}\right\} + O(\sqrt{x}). \tag{2}$$

We might then believe that the first sum on the right of (2) is the main term, and the second sum is a remainder. As is well known this argument is false since

$$\sum_{d \mid P(\sqrt{x})} \mu(d) \frac{x}{d} = x \prod_{p < \sqrt{x}} \left(1 - \frac{1}{p}\right) \sim \frac{2e^{-\gamma}x}{\log x},$$

Sieve Methods, Exponential Sums, and their Applications in Number Theory
Greaves, G.R.H., Harman, G., Huxley, M.N., Eds. ©Cambridge University Press, 1996

whereas, by the prime number theorem, $\pi(x) \sim x/\log x$. Thus it seems that the sieve of Eratosthenes-Legendre cannot prove the prime number theorem. Indeed, it does not even give "good" upper and lower bounds. One might therefore conclude that this sieve is weak. A less hasty judgement would be that we are asking the wrong questions and supplying inadequate information for the sieve to perform in a satisfactory manner.

Now let
$$\mathcal{A} \subseteq \mathbb{Z} \cap (0, x], \qquad \mathcal{A}_d = \{n : nd \in \mathcal{A}\}.$$

The formula of Legendre may then be generalized to
$$\sum_{p \in \mathcal{A}} 1 = \sum_{d \mid P(\sqrt{x})} \mu(d)|\mathcal{A}_d| + O(\sqrt{x}). \tag{3}$$

An alternative formulation would include a weight $f(p)$ with $|f(p)| \leq 1$ to obtain
$$\sum_{p \leq x} f(p) = \sum_{\substack{d \mid P(\sqrt{x}) \\ dn \leq x}} \mu(d)f(dn) + O(\sqrt{x}). \tag{4}$$

At first sight we have gained nothing since $\mu(\cdot)$ is an "unknown function" just as much as the characteristic function of the set of primes. However, in the 1930's Vinogradov [22] devised a method for tackling the sum
$$\sum_{\substack{d \mid P(\sqrt{x}) \\ dn \leq x}} \mu(d)f(dn).$$

The first step is to split into two cases: $d \leq M$ (Type I), $d > M$ (type II*). For the type I sum we can make use of the fact that for each d the variable n ranges over a large number ($\geq x/M$) of consecutive integers. If f is at all reasonable we can then give a non-trivial estimate for this sum. The novelty of Vinogradov's approach was a difficult argument which decomposed the type II* sum into $\ll \log^B x$ sums (for some B) like
$$\sum_{\substack{A \leq m < 2A \\ mn \leq x}} a_m b_n f(mn) \qquad \text{(called a type II sum)}.$$

Here A is neither too small nor too big, and the coefficients a_m, b_n are bounded by the divisor function. One common way of tackling such a sum is to apply Cauchy's inequality to remove one of these "unknown" coefficients.

For Vinogradov's proof of the ternary Goldbach conjecture for all large n he needed a "minor arc" estimate for the exponential sum
$$\sum_{p \leq x} e(\alpha p), \quad \text{where} \quad e(\theta) = \exp(2\pi i \theta).$$

The bound he obtained was essentially

$$\ll x \log^C x \left(\frac{1}{x^{\frac{1}{5}}} + \frac{1}{q^{\frac{1}{2}}} + \frac{q^{\frac{1}{2}}}{x^{\frac{1}{2}}} \right) \tag{5}$$

if

$$|\alpha - a/q| < q^{-2}, \qquad (a, q) = 1.$$

(In fact his result had a different shape depending on the size of q.) This replaced a bound of Hardy and Littlewood which was conditional on the GRH.

In the following years there were several other applications of Vinogradov's method. In [23] he proved that for all irrational γ and any real ξ there are infinitely many primes p with

$$\|\gamma p + \xi\| < p^{-\frac{1}{5} + \epsilon}$$

where $\| \cdot \|$ denotes distance to the nearest integer. Piatetski-Shapiro [18] showed that

$$\sum_{\substack{p=[n^c] \\ n \le x}} 1 \sim \frac{x}{c \log x} \quad \text{for} \quad c < \tfrac{12}{11}. \tag{6}$$

Vinogradov's method may be summarised thus: If type I & II sums of certain "shapes" can be estimated then the sieve of Eratosthenes-Legendre gives the "correct" formula for a sum over primes. We now see where we went wrong in our first application of the Eratosthenes-Legendre sieve. From Vinogradov's pioneering work we know that two types of information are needed on bilinear forms. It is also implicit that the prime number theorem must be assumed. One should compare this situation with the not altogether accurate statement that "sieve methods alone cannot give primes". The correct statement is "sieves cannot give primes with type I arithmetical information alone".

One barrier to the widespread use of Vinogradov's method was the obscure combinatorial argument used to reduce the type II* sum into type II sums. In the 1970's Vaughan made a major simplification to this subject with his famous identity. If we write

$$\Lambda(n) = \begin{cases} \log p & \text{if} \quad n = p^k \\ 0 & \text{otherwise} \end{cases}$$

as usual for the von Mangoldt function, then Vaughan was led to consider the following identity when working on the Bombieri-Vinogradov theorem:

$$\sum_{n=1}^{\infty} \frac{\Lambda(n)}{n^s} = -\frac{\zeta'}{\zeta}(s)$$

$$= F(s) - \zeta(s)F(s)G(s) - \zeta'(s)G(s)$$

$$+ \left(-\frac{\zeta'}{\zeta}(s) - F(s)\right)\left(1 - \zeta(s)G(s)\right),$$

where

$$F(s) = \sum_{m \leq U} \frac{\Lambda(m)}{m^s}, \qquad G(s) = \sum_{d \leq V} \frac{\mu(d)}{d^s}.$$

Since the functions n^{-s} are linearly independent we thus obtain Vaughan's identity

$$\Lambda(n) = a_1(n) + a_2(n) + a_3(n) + a_4(n)$$

with

$$a_1(n) = \begin{cases} \Lambda(n) & \text{if } n \leq U \\ 0 & \text{if } n > U, \end{cases} \qquad a_2(n) = - \sum_{\substack{mdr=n \\ m \leq U, d \leq V}} \Lambda(m)\mu(d),$$

$$a_3(n) = \sum_{\substack{hd=n \\ d \leq V}} \mu(d)\log h, \qquad a_4(n) = - \sum_{\substack{mk=n \\ m > U, k > 1}} \Lambda(m) \sum_{\substack{d|k \\ d \leq V}} \mu(d).$$

Hence, if $|f(n)| \leq 1$, the choice $U = V$, together with partial summation to remove the $\log h$ factor arising from $a_3(n)$, gives

$$\sum_{n \leq x} \Lambda(n)f(n) = O(U) + \text{sums of type I \& II}.$$

Here the sums are as follows:

$$\text{Type I:} \sum_{\substack{m \leq M \\ mn \leq x}} a_m f(mn) \qquad \text{Type II:} \sum_{\substack{K \leq m < 2K \\ mn \leq x}} a_m b_n f(mn)$$

where

$$M \leq \max\left(x^{1-\beta}, x^{2\beta}\right), \qquad x^\beta \leq K \leq x^{1-\beta},$$

for some $\beta \in \left(0, \frac{1}{2}\right)$. For example, to obtain (5) we must take $\beta = \frac{2}{5}$.

Subsequently Heath-Brown produced what he called a "Generalized Vaughan Identity" [9]. This allowed the introduction of further arithmetical information, namely

$$\text{Type I}_j \text{ sum:} \sum a_m f(mn_1 \ldots n_j)$$

$$\text{Type II}_3 \text{ sum:} \sum a_m b_n c_r f(mnr) \qquad \text{Type I/II sum:} \sum a_m b_n f(lmn).$$

Heath-Brown [10] also re-opened interest in the Linnik Identity ([17], eq.
(0.6.13), applied in [6] and [11], for example).

The discovery of these clearer (but no stronger !) identities led to many
advances and enabled elementary proofs of otherwise "deep" results to be
found (e.g. [21]). A combination of these identities with upper and lower
bound sieve methods was also very effective. For example, if we write p_n
for the nth prime, then Heath-Brown and Iwaniec [12] obtained

$$p_{n+1} - p_n \ll p_n^{\frac{11}{20}+\epsilon},$$

improving work of Iwaniec and Jutila [15] which had improved upon Huxley's result [14] obtained without sieves or sieve identities.

We now reconsider Eratosthenes-Legendre. Let

$$\mathcal{A} \subseteq \mathcal{B} = \mathbb{Z} \cap \left(\tfrac{1}{2}x, x\right], \qquad S(\mathcal{A}, z) = \left|\{n \in \mathcal{A}; p|n \Rightarrow p \geq z\}\right|,$$

(so the number of primes in $\mathcal{A}$ is $S(\mathcal{A}, \sqrt{x})$, the number of primes between $\tfrac{1}{2}x$ and x is $S(\mathcal{B}, \sqrt{x})$). In [8] the author has proved the following
result, which clearly shows what is possible to obtain given the correct information input into the sieve of Eratosthenes-Legendre (the author had
implicitly given a result of this shape some years earlier [7] but did not
use the Eratosthenes-Legendre sieve then; Balog [4] has given another proof
using a lemma of Richert).

Lemma 1. (Vinogradov made easy) *Suppose that for any two sequences of
complex numbers a_m, b_n with $|a_m| \leq \tau(m)$, $|b_n| \leq \tau(n)$, we have for some
$\lambda > 0$, $\alpha > 0$, $0 < \beta \leq \tfrac{1}{2}$, $M \geq 1$, that*

$$\sum_{\substack{mn \in \mathcal{A} \\ m \leq M}} a_m = \lambda \sum_{\substack{mn \in \mathcal{B} \\ m \leq M}} a_m + O(Y) \qquad \text{(Type I)} \quad (7)$$

$$\sum_{\substack{mn \in \mathcal{A} \\ x^\alpha \leq m \leq x^{\alpha+\beta}}} a_m b_n = \lambda \sum_{\substack{mn \in \mathcal{B} \\ x^\alpha \leq m \leq x^{\alpha+\beta}}} a_m b_n + O(Y) \quad \text{(Type II)}. \ (8)$$

Then, if

$$|c_r| \leq 1, \qquad x^\alpha < M,$$
$$R < \min(x^{1-\alpha}, M), \qquad M \geq x^{1-\alpha} \ \text{ if } \ r > x^{\alpha+\beta},$$

we have

$$\sum_{r \sim R} c_r S(\mathcal{A}_r, x^\beta) = \lambda \sum_{r \sim R} c_r S(\mathcal{B}_r, x^\beta) + O(Y \log^3 x). \qquad (9)$$

Here $r \sim R$ indicates $R \leq r < 2R$.

Proof. Write $z = x^\beta$. We have

$$S(\mathcal{A}_r, z) = \sum_{\substack{d \mid P(z) \\ drn \in \mathcal{A}}} \mu(d).$$

Write

$$\psi(m) = \sum_{mn \in \mathcal{A}} 1 - \lambda \sum_{mn \in \mathcal{B}} 1.$$

Thus

$$S(\mathcal{A}_r, z) = \lambda \sum_{\substack{d \mid P(z) \\ drn \in \mathcal{B}}} \mu(d) + \sum_{d \mid P(z)} \mu(d)\psi(dr)$$

$$= \lambda S(\mathcal{B}_r, z) + \Sigma_1(r) + \Sigma_2(r)$$

where

$$\Sigma_1(r) = \sum_{\substack{d \mid P(z) \\ dr \leq M}} \mu(d)\psi(dr), \qquad \Sigma_2(r) = \sum_{\substack{d \mid P(z) \\ dr > M}} \mu(d)\psi(dr).$$

Clearly Σ_1 is a type I sum which can be estimated by (7). For Σ_2 we note that $d \mid P(z)$ implies that every prime factor of d is less than z. For a given d considered by the sum we have $dr > M > x^\alpha$. Thus there is a positive integer t and primes $p_1 > p_2 > \cdots > p_t$ such that, on writing $\pi_t = p_1 \ldots p_t$, we have

$$d = \pi_t d_t, \qquad d_t \mid P(p_t), \qquad \pi_t r > x^\alpha \geq \pi_{t-1} r. \qquad (10)$$

It follows that the sum $\Sigma_2(r)$ can be split into subsums $\Sigma_2(r, t)$ of the kind

$$(-1)^t \sum_{p_1, \ldots, p_t} \sum_{\substack{d \mid P(p_t) \\ \pi_t dr > M}} \mu(d)\psi(dr\pi_t)$$

where the first summation is over $p_1, \ldots, p_t$ with $\pi_t r > x^\alpha \geq \pi_{t-1} r$. In particular, $\pi_t r = p_t \pi_{t-1} r \leq x^{\alpha+\beta}$, which will be required for the application of (8). Since d has $\ll \log x / \log \log x$ prime factors, this is the maximum number of sums which can arise. Before we can apply (8) we need to disentangle dependencies between the summation ranges, namely $\pi_t r d > M$ and $d \mid P(p_t)$. The first involves only a simple application of the Perron formula; the second requires a little more work to consider the highest prime factor of d, say q, and the relation between q and p_t is then covered by another application of the Perron formula (see [8]).

We have thus arrived at the following conclusion: the sieve of Erat-osthenes-Legendre will give a formula for $S(\mathcal{A}, x^\beta)$ (so we need $\beta \geq \frac{1}{2}$ to get primes), or sums involving $S(\mathcal{A}_r, x^\beta)$. Hopefully the process in the above lemma is more transparent than Vinogradov's method. Of course we have not yet finished our task since the requirement $\beta \geq \frac{1}{2}$ is far too stringent for many applications. We now show how to use the sums $S(\mathcal{A}_r, x^\beta)$ as "building blocks" to make up or approximate $S(\mathcal{A}, x^{1/2})$. To do this we need the elementary Buchstab identity

$$S(\mathcal{A}, z_1) = S(\mathcal{A}, z_2) - \sum_{z_2 \leq p < z_1} S(\mathcal{A}_p, p).$$

This is equivalent to taking out the highest prime factor in the proof of the above lemma. Now say $\alpha = \frac{1}{3} = \beta$, $M = x^{2/3}$ in the lemma. Then

$$S(\mathcal{A}, \sqrt{x}) = S(\mathcal{A}, \sqrt[3]{x}) - \sum_{\sqrt[3]{x} \leq p < \sqrt{x}} S(\mathcal{A}_p, p)$$

$$= \Sigma_A + \Sigma_B,$$

say. Now Σ_A can be estimated directly from the lemma, while for σ_B we note that as $p \geq \sqrt[3]{x} \geq \sqrt{x/p}$ we have

$$\Sigma_B = \sum_{\sqrt[3]{x} \leq p < \sqrt{x}} S(\mathcal{A}_p, \sqrt[3]{x}),$$

which can be estimated directly from (8). Hence we obtain

$$S(\mathcal{A}, x^{\frac{1}{2}}) = \lambda S(\mathcal{B}, x^{\frac{1}{2}}) + O(Y \log^3 x)$$

$$= \lambda(\pi(x) - \pi(\tfrac{1}{2}x)) + O(Y \log^3 x).$$

We have obtained this with only the Eratosthenes-Legendre sieve and the right information input. Usually the error term for $\pi(x)$ (given our current state of knowledge) will be worse than the error term $Y \log^3 x$.

The choice $\alpha = \beta = \frac{1}{3}$ made things easy in the above. With further decompositions we can obtain the same conclusion as arises from Vaughan's identity. Now consider the example of Vinogradov's result for $\|\gamma p\| < p^{-\theta+\epsilon}$. From our knowledge of exponential sums we can take $\alpha = \theta$, $\beta = 1 - 3\theta$ in Vaughan's identity or our own work. Hence we obtain the following result of Vaughan [20]:

$$\sum_{\substack{p \in \mathcal{B} \\ \|\gamma p\| < x^{-\theta}}} 1 = 2x^{-\theta}\pi(x)(1 + o(1)).$$

for $\theta < \frac{1}{4}$ and infinitely many x (depending on the convergents to γ). The proof using Vaughan's identity is shorter than the one using the process we have described, but we can now make use of our control over the signs of the sums involved after a Buchstab decomposition. We have problems if we want to take $\theta > \frac{1}{4}$ since the method "breaks up" around $\sqrt{x}$. That is, we can't estimate Type II sums for $x^{1-2\theta} < K < x^{2\theta}$. However, we can still estimate

$$\sum_{n \le x^{1-\theta}} a_n S(\mathcal{A}_n, x^{1-3\theta}).$$

Now let $\tau = 1 - 3\theta$ and put the building blocks together like this ($X = \sqrt{x}$ in the following, and $z = x^\tau$)

$$S(\mathcal{A}, X) = S(\mathcal{A}, z) - \sum_{z \le p < X} S(\mathcal{A}_p, p)$$

$$= S(\mathcal{A}, z) - \sum_{\substack{z \le p < x^\theta \\ \text{or } x^{1-2\theta} < p < X}} S(\mathcal{A}_p, z) - \sum_{x^\theta \le p \le x^{1-2\theta}} S(\mathcal{A}_p, p)$$

$$+ \sum_{\substack{z \le p < q < \min(p, \sqrt{x/p}) \\ p \le x^\theta \text{ or } p > x^{1-2\theta}}} S(\mathcal{A}_{pq}, q).$$

Let ∇ be the set of $(p, q) \in \mathbb{Z}^2$ for which we have no type II estimates with $z \le q < p < X$. Then

$$S(\mathcal{A}, X) = \lambda S(\mathcal{B}, X) - \sum_{(p,q) \in \nabla} \left(\lambda S(\mathcal{B}_{pq}, q) - S(\mathcal{A}_{pq}, q) \right) + O\left(\lambda x^{1-\frac{\eta}{2}} \right)$$

$$\ge \lambda \left(S(\mathcal{B}, X) - \sum_{(\nabla)} S(\mathcal{B}_{pq}, q) \right) + O\left(\lambda x^{1-\frac{\eta}{2}} \right).$$

We thus have a lower bound for $S(\mathcal{A}, X)$ non-trivial if ∇ is not too big. Clearly further decompositions are possible (see [8]). We can easily obtain

$$S(\mathcal{A}, X) \ge \left(1 - 80(\theta - \tfrac{1}{4})^2 \right) \lambda S(\mathcal{B}, X) \quad \text{for} \quad \tfrac{1}{4} \le \theta \le \tfrac{2}{7},$$

and thus obtain a "smooth transition" from identity to lower bound. With more work we get

$$S(\mathcal{A}, X) \ge \tfrac{1}{10} \lambda S(\mathcal{B}, X) \quad \text{for} \quad \theta \le \tfrac{7}{22} = 0.3\dot{1}\dot{8}$$

so the inequality

$$\|\gamma p\| < p^{-\frac{7}{22}}$$

has infinitely many solutions in primes p for irrational γ. To obtain this result we needed to reverse the rôles of the variables at certain stages of the argument. For example, we needed to change a sum

$$\sum_{\substack{p \sim P \\ q}} S(\mathcal{A}_{pq}, q) \quad \text{into} \quad \sum_{q,r} S(\mathcal{A}^*_{rq}, \sqrt{2P})$$

where $\mathcal{A}^*$ for all practical purposes behaves the same as $\mathcal{A}$.

To generalise the above, let Δ represent "arithmetical information" that we can estimate type I, II, I_j etc. sums of certain shapes (the type I_j sums arise by applying the sieve to j variables simultaneously), and introduce a level of difficulty parameter θ (in the above example $\|\gamma p\| < p^{-\theta}$, or for primes in short intervals $[x, x^{1-\theta}]$) where increasing θ makes the problem harder, that is Δ is "smaller". In general we will obtain

$$S(\mathcal{A}, X) \geq C(\Delta)\lambda S(\mathcal{B}, X)$$

where $C(\Delta) = 1$ for $\theta < \theta_0$, $C(\Delta)$ is often continuous (if not then it is piecewise continuous) with $C(\Delta) \to 0$ as $\theta \to \theta_1$ (this value is usually hard to calculate since $C(\Delta) = 1 - \chi$ where χ represents several messy integrals). Analogous results can be obtained for upper bounds to $S(\mathcal{A}, X)$. It must be noted that for several applications use must be made of the special nature of the coefficients which arise in our lemma, and sometimes a form of the sieve fundamental lemma has to be utilised. An alternative form of our lemma is then as follows.

Lemma 2. *Let the hypotheses of the first lemma be given, with the following additions and alterations. Let $\epsilon > 0$ be given, and u with*

$$\log^{1-\epsilon} x > u > \frac{\log \log x}{\log \log \log x}.$$

Suppose $R < Mx^{-\epsilon}$ and let $w = x^{1/u}$. Assume that (8) holds for all a_m of the form

$$a_m = \sum_{\substack{w \leq p < z \\ ph=m}} d_h p^{it}, \qquad |t| \leq x^2 \lambda^{-1}, \qquad |d_h| \ll \tau(h).$$

Then

$$\sum_{r \sim R} c_r S(\mathcal{A}_r, x^\beta) = \lambda \sum_{r \sim R} c_r S(\mathcal{B}_r, x^\beta)(1 + E) + O(Y \log^3 x),$$

where

$$E \ll \log x \, \exp\left(-\tfrac{1}{2} u\epsilon \log u\epsilon\right).$$

We now give some examples of other results obtainable by the method described here.

1) *The greatest prime factor of an integer in* $[x - x^{\frac{1}{2}+\epsilon}, x]$. Here we take

$$\mathcal{A} = \left\{ n : m \sim x^{1-\theta}, \ m \text{ "smooth"}, \ mn \in [x - x^{\frac{1}{2}+\epsilon}, x] \right\}.$$

This work is discussed by Heath-Brown and Jia in work to appear. The sums we can estimate are as follows:

Type II sum	$\displaystyle\sum_{\substack{u \sim U \\ uv \in \mathcal{A}}} a_u b_v$	if	$x^{\theta - \frac{1}{2}} \leq U \leq x^{\frac{1}{2}}$
Type I or I/II	$\displaystyle\sum_{\substack{u \sim U, v \sim V \\ nuv \in \mathcal{A}}} a_u b_v$	if	$V \leq x^{\frac{1}{2}}, \quad VU^2 \leq x^{\theta}$
Type I_2	$\displaystyle\sum_{\substack{m \sim M, n \sim N \\ mnu \in \mathcal{A}}} a_u$	if	$MN \geq x^{\frac{1}{2}}$.

This gives the conclusion $\theta_0 = \frac{5}{6}$, $\theta_1 > \frac{17}{18}$, so there are integers in the given interval having a prime factor exceeding $x^{19/20}$ for all large x.

2) *Primes in short intervals* $[x - x^{1-\theta}, x]$. Here the arithmetic input is very messy. (For large θ we cannot obtain estimates for type II sums, instead we must use type II_3 and II_4 sums and make use of the special form of the coefficients in (8)). For this problem we obtain $\theta_0 = \frac{5}{12}$ (Huxley [14], Heath-Brown [11] by sieve identities) and $\theta_1 > 0.465$ (Baker and Harman [1], see also the work by Baker, Harman and Pintz in this volume).

3) *Primes in "almost all" short intervals* $[x - x^{1-\theta}, x]$. Again the arithmetic input is messy (see Baker, Harman and Pintz in this volume). One obtains $\theta_0 = \frac{5}{6}$, and $\theta_1 > \frac{17}{18}$ (Wong [26], improving work by Watt [24]), or even (Jia in work to appear) $\theta_1 > \frac{19}{20}$.

4) *Primes of the form* $[n^{\theta}]$. Here

$$\mathcal{A} = \left\{ [n^{\theta}] : \tfrac{1}{2}x \leq n^{\theta} < x \right\}.$$

For $1 < \theta < \frac{13}{11}$ let $\gamma = \theta^{-1}$. Then we can estimate Type I sums if $M < x^{3\gamma - 2}$ and Type II sums if

$$x^{1-\gamma} \leq M \leq x^{5\gamma - 4} \quad \text{or} \quad x^{3-3\gamma} \leq M \leq x^{3\gamma - 2} \quad \text{or} \quad x^{5-5\gamma} \leq M \leq x^{\gamma}.$$

We obtain the conclusion that $\theta_0 = \frac{15}{13}$ [13] (or even $\frac{6121}{5302}$ [19]), and $\theta_1 > \frac{20}{17}$ [3], now improved to $\theta_1 > \frac{13}{11}$ in [16].

5) *An extended Bombieri-Vinogradov Theorem.* Fix a and write

$$\pi(x; q, a) = \sum_{\substack{p \leq x \\ p \equiv a \,(\text{mod } q)}} 1.$$

In [2] we obtain, for any $A > 0$, $\epsilon > 0$,

$$\sum_{\substack{\sqrt{x} \leq q \leq x^{\theta} \\ (q,a)=1}} \left| \pi(x; q, a) - u(a, q) \frac{x}{\phi(q) \log x} \right| \ll \log^{-A} x \sum_{q \leq x^{\theta}} \frac{x}{\phi(q) \log x},$$

with

$$C_1(\theta) < u(a, q) < C_2(\theta), \quad C_1\left(\tfrac{1}{2}\right) = 1 - \epsilon, \quad C_2\left(\tfrac{1}{2}\right) = 1 + \epsilon,$$
$$C_1(\theta) > 0 \quad \text{for} \quad \theta \leq 0.52,$$
$$C_2(\theta) < 1.015 \quad \text{for} \quad \theta \leq 0.51, \qquad C_2(\theta) < 2 \quad \text{for} \quad \theta < 0.53,$$
$$\int_{\frac{1}{2}}^{\frac{3}{5}} C_2(\theta) < 0.241.$$

This result has the following applications: the greatest prime factor of $p + a$ exceeds $p^{0.676}$ infinitely often; the greatest prime factor of $p + a$ does not exceed $p^{0.299}$ infinitely often.

6) *Non-linear Diophantine approximation with primes.* In [25] it is shown that the inequality $\|\gamma p^k\| < p^{-\rho(k)}$ (where $k \geq 4$ is an integer) has infinitely many solutions in primes p for irrational γ with

$$\rho(n) = \begin{cases} \frac{1}{21} & \text{if} \quad k = 4 \\ (0.815)2^{-k} & \text{if} \quad k \geq 5. \end{cases}$$

7) *The equidistribution of roots of a quadratic congruence to prime moduli.* Let $f(x) = ax^2 + 2bx + c \in \mathbb{Z}[x]$, with $D = ac - b^2 > 0$. Then, for each fixed $h \neq 0$ we have

$$\sum_{p \leq x} \sum_{p|f(\nu)} e\left(\frac{h\nu}{p}\right) = o\big(\pi(x)\big).$$

This corresponds to $\alpha = \epsilon$, $\beta = \frac{1}{3} - 2\epsilon$, $M = x^{\frac{1}{2} - \epsilon}$ in our description of the sieve, and the $o(\cdot)$ reflects the smooth transition from an identity to upper and lower bounds [6].

To finish we give some examples of the messy integrals involved. Let $\omega(u)$ denote Buchstab's function

$$\omega(u) = u^{-1} \quad \text{if} \quad u \le 2, \qquad \omega(u) = \frac{1 + \log(u-1)}{u} \quad \text{if} \quad 2 \le u \le 3,$$

$$\omega(u) \to e^{-\gamma} = 0.56\ldots \quad \text{as} \quad u \to \infty.$$

The significance of ω is that the probability a number around x in size has all its prime factors $> x^{1/u}$ is $u\omega(u)/\log x$. Thus the integrals which arise have the form

$$\int_{\mathcal{D}} \omega\left(\frac{1 - \alpha - \beta - \gamma - \delta}{\delta}\right) \frac{d\alpha\, d\beta\, d\gamma\, d\delta}{\alpha\beta\gamma\delta^2},$$

$$\int_{\mathcal{E}} \omega\left(\frac{\alpha - \gamma}{\gamma}\right) \omega\left(\frac{1 - \alpha - \beta - \gamma - \delta}{\delta}\right) \frac{d\alpha\, d\beta\, d\gamma\, d\delta}{\beta\gamma^2\delta^2},$$

or

$$\int_{(\alpha,\beta)\in\mathcal{E}} \frac{d\alpha\, d\beta}{\alpha\beta} \min\left(\frac{1}{\beta}\omega\left(\frac{1-\alpha-\beta}{\beta}\right), \int_{(\alpha,\beta,\gamma,\delta)\in\mathcal{D}} \omega\left(\frac{1-\alpha-\beta-\gamma-\delta}{\delta}\right) \frac{d\gamma\, d\delta}{\gamma\delta^2}\right).$$

Here $\mathcal{D}, \mathcal{E}$ are complicated subsets of $[\,\tau, \frac{1}{2}\,]^k$ (assuming we can give a formula for $S(\mathcal{A}, x^\tau)$) corresponding to combinations of k variables $p_1, \ldots, p_k$ for which we can neither give a formula for

$$\sum S(\mathcal{A}_{p_1 \ldots p_k}, p_k)$$

nor apply the Buchstab identity twice more. The "minimum" occuring in the last integral arises since it may not always be more efficient to apply the Buchstab decomposition twice more.

Conclusion. Contrary to some popular beliefs, the sieve of Eratosthenes-Legendre is very powerful when used correctly. One major outstanding problem is to characterise all sets of arithmetical information which will lead to non-trivial lower bounds from the sieve process described here.

References

1 R.C. Baker and G. Harman: The difference between consecutive primes. *Proc. London Math. Soc.* (3) **72** (1996), 261–280.
2 R.C. Baker and G. Harman: The Brun-Titchmarsh theorem on average. In *Analytic Number Theory: Proceedings of a Conference in Honor of Heini Halberstam* Vol. I, 39–103 (B.C. Berndt, H.G. Diamond and A.J. Hildebrand, eds.). Birkhäuser (1996).

3 R.C. Baker, G. Harman and J. Rivat: Primes of the form $[n^c]$. *J. Number Theory* **50** (1995), 261–277.

4 A. Balog: On the distribution of p^θ mod 1. *Acta Math. Hungar.* **45** (1985), 179–199.

5 E. Bombieri, J.B. Friedlander and H. Iwaniec: Primes in arithmetic progressions to large moduli III. *J. Amer. Math. Soc.* **2** (1989), 215–224.

6 W. Duke, J.B. Friedlander and H. Iwaniec: Equidistribution of roots of a quadratic congruence to prime moduli. *Annals of Math.* **141** (1995), 423–441.

7 G. Harman: On the distribution of αp modulo one. *J. London Math. Soc.* (2) **27** (1983), 9–18.

8 G.Harman: On the distribution of αp modulo one II. *Proc. London Math. Soc.* (3) **72** (1996), 241–260.

9 D.R. Heath-Brown: Prime numbers in short intervals and a generalized Vaughan identity. *Canadian J. Math.* **34** (1982), 1365–1377.

10 D.R. Heath-Brown: Sieve identities and gaps between primes. *Astérisque* **94** (1982), 61–65.

11 D.R. Heath-Brown: The number of primes in a short interval. *J. Reine Angew. Math.* **389** (1988), 22–63.

12 D.R. Heath-Brown and H. Iwaniec: On the difference between consecutive primes. *Invent. Math.* **55** (1979), 49–69.

13 L. Hong-Quan and J. Rivat: On the Piatetski-Shapiro prime number theorem. *Bull. London Math. Soc.* **24** (1992), 143–147.

14 M.N. Huxley: On the difference between consecutive primes. *Invent. Math.* **15** (1972), 155–164.

15 H. Iwaniec and M. Jutila: Primes in short intervals. *Arkiv Mat.* **17** (1979), 167–176.

16 Chaohua Jia: On the Piatetski-Shapiro prime number theorem II. *Sc. China Ser. A* **36** (1993), 913–926.

17 Yu.V.Linnik: *The Dispersion Method in Binary Additive Problems* (Trans. Math. Mono. 4). Amer. Math. Soc. (1963).

18 I. Piatetski-Shapiro: On the distribution of prime numbers in sequences of the form $[f(n)]$. *Math. Sb.* **33** (1953), 559–566.

19 J. Rivat: *Autour d'une théorème de Piatetski-Shapiro*. Thèse, Université de Paris Sud (1992).

20 R.C. Vaughan: On the distribution of αp modulo 1. *Mathematika* **24** (1977), 135–141.

21 R.C. Vaughan: An elementary method in prime number theory. *Acta Arith.* **37** (1980), 111–115.

22 I.M. Vinogradov: A new estimation of a certain sum containing primes. *Mat. Sb.* **44** (1937), 783–791.

23 I.M. Vinogradov: *The Method of Trigonometric Sums in the Theory of Numbers* (translated from the Russian by K.F. Roth and A. Davenport). Wiley-Interscience (1954).

24 N. Watt: Intervals almost all containing primes. *Acta Arith.* **72** (1995), 131–167.

25 K.C. Wong: On the distribution of αp^k modulo one. *Glasgow Math. J.*, to appear.

26 K.C. Wong: Primes in almost all short intervals (preprint). (1995).

10. On Hypothesis K* in Waring's Problem

C. Hooley

1. Introduction

Enunciated by Hardy and Littlewood in the memoir *Partitio Numerorum* VI [2] in order to further their study of Waring's problem, *Hypothesis K* asserted that the number $r_k(n)$ of representations of a positive integer n as the sum of k non-negative kth powers is limited by the bound

$$r_k(n) = O\left(n^\epsilon\right). \tag{1}$$

Although Mahler [5] shewed that this relation was false for $k = 3$ by using properties of cubic surfaces, the belief in this aspect of Hardy and Littlewood's procedures was not seriously undermined because they had applied their hypothesis through the sole agency of the consequential equation

$$S_k(x) = \sum_{n \leq x} r_k^2(n) = O\left(x^{1+\epsilon}\right), \tag{2}$$

which we term *Hypothesis K** for convenience and which may well be true even when (1) is false. Indeed, subsequent researches such as those due to Hua produced important upper bounds for $S_k(x)$, which, though far weaker than (2), enabled significant progress to be made in the elucidation of Waring's problem. Yet the bounds achieved for $S_k(x)$ were insufficiently strong for the solution of the more significant problems such as the production of an asymptotic formula for the number of representations of a large number as the sum of seven cubes, albeit important work due to Vaughan has resulted in the achievement of the asymptotic formula for the eight cubes question by a narrow but decisive margin.

In a memoir published some ten years ago ([3], to which hereafter we refer by the symbol I for brevity) we studied the effect on Waring's problem of assuming the Riemann hypothesis for certain Hasse-Weil L-functions defined over cubic three-folds, proving in particular the bound

$$S_3(x) = O\left(x^{\frac{20}{19}+\epsilon}\right) \tag{3}$$

Sieve Methods, Exponential Sums, and their Applications in Number Theory
Greaves, G.R.H., Harman, G., Huxley, M.N., Eds. ©Cambridge University Press, 1996

that was shewn to have a number of useful implications for the theory. But we signalled there our intention of returning to the analysis of $S_3(x)$ not only because of the likelihood of an asymptotic formula of the type

$$S_k(x) \sim A(k)x \quad (x \to \infty)$$

but also on account of our belief — expressed informally during several lectures on the subject — that the attainment of *Hypothesis K^** was within the potential of the method in I. Therefore, having predicted in [4] the value of $A(k)$ above by developing various methods for producing lower bounds of the type

$$S_k(x) > A'(k)x \quad \bigl(x > x_0(k)\bigr),$$

we now revert to the main theme of I and substantiate *Hypothesis K^** itself on the same assumptions as before.

Much of the previous method is retained, including such innovations as the *double Kloostermann refinement* that are described in the Introduction to I. But now we avoid the partial summation introduced in equation (72) therein that was instrumental in dulling the effect of the double Kloostermann refinement on the senior arcs and making it necessary to bring in the junior arcs on which a normal Kloostermann refinement was used. This is engineered by exploiting so systematically the positivity of the integrands in the process that the factors attached to the exponential sums $Q(\mathbf{m}; k)$ become essentially independent of k in the penultimate part of the relevant analysis. Thus, save in the treatment of the terms corresponding to determinations of $\mathbf{m}$ for which the discriminant $\Delta(\mathbf{m})$ vanishes, all Farey arcs are now both major and senior, in the respective senses of Hardy and Littlewood [1] and of I.

There may well be applications of these methods to both the conditional and unconditional study of the representation of large numbers by general ternary cubic forms, the theory of which has been much neglected save in the case of norm forms. But any account of this matter must await a more detailed examination of all the issues involved.

After completing the proof, we briefly indicate some consequences of our theorem that were not covered by our earlier result (3).

Finally, Dr. Heath-Brown has informed me that he has independently proved (2) for $k = 3$ by a different method involving the assumption of essentially the same hypothesis; an account of this will appear in due course.

On the whole, we retain the notation in I because we make frequent references to the results and work in the former publication; the main difference concerns the conventions related to x and x_1 in equation (5) below.

2. The revised initial treatment

The starting point of our analysis being the inequality

$$S(x) = S_3(x) = \sum_{n \leq x} r_3^2(n) = O\{R^*(8x)\} \qquad (4)$$

expressed in I(5), we replace x by x_1 until §6 because the former symbol is now best reserved for a certain free variable occurring in the treatment. Accordingly, assuming

$$X_1 = x_1^{\frac{1}{3}}, \quad X = x^{\frac{1}{3}}, \text{ and } \quad X_1 \leq X \leq 2X_1, \qquad (5)$$

we have

$$R^*(x_1) = \int_0^1 |F(\theta, X_1)|^6 \, d\theta \leq \frac{1}{X_1} \int_{X_1}^{2X_1} R^*(x) \, dx = R^\dagger(X_1), \qquad (6)$$

say, because it is clear from its mode of formation that $R^*(x)$ is a non-decreasing function. Yet, although we already diverge from the path taken in I by forming the average $R^\dagger(X_1)$ that is an essential component of the new method, the initial treatment of the integrand $R^*(x)$ is similar to that in I save for some slight changes in its environment.

The order M of the Farey's series appearing in the work is taken to be close to $\sqrt{x_1}$, a choice more traditional for this sort of problem than the one adopted in I. Specifically, if 2^{M_1} be the least power of 2 that is not less than $\sqrt{x_1}$, we find it convenient to set

$$M = 2^{M_1} - 1 \qquad (7)$$

and then to subdivide the integers k not exceeding M into disjoint sets of the type

$$Y \leq k < 2Y, \qquad (8)$$

where $Y = Y_j$ is of the form 2^{j-1} and $1 \leq j \leq M_1$. Then, after following the analysis of I almost verbatim from (70) therein to the beginning of §4, we reach W in (19) and redefine it to suit our present design. In fact, on the assumption that (5) holds and k is in a typical range (8), we set

$$\lambda = \lambda_k = \lambda_{k,Y} = k/(2Y) < 1 \qquad (9)$$

and then define $W = W(x_1, k, Y)$ by

$$W = \max\left(\lambda^3 X_1^2 Y |\phi| \log^4 x_1, \, Y X_1^{-1} \log^4 x_1\right), \qquad (10)$$

the new value of which does not materially affect the analysis of the sums $R_1^*(x)$ and $R_3^*(x)$ that arise from the sums

$$F_2(h, k; \phi) = F_2(h, k; \phi; X, X_1, Y), \quad F_3(h, k; \phi) = F_3(h, k; \phi; X, X_1, Y)$$

in I(21). Indeed, if we note that still $|m/k| > 6X^2|\phi|$ when $|m| > W$ and that now [‡]

$$F_3(h, k; \phi) = O\left(X \sum_{|m| > \frac{1}{4}kX^{-1}\log^4 x} e^{-A_0(|m|X/k)^{\frac{1}{3}}} \right),$$

we see that
$$R_1^*(x) = O(x) \quad \text{and} \quad R_3^*(x) = O(1)$$

as before. Hence, by (6) and I (23), we end the initial part of treatment by deducing that

$$R^\dagger(X_1) = O(X_1) + \frac{1}{X_1} \int_{X_1}^{2X_1} R_2^*(x)\, dx = O(X_1) + R_2^\dagger(X_1), \qquad (11)$$

say, at which point we cease to emulate the pattern set in I.

3. Treatment of $R_2^\dagger(X_1)$; first stage

We make a thorough use of the positivity of the function $G_2(\phi, k)$ from which $R_2^\dagger(X_1)$ springs. First, by (11), we infer that

$$R_2^\dagger(X_1) = O\left(\sum_{k \leq M} \frac{1}{X_1} \int_{X_1}^{2X_1} \int_{-1/Mk}^{1/Mk} G_2(\phi, k)\, d\phi\, dX \right)$$

$$= O\left(\sum_{1 \leq j \leq M_1} \sum_{Y_j \leq k < 2Y_j} \frac{1}{X_1} \int_{X_1}^{2X_1} \int_{-1/Mk}^{1/Mk} G_2(\phi, k)\, d\phi\, dX \right)$$

$$= O\left(\sum_{1 \leq j \leq M_1} P(X_1, Y_j) \right), \qquad (12)$$

say, and then consider the contribution $B(X_1, k) = B(k)$ to $P(X_1, Y)$ due to a typical value of k in $(Y, 2Y)$. Confirming that $G_2(\phi, k)$ is a non-negative

---

[‡] There is a slight mistake in the corresponding equation in I; the lower limit for the summation over m should be $\frac{1}{2}kX^{-1}\log^4 x$.

function that is still defined even when the condition on X in (5) is no longer imposed, we then have

$$
B(k) \leq \frac{1}{X_1} \int_0^{4\lambda X_1} \frac{1}{\lambda^4} \left(\frac{k}{Y}\right)^{\frac{5}{2}} \int_{-1/\lambda^3 MY}^{1/\lambda^3 MY} G_2(\phi, k) \, d\phi \, dX
$$

$$
= \frac{2}{X_1 Y^{\frac{5}{2}}} \int_0^{4\lambda X_1} \frac{1}{\lambda^4} \int_0^{1/\lambda^3 MY} k^{\frac{5}{2}} G_2(\phi, k) \, d\phi \, dX
$$

$$
= \frac{2}{X_1 Y^{\frac{5}{2}}} \int_0^{4\lambda X_1} \frac{1}{\lambda^4} C(X, k) \, dX, \tag{13}
$$

say, because of (9) and because I(33) states that

$$
G_2(\phi, k) = \frac{1}{k^6} \sum_{\|\mathbf{m}\| \leq W} H(\phi, \mathbf{m}/k; X) Q(\mathbf{m}; k) \tag{14}
$$

is an even function of ϕ; here we retain the convention that all the components of $\mathbf{m}$ are non-zero. Next set

$$
X_2 = X_1^2 Y \log^4 x_1, \quad X_3 = Y X_1^{-1} \log^4 x_1, \quad X_4 = \left(X_1^2 \log^4 x_1\right)/M \tag{15}
$$

for conciseness and substitute (14) in the integral defining $C(X, k)$ to get

$$
C(X, k)
$$

$$
= \sum_{\|\mathbf{m}\| \leq Y X_1^{-1} \log^4 x_1} \frac{Q(\mathbf{m}; k)}{k^{\frac{7}{2}}} \int_0^{1/\lambda^3 MY} H\left(\phi, \frac{\mathbf{m}}{k}; X\right) d\phi
$$

$$
+ \sum_{Y X_1^{-1} \log^4 x_1 < \|\mathbf{m}\| \leq (X_1^2 \log^4 x_1)/M} \frac{Q(\mathbf{m}; k)}{k^{\frac{7}{2}}} \int_{\|\mathbf{m}\|/\lambda^3 X_2}^{1/\lambda^3 MY} H\left(\phi, \frac{\mathbf{m}}{k}; X\right) d\phi
$$

$$
= \sum_{\|\mathbf{m}\| \leq X_3} \frac{Q(\mathbf{m}; k)}{k^{\frac{7}{2}}} H_1(k, \mathbf{m}; X) + \sum_{X_3 < \|\mathbf{m}\| \leq X_4} \frac{Q(\mathbf{m}; k)}{k^{\frac{7}{2}}} H_2(k, \mathbf{m}; X), \tag{16}
$$

say, after changing the order of integration and summation.

To apply (16) we recall the genesis of $H(\phi, \mathbf{m}/k; X)$ in I(33) from the integral

$$
J(\phi, \mathbf{m}/k; X) = 2 \int_0^X \gamma(t/X) \cos 2\pi \left(\phi t^3 + mt/k\right) dt,
$$

which through the substitution $t = \lambda t'$ is shewn to be the same as

$$\lambda J(\lambda^3 \phi, m/2Y; X/\lambda)$$

by (9). Hence, if we use the consequential determination

$$\lambda^6 H(\lambda^3 \phi, \mathbf{m}/2Y; X/\lambda)$$

of the integrands in H_1 and H_2, we obtain

$$H_1 = \lambda^3 \int_0^{1/MY} H(\phi, \mathbf{m}/2Y; X/\lambda) \, d\phi,$$

$$H_2 = \lambda^3 \int_{\|\mathbf{m}\|/X_2}^{1/MY} H(\phi, \mathbf{m}/2Y; x/\lambda) \, d\phi$$

by means of a further transformation $\phi = \lambda^{-3} \phi'$. These values are then absorbed in (13) via (16) to yield

$$B(k)$$

$$\leq \frac{2}{Y^{\frac{5}{2}}} \sum_{\|\mathbf{m}\| \leq X_3} \frac{Q(\mathbf{m}; k)}{k^{\frac{7}{2}}} \int_0^{1/MY} \frac{1}{\lambda X_1} \int_0^{4\lambda X_1} H\left(\phi, \frac{\mathbf{m}}{2Y}; \frac{X}{\lambda}\right) dX \, d\phi$$

$$+ \frac{2}{Y^{\frac{5}{2}}} \sum_{X_3 < \|\mathbf{m}\| \leq X_4} \frac{Q(\mathbf{m}; k)}{k^{\frac{7}{2}}} \int_{\|\mathbf{m}\|/X_2}^{1/MY} \frac{1}{\lambda X_1} \int_0^{4\lambda X_1} H\left(\phi, \frac{\mathbf{m}}{2Y}; \frac{X}{\lambda}\right) dX \, d\phi$$

$$= \frac{2}{Y^{\frac{5}{2}}} \sum_{\|\mathbf{m}\| \leq X_3} \frac{Q(\mathbf{m}; k)}{k^{\frac{7}{2}}} \int_0^{1/MY} \frac{1}{X_1} \int_0^{4X_1} H\left(\phi, \frac{\mathbf{m}}{2Y}; X\right) dX \, d\phi$$

$$+ \frac{2}{Y^{\frac{5}{2}}} \sum_{X_3 < \|\mathbf{m}\| \leq X_4} \frac{Q(\mathbf{m}; k)}{k^{\frac{7}{2}}} \int_{\|\mathbf{m}\|/X_2}^{1/MY} \frac{1}{X_1} \int_0^{4X_1} H\left(\phi, \frac{\mathbf{m}}{2Y}; X\right) dX \, d\phi$$

$$= \frac{2}{Y^{\frac{5}{2}}} \sum_{\|\mathbf{m}\| \leq X_3} \frac{Q(\mathbf{m}; k)}{k^{\frac{7}{2}}} H_3(\mathbf{m}) + \frac{2}{Y^{\frac{5}{2}}} \sum_{X_3 < \|\mathbf{m}\| \leq X_4} \frac{Q(\mathbf{m}; k)}{k^{\frac{7}{2}}} H_4(\mathbf{m})$$

$$= B_1(k) + B_2(k), \tag{17}$$

say, which inequality is especially suitable for further exploitation because the coefficients of $Q(\mathbf{m}; k)$ within it are independent of k in a given range (8).

The two types of these coefficients are assessed through the bound

$$H(\phi, \mathbf{m}/2Y; X) = O\left(X_1^6\right)$$

that is trivial for $0 \le X \le 4X_1$ and the universal bound

$$H(\phi, \mathbf{m}/2Y; X) = O\left(\frac{Y^{\frac{3}{2}}}{|\phi|^{\frac{3}{2}} |m_1 \ldots m_6|^{\frac{1}{4}}}\right)$$

that flows from Lemma 2 in I. The integral

$$\frac{1}{X_1} \int_0^{4X_1} H(\phi, \mathbf{m}/2Y; X) \, dX$$

being subject to the same limits, it follows that

$$H_3(\mathbf{m}) = O\left(X_1^6 \int_0^{1/X_1^3} d\phi\right) + O\left(\frac{Y^{\frac{3}{2}}}{|m_1 \ldots m_6|^{\frac{1}{4}}} \int_{1/X_1^3}^\infty \frac{d\phi}{\phi^{\frac{3}{2}}}\right)$$

$$= O\left(X_1^3\right) + O\left(\frac{Y^{\frac{3}{2}} X_1^{\frac{3}{2}}}{|m_1 \ldots m_6|^{\frac{1}{4}}}\right)$$

$$= O\left(\frac{X_1^{\frac{3}{2}+\epsilon} Y^{\frac{3}{2}}}{|m_1 \ldots m_6|^{\frac{1}{4}}}\right) \tag{18}$$

in view of (17) and the limit of summation in $B_1(k)$; similarly

$$H_4(\mathbf{m}) = O\left(\frac{Y^{\frac{3}{2}}}{|m_1 \ldots m_6|^{\frac{1}{4}}} \int_{\|\mathbf{m}\|/X_2}^\infty \frac{d\phi}{\phi^{\frac{3}{2}}}\right)$$

$$= O\left(\frac{Y^{\frac{3}{2}} X_2^{\frac{1}{2}}}{|m_1 \ldots m_6|^{\frac{1}{4}} \|\mathbf{m}\|^{\frac{1}{2}}}\right) = O\left(\frac{X_1 Y^2}{|m_1 \ldots m_6|^{\frac{1}{4}} \|\mathbf{m}\|^{\frac{1}{2}}}\right) \tag{19}$$

by (15).

Lastly, before commencing the requisite summations over k and $\mathbf{m}$, we split each sum $B_i(k)$ into two parts $B_{i,1}(k)$ and $B_{i,2}(k)$ that answer, respectively, to determinations of $\mathbf{m}$ for which $\Delta(\mathbf{m}) \ne 0$ and $\Delta(\mathbf{m}) = 0$ (see I(34) for the definition of $\Delta(\mathbf{m})$), the donation of all sums of type $B_{i,l}(k)$ to $P(X_1, Y)$ in (12) being denoted by $P_{i,l}(X_1, Y)$ so that

$$P(X_1, Y) = O\{P_{1,1}(X_1, Y)\} + O\{P_{2,1}(X_1, Y)\}$$
$$+ O\{P_{1,2}(X_1, Y)\} + O\{P_{2,2}(X_1, Y)\}. \tag{20}$$

4. Summations over k and $\mathbf{m}$

So far our work has been free from any hypothesis. But now we assume Hypothesis HW of I and remember the consequent formula

$$\sum_{k \leq y} \frac{Q(\mathbf{m}; k)}{k^{\frac{7}{2}}} = O\left(X^{\epsilon} y^{\frac{1}{2}} \prod_{1 \leq j \leq 6} \varpi(m_j)\right)$$

that is valid for $y, \|\mathbf{m}\| \leq X_1^{A_1}$, and $\Delta(\mathbf{m}) \neq 0$ when expressed in the notation of I, Lemma 9. Then, by (12) and (17) and then by (18), we have

$$P_{1,1}(X_1, Y) = \frac{2}{Y^{\frac{5}{2}}} \sum_{\substack{\|\mathbf{m}\| \leq X_3 \\ \Delta(\mathbf{m}) \neq 0}} H_3(\mathbf{m}) \sum_{Y \leq k < 2Y} \frac{Q(\mathbf{m}; k)}{k^{\frac{7}{2}}}$$

$$= O\left(\frac{X_1^{\epsilon}}{Y^2} \sum_{\|\mathbf{m}\| \leq X_3} |H_3(\mathbf{m})| \prod_{1 \leq j \leq 6} \varpi(m_j)\right)$$

$$= O\left(\frac{X_1^{\frac{3}{2}+\epsilon}}{Y^{\frac{1}{2}}} \prod_{1 \leq j \leq 6} \sum_{0 < m_j \leq X_3} \frac{\varpi(m_j)}{m_j^{\frac{1}{4}}}\right)$$

and consequently deduce from I, Lemma 12, and from (15) that

$$P_{1,1}(X_1, Y) = O\left(\frac{X_1^{\frac{3}{2}+\epsilon} X_3^{\frac{9}{2}}}{Y^{\frac{1}{2}}}\right) = O\left(\frac{X_1^{\epsilon} Y^4}{X_1^3}\right); \tag{21}$$

note here that the argument is valid for $Y < X_1 \log^{-4} x_1$ but gives a trivial result. Similarly, by (17), (19), and an obvious variant of Lemma 12 in I, we have

$$P_{2,1}(X_1, Y) = \frac{2}{Y^{\frac{5}{2}}} \sum_{\substack{X_3 < \|\mathbf{m}\| \leq X_4 \\ \Delta(\mathbf{m}) \neq 0}} H_4(\mathbf{m}) \sum_{Y \leq k < 2Y} \frac{Q(\mathbf{m}; k)}{k^{\frac{7}{2}}}$$

$$= O\left(\frac{X_1^{\epsilon}}{Y^2} \sum_{\|\mathbf{m}\| \leq X_4} |H_4(\mathbf{m})| \prod_{1 \leq j \leq 6} \varpi(m_j)\right)$$

$$= O\left(X_1^{1+\epsilon} \sum_{\|\mathbf{m}\| \leq X_4} \frac{1}{|m_1 \ldots m_6|^{\frac{1}{4}} \|\mathbf{m}\|^{\frac{1}{2}}} \prod_{1 \leq j \leq 6} \varpi(m_j)\right)$$

$$= O\left(X_1^{1+\epsilon} \sum_{0 < m_1 \leq X_4} \frac{\varpi(m_1)}{m_1^{\frac{3}{4}}} \prod_{2 \leq j \leq 6} \sum_{0 < m_j \leq X_4} \frac{\varpi(m_j)}{m_j^{\frac{1}{4}}}\right)$$

$$= O\left(X_1^{1+\epsilon} X_4^4\right) = O\left(\frac{X_1^{9+\epsilon}}{M^4}\right) \tag{22}$$

with (15) in mind.

Strictly speaking, we see that the estimations of

$$P_{1,2}(X_1,Y), \quad P_{2,2}(X_1,Y)$$

are really covered by §10 of I, since the only essential difference between our practices in this matter is a change in the order of summation over $\mathbf{m}$ and integration over ϕ. Nevertheless, to provide coherence in exposition we provide a brief direct description of the estimations within the framework already set up. It being most convenient here to reverse the order of summations over k and $\mathbf{m}$, we observe that

$$B_{1,2}(k) = O\left(\frac{X_1^{\frac{3}{2}+\epsilon}}{Y^{\frac{1}{2}}} \sum_{\substack{\|\mathbf{m}\| \leq X_3 \\ \Delta(\mathbf{m})=0}} \prod_{1 \leq j \leq 6} \frac{(k,m_j)^{\frac{1}{4}}}{m_j^{\frac{1}{4}}}\right)$$

by (17), (18), and the relatively crude assessment of $Q(\mathbf{m};k)$ contained in Lemma 8 in I. Then, categorizing as in I the solutions of $\Delta(\mathbf{m}) = 0$ by considering sets typified by

(i) $\mathbf{m} = l(m_1'^{\,2}, \ldots, m_6'^{\,2})$,

(ii) $m_1 = m_2$, $m_3 = m_4$, $m_5 = m_6$,

(iii) $m_1 = m_2$, $(m_2, \ldots, m_6) = l(m_3'^{\,2}, \ldots, m_6'^{\,2})$,

we see the contribution to $B_{1,2}(k)$ due to the first lot of $\mathbf{m}$ is

$$O\left(\frac{X_1^{\frac{3}{2}+\epsilon}}{Y^{\frac{1}{2}}} \sum_{0 < lm_1'^{\,2}, \ldots, lm_6'^{\,2} \leq X_3} \frac{1}{l^{\frac{3}{2}}} \prod_{1 \leq j \leq 6} \frac{(k, lm_j'^{\,2})^{\frac{1}{4}}}{m_j'^{\frac{1}{2}}}\right)$$

$$= O\left\{\frac{X_1^{\frac{3}{2}+\epsilon}}{Y^{\frac{1}{2}}} \sum_{l \leq X_3} \left(\sum_{0 < m' \leq (X_3/l)^{\frac{1}{2}}} \frac{(k, m')^{\frac{1}{2}}}{m'^{\frac{1}{2}}}\right)^6\right\}$$

$$= O\left(\frac{X_1^{\frac{3}{2}+\epsilon} X_3^{\frac{3}{2}}}{Y^{\frac{1}{2}}} \sum_{l=1}^{\infty} \frac{1}{l^{\frac{3}{2}}}\right) = O(X_1^\epsilon Y).$$

Also the effect on $B_{1,2}(k)$ due to the second lot of $\mathbf{m}$ is

$$O\left(\frac{X_1^{\frac{3}{2}+\epsilon}}{Y^{\frac{1}{2}}} \sum_{0 < m_1, m_3, m_5 \leq X_3} \prod_{j=1,3,5} \frac{(k, m_j)^{\frac{1}{2}}}{m_j^{\frac{1}{2}}}\right)$$

$$= O\left(\frac{X_1^{\frac{3}{2}+\epsilon} X_3^{\frac{3}{2}}}{Y^{\frac{1}{2}}}\right) = O(X_1^\epsilon Y),$$

while that due to the third lot is similar by an amalgam of the arguments used in the two previous instances. Thus, by a summation over k, we find that

$$P_{1,2}(X_1, Y) = O\left(X^\epsilon Y^2\right). \tag{23}$$

The estimation of $P_{2,2}(X_1, Y)$ is sufficiently close to that of $P_{1,2}(X_1, Y)$ for us to limit our description to case (i) above. Indeed, by (17) and (19), the contribution in this instance to $B_{2,2}(k)$ is

$$O\left(X_1 \sum_{\substack{lm_1'^2 \leq X_4 \\ 0 < m_2', \ldots, m_6' \leq m_1'}} \frac{1}{l^2} \frac{(k, lm_1'^2)^{\frac{1}{4}}}{m_1'^{\frac{3}{2}}} \prod_{2 \leq j \leq 6} \frac{(k, lm_j'^2)^{\frac{1}{4}}}{m_j'^{\frac{1}{2}}}\right)$$

$$= O\left\{X_1 \sum_{l \leq X_4} \frac{1}{l^{\frac{1}{2}}} \sum_{0 < m_1' \leq (X_4/l)^{\frac{1}{2}}} \frac{(k, m_1')^{\frac{1}{2}}}{m_1'^{\frac{3}{2}}} \left(\sum_{0 < m' \leq m_1'} \frac{(k, m')^{\frac{1}{2}}}{m'^{\frac{1}{2}}}\right)^5\right\}$$

$$= O\left(X_1^{1+\epsilon} \sum_{l \leq X_4} \frac{1}{l^{\frac{1}{2}}} \sum_{0 < m_1' \leq (X_4/l)^{\frac{1}{2}}} (k, m_1')^{\frac{1}{2}} m_1'\right)$$

$$= O\left(X_1^{1+\epsilon} X_4 \sum_{l=1}^{\infty} \frac{1}{l^{\frac{3}{2}}}\right) = O\left(\frac{X_1^{3+\epsilon}}{M}\right),$$

the summation of which over k gives

$$O\left(\frac{X_1^{3+\epsilon} Y}{M}\right)$$

as the impact on $P_{2,2}(X_1, Y)$ of the triplets $\mathbf{m}$ in question. The other classes have a similar influence so that

$$P_{2,2}(X_1, Y) = O\left(\frac{X_1^{3+\epsilon} Y}{M}\right). \tag{24}$$

Therefore, in summation, we get

$$P(X_1, Y) = O\left(\frac{X_1^\epsilon Y^4}{X_1^3}\right) + O\left(\frac{X_1^{9+\epsilon}}{M^4}\right) + O\left(X_1^\epsilon Y^2\right) + O\left(\frac{X_1^{3+\epsilon} Y}{M}\right)$$

from (20), (21), (22), (23), and (24), whence

$$P(X_1, Y) = O\left(X_1^{3+\epsilon}\right)$$

because of the definitions of Y and M that relate to (7) and (8). Thus we conclude from (12) that

$$R_2^\dagger(X_1) = O\left(X_1^{3+\epsilon}\right). \tag{25}$$

5. The theorem on $S(x)$

Equations (4), (6), (11), and (25) yield at once our

Theorem. *Let $r(n)$ be the number of representations of n as the sum of three non-negative cubes. Then, as $x \to \infty$,*

$$\sum_{n \le x} r^2(n) = O\left(x^{1+\epsilon}\right)$$

if Hypothesis HW *(as stated in* I*) be true for the Hasse-Weil L-functions defined over the cubic varieties* $\mathcal{V}(\mathbf{m})$.

6. Applications of the theorem

The results on Waring's problem that were enumerated in Theorems 2–7 of I can now be derived with improved exponents. We also have the following conditional theorem that was beyond the reach of the previous method for $l \ge 4$.

The number of representations of a large number N as the sum of six non-negative cubes and a non-negative l-th power is asymptotically equivalent to

$$\frac{l}{l+1}\Gamma^6\left(\tfrac{4}{3}\right)N\mathfrak{S}(N),$$

where $\mathfrak{S}(N)$ is the singular series. All large numbers are expressible in the proposed form.

Lastly, let $\rho(x)$ be the number of positive integers not exceeding x that are expressible as a sum of three non-negative cubes. Then, by using the Cauchy-Schwarz inequality as in the proof of Theorem 8 in I, we have

$$\rho(x) > x^{1-\epsilon} \quad (x > x_0(\epsilon))$$

on Hypothesis HW. Almost best possible, this result should be contrasted with several statements made in [4].

References

All references given in the previous paper [3] cited below are still relevant. But, including [3], we draw particular attention to

1. G.H. Hardy and J.E. Littlewood: A new solution of Waring's problem. *Quarterly Journal of Mathematics* **48** (1920), 272–293.
2. G.H. Hardy and J.E. Littlewood: Some problems of *"Partitio Numerorum"*, VI: Further researches in Waring's problem. *Math. Zeitschrift* **23** (1925), 1–37.
3. C. Hooley: On Waring's problem. *Acta Math.* **157** (1986), 49–97.
4. C. Hooley: On some topics connected with Waring's problem. *J. Reine Angew. Math.* **369** (1986), 110–153.
5. K. Mahler: Note on hypothesis K of Hardy and Littlewood. *J. London Math. Soc.* **11** (1936), 136–138.

11. Moments of Differences between Square-free Numbers

M. N. Huxley

1. General Discussion

A square-free number is a positive integer with no repeated prime factor (but see [6], Art. 17.8). The square-free numbers have asymptotic density $1/\zeta(2) = 6/\pi^2$, so that two consecutive integers are both square-free infinitely often. Mirsky [10] proved more. To state a version of Mirsky's result, we let $s_1, \ldots, s_M$ be the finite sequence of square-free numbers not exceeding N. Let $D(h)$ be the number of i for which $s_{i+1} - s_i = h$.

Proposition 1. (Mirsky) *We have*

$$D(h) = \alpha(h)N + O\left(\frac{hN}{\log N \log\log N}\right)$$

for

$$h < \frac{\log N \log\log\log N}{(\log\log N)^2}.$$

Mirsky actually states Proposition 1 with error $O(N^{1+\epsilon-1/(h+3)})$. His argument is easily modified to give the result above. The constant $\alpha(h)$ is independent of N, and tends to zero faster than exponentially in h; the square-free numbers do not mimic a Poisson process of independent random variables. Two conjectures about the differences between consecutive square-free numbers could in principle be settled by extending the range for h and the accuracy in Proposition 1.

Conjecture 1. *The inequality*

$$s_{i+1} - s_i = O\left(N^\delta\right) \qquad\qquad (C_1(\delta))$$

holds for every $\delta > 0$ and every N.

Conjecture 2. *For every $\gamma \geq 0$, the asymptotic equality*

$$\sum_{i=1}^{M-1} (s_{i+1} - s_i)^\gamma \sim \beta(\gamma)N \qquad\qquad (C_2(\gamma))$$

holds with some constant $\beta(\gamma)$ as N tends to infinity.

Sieve Methods, Exponential Sums, and their Applications in Number Theory
Greaves, G.R.H., Harman, G., Huxley, M.N., Eds. ©Cambridge University Press, 1996

The case $C_2(\gamma)$ of Conjecture 2 implies $C_1(1/\gamma)$, and Filaseta [2] has shown that $C_2(\gamma)$ follows from $C_1(\delta)$ for some small δ depending only on γ. The most recent result towards Conjecture 1 is due to Filaseta and Trifonov [3].

Proposition 2. (Filaseta and Trifonov)

$$s_{i+1} - s_i = O(N^{\frac{1}{6}} \log N). \tag{1.1}$$

These conjectures and propositions have analogues for the sequences of k-free numbers, the positive integers with no prime factor occurring to the kth power or a higher power, for $k \geq 2$. The exponent in Proposition 2 is $1/(2k+1)$ for the k-free numbers.

Erdős established Conjecture 2 for $\gamma \leq 2$, and Hooley [7] increased the range to $\gamma \leq 3$. Hooley subsequently announced the further extension $\gamma \leq \frac{250}{79} = 3.164\ldots$, deeming the improvement insufficient to justify publication. Filaseta [2] obtained Conjecture 2 for $\gamma < \frac{29}{9} = 3.222\ldots$, and with Trifonov [4] has extended the range again to $\gamma < \frac{43}{13} = 3.307\ldots$. Meanwhile Graham [5] considered Conjecture 2 for general k-free numbers, obtaining the range

$$\gamma < 2k - 2 + \frac{4}{k+1}$$

for $k \geq 3$. Graham's result corresponds to $\gamma < \frac{10}{3} = 3.333\ldots$ at $k = 2$, and his method, with some modification, works for $k = 2$ also.

In this paper we obtain Conjecture 2 for $\gamma < \frac{11}{3} = 3.666\ldots$. Besides Proposition 2 and the results of [4], we consider triples of integers of the form $p^2 q$ by vector methods. The same argument gives Conjecture 2 for k-free numbers for

$$\gamma < 2k - 1 + \frac{2}{k+1}.$$

The constant $\beta(\gamma)$ in Conjecture 2 is defined as the sum of an infinite series whose convergence is usually proved indirectly. Our first lemma gives a bound for $\alpha(r)$ in Proposition 1.

Lemma 1. *For large* r

$$\log(\alpha(r+1)) \leq -\tfrac{5}{4} r \log \log r + O(r).$$

Proof. In this lemma we consider the infinite sequence $\mathcal{S}$ of all square-free numbers s_i. Let $\mathcal{A}$ be the subsequence of those s_i with $s_{i+1} - s_i = r+1$. For n in $\mathcal{A}$, let $q_i(n)$ be the least prime number p with $p^2|(n+i)$, for $i = 1, \ldots, r$. We classify $\mathcal{A}$ according to the sequence $q_1(n), \ldots, q_r(n)$. Let $q_1, \ldots, q_r$ be a list of primes with repeats allowed. Let $q'_1, \ldots, q'_s$ be $q_1, \ldots, q_r$ arranged in ascending order without repeats, and let q be the product $q'_1 \ldots q'_s$. The asymptotic density of the numbers in $\mathcal{A}$ with $q_i(n) = q_i$ for $i = 1, \ldots, r$ is at most $1/q^2$. We ask: given $q'_1, \ldots, q'_s$, how many possible sequences $q_1, \ldots, q_r$ can occur? We define an integer t by $q'_i \leq \sqrt{r}$ for $i \leq t$, $q'_i > \sqrt{r}$ for $i > t$, and we put $s = t + u$. If we know the residue class of n mod $q_i'^2$ for $i \leq t$, then we know where these primes occur in the list $q_1, \ldots, q_r$. In particular, the prime 2 occurs either $[r/4]$ or $[r/4] + 1$ times, depending on $n \pmod 4$, so

$$\frac{r}{4} - 1 + s \leq r \leq r \sum \frac{1}{p^2} + s \leq \frac{3r}{8} + s,$$

so that s is bounded in terms of r by

$$\frac{3r}{4} + 1 \geq s \geq \frac{5r}{8}. \tag{1.2}$$

The primes q'_i with $i > t$ are greater than $\sqrt{r}$, and they occur only once. For given $q'_1, \ldots, q'_s$, there are at most

$$q_1'^2 \ldots q_t'^2 u^u$$

possibilities for the sequence $q_1, \ldots, q_r$. Hence the density of those n in $\mathcal{A}$ with

$$\{q_1(n), \ldots, q_r(n)\} = \{q'_1, \ldots, q'_s\}$$

as sets is

$$\leq \frac{u^u}{q_{t+1}'^2 \ldots q_s'^2}.$$

To estimate this density, we write $p_1, p_2, \ldots$ for the prime numbers in order. Then $q'_i \geq p_i$, so for large i

$$\sum_{q'_i} \frac{1}{q_i'^2} \leq \left(1 + O\left(\frac{1}{i}\right)\right) \frac{1}{p_i \log i} \leq \left(1 + O\left(\frac{1}{i}\right)\right) \frac{1}{i \log^2 i}.$$

We use this for $i > t$. There are at most $2^{\sqrt{r}}$ choices for $q'_1, \ldots, q'_t$. The total density of those n in $\mathcal{A}$ with fixed values of t and u is

$$\leq \delta(t, u) = 2^{\sqrt{r}} \prod_{i=t+1}^{s} \frac{u}{i \log^2 i} \left(1 + O\left(\frac{1}{i}\right)\right),$$

with

$$\log \delta(t, u) = u \log s + O(\sqrt{r}) + O(\log s) - \sum_i (\log i + 2 \log \log i)$$

$$= s \log s + O(\sqrt{r} \log r) - \int_t^s (\log x + 2 \log \log x) \, dx$$

$$= -2s \log \log r + O(s).$$

The lemma follows on summing over t and u with $s = t + u$ in the range (1.2). $\square$

The argument is delicate, and the steps have to be taken in the right order. We shall prove Conjecture 2 in an equivalent form.

Conjecture 2'. *Let $s_1, \ldots, s_M$ be the square-free numbers n in the range $N/2 \le n \le N$. Then for every $\gamma \ge 0$, the asymptotic inequality*

$$\sum_{i=1}^{M-1} (s_{i+1} - s_i)^\gamma \sim \tfrac{1}{2}\beta(\gamma)M$$

holds as N tends to infinity, with some constant $\beta(\gamma)$.

We use Mirsky's asymptotic formula (Proposition 1) for $s_{i+1} - s_i = h+1$ with

$$h \le H_0 = \left(\frac{\log N}{\log \log N}\right)^{1/(\gamma+2)}.$$

These terms contribute

$$\tfrac{1}{2}\beta(\gamma)N + O\left(\frac{N}{(\log \log N)^2}\right)$$

to the sum in Conjecture 2', with $\beta(\gamma)$ given by

$$\beta(\gamma) = \sum_{h=0}^{\infty} (h+1)^\gamma \alpha(h),$$

which converges by Lemma 1. The main task is to estimate the number of gaps greater than H_0.

Our second lemma is a common ingredient of all treatments of gaps between square-free numbers.

Lemma 2. (Erdős) *There are constants $c_3 \ge 1$ and $c_4 \ge \frac{1}{4}$ such that if $n+1, \ldots, n+H$ are consecutive integers which are not square-free, and $H \ge c_3$, then there are at least $c_4 H$ integers h in $1 \le h \le H$ for which the smallest prime p with $p^2 \,|\, (n+h)$ has*

$$p \ge P_0(H) = \tfrac{1}{4}H \log H.$$

For N sufficiently large we have $H_0 \ge 2c_3$. Let H be a power of two with $H \ge \frac{1}{2}H_0 \ge c_3$. We classify the gaps with $s_{i+1} - s_i \ge H+1$ into large-prime gaps and small-prime gaps. Large-prime gaps contain a number $p^2 q$ with

$$p \ge P_1(H) = H^\gamma \log H.$$

The number of possible p^2q is

$$O\left(\sum_{p\geq P_1}\frac{N}{p^2}\right) = O\left(\frac{N}{P_1\log P_1}\right) = O\left(\frac{N}{H^\gamma\log^2 H}\right).$$

This is an upper bound for the number of large-prime gaps greater than H. We sum H through powers of two to see that large-prime gaps longer than H_0 contribute

$$O\left(\sum_{H\geq\frac{1}{2}H_0}H^\gamma\cdot\frac{N}{H^\gamma\log^2 H}\right) = O\left(\frac{N}{\log H_0}\right) = O\left(\frac{N}{\log\log N}\right).$$

In small-prime gaps, each set of H consecutive integers $n+1,\ldots,n+H$ contains at least $H/4$ numbers of the form p^2q with p prime and

$$P_0(H) \leq p < P_1(H).$$

For any n, let $F(H,n)$ be the number of solutions of $n < p^2q \leq n+H$ with p prime and $p \geq P_0(H)$, and let $F(H,P,n)$ be the number of solutions with p in the range $P \leq p < 2P$; here P is a power of two. A gap of length greater than H contains a number n with $F(H,n) \geq H/4$. There are at most $2\gamma\log H$ different powers of two, P, with $P_0(H)/2 \leq P < P_1(H)$, so

$$\sum_P{}' F(H,P,n) \geq \frac{H}{8},$$

where $\sum'$ denotes a sum over powers of two with

$$F(H,P,n) \geq \frac{H}{16\gamma\log H}. \tag{1.3}$$

We can restrict P further if we wish: if $P \geq P_2(H) = H^{\gamma-1}\log^2 H$, then the number of disjoint intervals for which (1.3) holds is so small that they contribute $O(N/\log\log N)$ in Conjecture 2.

If H_0 is so large that (1.3) implies $F(H,P,n) \geq 6$, then (1.3) implies that there are

$$\geq \frac{H^3}{12\times 2^{12}\gamma^3\log^3 H}$$

sextuples of integers p_1, p_2, p_3, q_1, q_2, q_3 with p_i primes lying in the range $P \leq p < 2P$, and with

$$n < p_3^2q_3 < p_2^2q_2 < p_1^2q_1 \leq \min(N, n+H).$$

Since $P \geq P_0(H) \geq H$, the primes p_1, p_2 and p_3 are necessarily distinct. We use the following lemma to remove all sextuples with large common factors between the q_i.

Lemma 3. *Let r, D, H, M be positive integers with $r \geq 2$. The number of r-tuples of distinct integers $n_1, \ldots, n_r$ with $M + 1 \leq n_i \leq M + H$ for $i = 1, \ldots, r$, and highest common factor satisfying $(n_1, \ldots, n_r) > D$ is 0 for $D \geq H$, and*

$$\leq \frac{2H^r}{(r-1)D^{r-1}}$$

for $D < H$.

Proof. Let $d = (n_1, \ldots, n_r)$. Then $n_1 \equiv n_2 \pmod{d}$, so $d \leq H - 1$. Let

$$K = \left[\frac{(M+H)}{d}\right] - \left[\frac{M}{d}\right] \leq \frac{H}{d} + 1$$

be the number of multiples of d in $M + 1, \ldots, M + H$. The number of r-tuples of distinct integers taken from $M + 1, \ldots, M + H$, all multiples of d, is

$$K(K-1)\ldots(K-r+1) \leq K(K-1)^{r-1} \leq \frac{2H}{d}\left(\frac{H}{d}\right)^{r-1}.$$

This number is an upper bound for the number of r-tuples whose highest common factor is d. Hence the number of r-tuples in the lemma is

$$\leq \sum_{d=D+1}^{H-1} \frac{2H^r}{d^r} \leq 2H^r \int_D^\infty \frac{dx}{x^r} = \frac{2H^r}{(r-1)D^{r-1}}. \qquad \square$$

For $D = 2^9(\gamma \log H)^{3/2}$, there are at most

$$\frac{H^3}{D^2} = \frac{H^3}{2^{18}\gamma^3 \log^3 H}$$

sextuples with $(q_1, q_2, q_3) > D$. For $D' = 2D^2 = 2^{19}\gamma^3 \log^3 H$, there are, for each pair i and j, at most

$$H \cdot \frac{2H^2}{D'} = \frac{H^3}{2^{18}\gamma^3 \log^3 H}$$

sextuples with $(q_i, q_j) > D'$. At least

$$S(H) = \frac{H^3}{3 \times 2^{16}\gamma^3 \log^3 H} \tag{1.4}$$

sextuples remain.

To explain our modified version of Hooley's combinatorics, we define two counting functions $T(H,P)$ and $S(H,P)$, where H and P are powers of two with

$$\tfrac{1}{2}H_0 \leq H \leq H_1 = c_5 N^{\frac{1}{5}} \log N, \tag{1.5}$$

where H_1 is the upper bound of Proposition 2, and

$$\tfrac{1}{2}P_0(H) \leq P \leq P_1(H). \tag{1.6}$$

Firstly, $T(H,P)$ is the maximum over $N/2 \leq n \leq N - H$ of the number of integers in $n+1, \ldots, n+H$ which can be written as $p^2 q$ with p prime in the range $P \leq p < 2P$. Secondly, $S(H,P)$ is the number of sextuples of positive integers p_1, p_2, p_3, q_1, q_2, q_3 with p_i prime numbers in the range $P \leq p_i < 2P$ with

$$\tfrac{1}{2}N \leq p_3^2 q_3 < p_2^2 q_2 < p_1^2 q_1 \leq \min(N, p_3^2 q_3 + H - 1),$$

and $(q_1, q_2, q_3) \leq D$, and $(q_i, q_j) \leq D'$ for each pair of suffices i, j.

Lemma 4. *For N sufficiently large we have*

$$\sum_{i=1}^{M-1} (s_{i+1} - s_i)^\gamma = \tfrac{1}{2}\beta(\gamma)N + O\left(\frac{N}{\log \log N}\right) \tag{1.7}$$

in Conjecture 2', provided that for each pair of powers of two, H and P, in the ranges (1.5) and (1.6) we have either

$$T(H,P) < \frac{H}{64\gamma \log H}, \tag{1.8}$$

or a bound for $S(H,P)$ of one of the following forms:

$$S(H,P) = O\left(\frac{N}{H^{\gamma-3} \log^6 H}\right), \tag{1.9}$$

or for some $\eta > 0$ and some $P' \leq P$, which may depend on H but not on P,

$$S(H,P) = O\left(\frac{N}{H^{\gamma-3} \log^5 H}\left(\frac{P'}{P}\right)^\eta\right), \tag{1.10}$$

or for some $\eta > 0$ and some $P' \geq P$, which may depend on H but not on P,

$$S(H,P) = O\left(\frac{N}{H^{\gamma-3} \log^5 P}\left(\frac{P}{P'}\right)^\eta\right). \tag{1.11}$$

The implied constants depend on γ and η.

Proof. We have to estimate the contribution of small-prime gaps. Since $T(H, P)$ is an upper bound for $F(H, P, n)$, values of H and P for which (1.8) holds cannot occur in (1.3). Consider the small-prime gaps for which $H \leq h < 2H$. Each gap contributes $O(H^\gamma)$ to the sum in (1.7), and at least $S(H)$ sextuplets to one of the sums $S(H, P)$. The contribution of these gaps to the right hand side of (1.7) is

$$O\left(\frac{H^\gamma}{S(H)} \sum_P S(H, P)\right) = O\left(H^{\gamma-3} \sum_P S(H, P) \log^3 H\right).$$

Ranges for P on which one of the bounds (1.9), (1.10) or (1.11) holds give $O(N/\log^2 H)$, which sums over powers of two in the range (1.5) to

$$O\left(\frac{N}{\log H_0}\right) = O\left(\frac{N}{\log \log N}\right). \qquad \square$$

If $p^2 q$ is counted in $F(H, P, n)$, then the point (p, q) is an integer point close to the plane curve $x^2 y = n$. There is a growing theory (see [8], [4], [9]) of bounds for the number of integer points close to a curve. Proposition 2 is established by demonstrating (1.8) for all P when $H \geq H_0$.

2. Vector methods

We approach $S(H, P)$ in two stages. First we fix primes p_1, p_2, p_3 with $P \leq p_i < 2P$, and ask how many integer vectors $\mathbf{q} = (q_1, q_2, q_3)$ give a sextuple $(p_1, p_2, p_3, q_1, q_2, q_3)$ counted in $S(H, P)$. Secondly, for K a power of two, we bound $S(H, P, K)$, the number of triplets of primes p_1, p_2, p_3 for which the number of such vectors $\mathbf{q}$ lies in the range K to $2K - 1$. By the Riesz interchange we have

$$S(H, P) \leq \sum_K 2K\, S(H, P, K), \qquad (2.1)$$

where K is summed through powers of two.

Let $\mathbf{p}$ be the vector

$$\mathbf{p} = (p_2^2 p_3^2,\ p_1^2 p_3^2,\ p_1^2 p_2^2). \qquad (2.2)$$

We put

$$p_1^2 q_1 - p_i^2 q_i = h_i \quad \text{when} \quad 1 \leq h_i \leq H. \qquad (2.3)$$

Then

$$\mathbf{p} \times \mathbf{q} = ((h_2 - h_3) p_1^2,\ h_3 p_2^2,\ -h_2 p_3^2).$$

Since N is large, we can suppose that $H \leq N/4$ by Proposition 2. We have

$$\sqrt{3}P^4 \leq |\mathbf{p}| \leq 16\sqrt{3}P^4, \qquad \frac{\sqrt{3}N}{32P^2} \leq |\mathbf{q}| \leq \frac{\sqrt{3}N}{P^2}.$$

By (2.3), $|\mathbf{p} \times \mathbf{q}| \leq 4\sqrt{3}HP^2$, so the angle θ between $\mathbf{p}$ and $\mathbf{q}$ satisfies

$$\sin\theta \leq \frac{128H}{\sqrt{3}N}. \tag{2.4}$$

Our next lemma is set in three-dimensional space, and it could be extended to higher dimensions if required.

Lemma 5. *Let $Q > 0$ and let $\mathbf{p}$ be a fixed vector in $\mathbb{R}^3$. Suppose that $\mathbf{q}^{(1)}, \ldots, \mathbf{q}^{(K)}$ are integer vectors with $|\mathbf{q}^{(i)}| \leq Q$ which make acute angles at most α with $\mathbf{p}$. Then either*

$$K \leq 12\pi Q^3 \sin^2\alpha, \tag{2.5}$$

or the vectors $\mathbf{q}^{(i)}$ lie in some two-dimensional subspace.
If the vectors $\mathbf{q}^{(i)}$ lie in a two-dimensional subspace, then either

$$K \leq 4\pi Q^2 \sin\alpha, \tag{2.6}$$

or the vectors $\mathbf{q}^{(i)}$ are multiples of some primitive integer vector $\mathbf{r}$ with $|\mathbf{r}| \leq Q/K$.

Proof. We consider the rays l_i through the origin in the directions of the vectors $\mathbf{q}^{(i)}$. Let $\mathbf{r}_j$ be the primitive integer vector in the direction of l_j. If $\mathbf{q}^{(i)}$ lies along l_j, then $\mathbf{q}^{(i)} = d_{ij}\mathbf{r}_j$ for some positive integer d_{ij}. Let w_j be the number of indices i for which $\mathbf{q}^{(i)}$ lies along l_j, and let Q_j be the lattice point corresponding to the vector $\mathbf{q}^{(i)}$ for which d_{ij} is maximum, so $d_{ij} \geq w_j$. If three rays l_i, l_j, l_k are not coplanar, then the tetrahedron $OQ_iQ_jQ_k$ has volume

$$\geq \tfrac{1}{6}w_iw_jw_k\left|(\mathbf{r}_i \times \mathbf{r}_j) \cdot \mathbf{r}_k\right| \geq \tfrac{1}{6}w_iw_jw_k \geq \tfrac{1}{18}(w_i + w_j + w_k).$$

Suppose that there are J rays l_j. If they are not all coplanar, then we can form $J - 2$ disjoint tetrahedra of the type $OQ_iQ_jQ_k$, with total volume

$$\geq \tfrac{1}{18}\sum w_j = \tfrac{1}{18}K.$$

These tetrahedra fit inside the intersection of a cone vertex at O, axis $\mathbf{p}$, semi-angle α, and the sphere centre O, radius Q. Comparing volumes, we have

$$\frac{K}{18} \leq \frac{2\pi}{3}Q^3(1 - \cos\alpha) = \frac{2\pi}{3}Q^3 \cdot \frac{1 - \cos^2\alpha}{1 + \cos\alpha} \leq \frac{2\pi}{3}Q^3 \sin^2\alpha,$$

which gives the bound (2.5).

When $J \geq 2$, but all the rays l_j are coplanar, there are $J - 1$ disjoint triangles of the type OQ_iQ_j with area

$$\geq \tfrac{1}{2}w_iw_j|\mathbf{r}_i \times \mathbf{r}_j| \geq \tfrac{1}{2}w_iw_j \geq \tfrac{1}{4}(w_i + w_j).$$

The total area of these triangles is

$$\geq \tfrac{1}{4}\sum w_j = \tfrac{1}{4}K,$$

and they fit inside a sector of a circle radius Q, so

$$\tfrac{1}{4}K \leq 2\alpha Q^2 \leq \pi Q^2 \sin\alpha,$$

which gives (2.6). $\square$

In Lemma 5 we take $Q = \sqrt{3}N/P^2$, $\sin\alpha = 128H/\sqrt{3}N$, so that (2.5) becomes

$$K \leq \frac{199\,608\sqrt{3}\pi H^2 N}{P^6}. \tag{2.7}$$

We distinguish three cases.

Case 1 comprises all triplets of primes for which (2.7) is false, and the vectors $\mathbf{q}^{(i)}$ are multiples of some primitive integer vector $\mathbf{r} = (r_1, r_2, r_3)$. We have

$$p_1^2 r_1 - p_i^2 r_i = g_i, \tag{2.8}$$

for some g_2, g_3 with $1 \leq g_i \leq H/K$, so that $K \leq H$.

Case 2 comprises all triplets of primes for which (2.7) is false and the vectors $\mathbf{q}^{(i)}$ span a space of dimension two.

Case 3 comprises all triplets of primes for which (2.7) holds. When

$$p_1^2 p_2^2 p_3^2 \leq N, \tag{2.9}$$

then for given h_2 and h_3 in (2.3), the values of q_1 lie in a residue class modulo $p_2^2 p_3^2$, and q_1 determines q_2 and q_3. The number of values of q_1 is at most

$$\left[\frac{N}{p_1^2 p_2^2 p_3^2}\right] + 1 \leq \frac{N}{P^6} + 1 \leq \frac{2N}{P^6},$$

so $K \leq 2H^2N/P^6$. All triplets of primes satisfying (2.9) are in Case 3.

In Cases 1 and 2 (2.9) is false. If two vectors $\mathbf{q}^{(i)}$ and $\mathbf{q}^{(j)}$ have the same values of h_2 and h_3, then

$$p_2^2 p_3^2 \left| \left(q_1^{(i)} - q_1^{(j)} \right), \right.$$

and so (2.9) holds. Hence in Cases 1 and 2 no two $\mathbf{q}$ vectors have the same h_2 and h_3, and $K \leq H^2$.

Lemma 6. *In Case* 2 *there is a primitive integer vector* **s** *in the span of the vectors* $\mathbf{q}^{(i)}$ *with*

$$\mathbf{s} = (p_2^2 t, \ p_1^2 t, \ u)$$

for some integers t, u *with*

$$1 \le |p_1^2 p_2^2 t - p_3^2 u| \le \frac{4\pi\sqrt{2}H^2}{K}, \tag{2.10}$$

$$|\mathbf{s}| \le \frac{1024\sqrt{3}\pi H N}{K P^2}. \tag{2.11}$$

Proof. We continue the argument of Lemma 5. The rays l_j lie in a plane through the origin, which has a primitive integer normal vector **n**. The cross-products $\mathbf{r}_i \times \mathbf{r}_j$ are integer multiples of **n**. Let k and l be a pair of indices for which the cross-product is shortest. Let

$$\mathbf{r}_k \times \mathbf{r}_l = a\mathbf{n}, \qquad |\mathbf{r}_k \times \mathbf{r}_l| = a|\mathbf{n}| = A.$$

The triangles OQ_iQ_j have area $\ge \frac{1}{2}w_iw_jA \ge \frac{1}{4}A(w_i + w_j)$, so we can sharpen (2.6) to

$$AK \le 4\pi Q^2 \sin\alpha \le \frac{512\sqrt{3}\pi H N}{P^4}.$$

Now we consider the two-dimensional vectors $\mathbf{h}^{(i)} = \left(h_2^{(i)}, h_3^{(i)}\right)$, which are obtained from $\mathbf{q}^{(i)}$ by a linear map T of rank 2 defined over the integers by

$$\begin{bmatrix} h_2 \\ h_3 \end{bmatrix} = \begin{bmatrix} p_1^2 & -p_2^2 & 0 \\ p_1^2 & 0 & -p_3^2 \end{bmatrix} \begin{bmatrix} q_1 \\ q_2 \\ q_3 \end{bmatrix}.$$

Let $Tr_j = \mathbf{g}_j = \left(g_2^{(j)}, g_3^{(j)}\right)$. Linear maps of rank 2 in two dimensions preserve ratios of areas, so $|\mathbf{g}_i \times \mathbf{g}_j| \ge |\mathbf{g}_k \times \mathbf{g}_l| = B$, say. The vectors $\mathbf{h}^{(i)}$ have length at most $Q' = \sqrt{2}H$ and make an angle at most $\alpha' = \pi/4$ with $\mathbf{p}' = (1,1)$. Estimating the areas of triangles as above, we use the argument of Lemma 5 to get

$$BK \le 4\pi Q'^2 \sin\alpha' \le 4\sqrt{2}\pi H^2.$$

We construct the three-dimensional vector $\mathbf{s} = g_2^{(k)}\mathbf{r}_l - g_2^{(l)}\mathbf{r}_k$, with

$$p_1^2 s_1 - p_2^2 s_2 = 0, \tag{2.12}$$

$$p_1^2 s_1 - p_3^2 s_3 = \pm B. \tag{2.13}$$

For some integer t we have $s_1 = p_2^2 t$, $s_2 = p_1^2 t$. We put $u = s_3$. The simultaneous equations (2.12), (2.13), and $\mathbf{n} \cdot \mathbf{s} = n_1 s_1 + n_2 s_2 + n_3 s_3 = 0$ for s_1, s_2, s_3 have the solution

$$\mathbf{s} = \pm \frac{B}{\mathbf{p} \cdot \mathbf{n}} \left(n_3 p_2^2, \ n_3 p_1^2, \ -n_1 p_2^2 - n_2 p_1^2 \right),$$

with $\mathbf{p}$ as in (2.2).

To compute $\mathbf{p} \cdot \mathbf{n}$, we write $\mathbf{r}$, $\mathbf{r}'$ for $\mathbf{r}^{(k)}$, $\mathbf{r}^{(l)}$ to simplify the notation, and we use

$$a\,\mathbf{p} \cdot \mathbf{n} = \mathbf{p} \cdot (\mathbf{r} \times \mathbf{r}') = \begin{vmatrix} p_2^2 p_3^2 & p_1^2 p_3^2 & p_1^2 p_2^2 \\ r_1 & r_2 & r_3 \\ r_1' & r_2' & r_3' \end{vmatrix}$$

$$= \frac{1}{p_1^2 p_2^2 p_3^2} \begin{vmatrix} p_1^2 p_2^2 p_3^2 & p_1^2 p_2^2 p_3^2 & p_1^2 p_2^2 p_3^2 \\ p_1^2 r_1 & p_2^2 r_2 & p_3^2 r_3 \\ p_1^2 r_1' & p_2 r_2' & p_3^2 r_3' \end{vmatrix}$$

$$= \frac{1}{p_1^2 p_2^2 p_3^2} \begin{vmatrix} p_1^2 p_2^2 p_3^2 & 0 & 0 \\ p_1^2 r_1 & -g_2 & -g_3 \\ p_1^2 r_1' & -g_2' & -g_3' \end{vmatrix}$$

$$= g_2 g_3' - g_2' g_3 = \pm B.$$

Since a is a positive integer, we deduce that

$$|\mathbf{s}| \leq 2P^2 |a\mathbf{n}| = 2AP^2 \leq \frac{1024\sqrt{3}\pi H N}{K P^2}.$$

The vector $\mathbf{s}$ that we have constructed may not be primitive, but when we divide by the highest common factor (t, u) to make it primitive, then the inequalities (2.10) and (2.11) still hold. $\square$

The treatment of Case 1 is unsatisfactory. We drop a dimension, and consider only $r = 2$ in (2.8). The angle between the vectors $\mathbf{p}' = (p_2^2, p_1^2)$ and (r_1, r_2) is at most α, since orthogonal projection does not increase angles. Suppose that for given $\mathbf{p}'$ there are L different integer vectors (r_1, r_2) with $r_i \leq N/KP^2$, each making an angle at most α with $\mathbf{p}'$. By Lemma 5 in two dimensions with $Q = \sqrt{2}\,N/KP^2$, either

$$L \leq 4\pi Q^2 \sin \alpha \leq \frac{1024\pi}{\sqrt{3}} \frac{H N}{K^2 P^4}, \tag{2.14}$$

or the vectors (r_1, r_2) are multiples of some primitive integer vector $\mathbf{s} = (t, u)$ with

$$t, u \leq \frac{\sqrt{2} N}{K L P^2}, \tag{2.15}$$

$$1 \leq |p_1^2 t - p_2^2 u| \leq \frac{H}{K L}. \tag{2.16}$$

Case 1(b) comprises triplets (p_1, p_2, p_3) of primes for which (2.14) holds, and Case 1(a) comprises the remaining triplets in Case 1. This subdivision is indirect: we declare a certain number K of solutions (this involves all three primes), and then subdivide using p_1 and p_2 only.

3. Counting arguments

We write $S_i(H, P, K)$ for the number of triplets of primes counted in $S(H, P, K)$ for which Case i holds, and $S_i(H, P)$ for the contribution of Case i to $S(H, P)$. The prime number theorem gives a trivial estimate

$$S_i(H, P, K) \ll \frac{P^3}{\log^3 P} \,. \tag{3.1}$$

From (2.1), (2.7), and (3.1) we have

$$S_3(H, P) \ll \frac{H^2 N}{P^3 \log^3 P} \,.$$

Using the lower bound for P in (1.6), we see that this expression satisfies (1.10) with $\eta = 3$ for $\gamma \le 4$.

In Cases 1 and 2 (2.9) is false, so $P \gg N^{1/6}$ and $\log P \asymp \log N$.

In Case 1 the primitive integer vector $\mathbf{r}$ is unique. If we know p_1, p_2, r_1 and r_2, then $p_3^2 r_3$ in (2.8) is one of at most H/K integers. Since (2.9) is false, for each such integer there are at most three prime factors large enough to be p_3. Hence

$$S_1(H, P, K) \le \frac{3H}{K} S'(H, P, K),$$

where $S'(H, P, K)$ is the number of ordered sets of integers p_1, p_2, r_1, r_2, with p_1 and p_2 distinct primes in the range $P \le p_i < 2P$, and

$$\frac{N}{16K} \le p_i^2 r_i \le \frac{N}{K}$$

with $1 \le p_1^2 r_1 - p_2^2 r_2 \le H/K$.

Let $S_1'(H, P, K)$ and $S_2'(H, P, K)$ be the contributions to $S'(H, P, K)$ from Case 1(a) and Case 1(b) respectively, and define

$$S_{11}(H, P, K), \ S_{12}(H, P, K), \ S_{11}(H, P), \ S_{12}(H, P)$$

similarly. For L a power of two, let $S_1'(H, P, K, L)$ and $S_2'(H, P, K, L)$ be the number of pairs of primes in Case 1(a) or Case 1(b) respectively for which there are between L and $2L - 1$ vectors (r_1, r_2). By the Riesz interchange

$$S_i'(H, P, K) \le \sum_L 2L \, S_i'(H, P, K, L),$$

where L is summed through powers of two. The prime number theorem
gives a trivial estimate

$$S_i'(H, P, K, L) \ll \frac{P^2}{\log^2 P}. \tag{3.2}$$

From (2.14) and (3.2) we have

$$S_2'(H, P, K) \ll \frac{P^2}{\log^2 P} \cdot \frac{HN}{K^2 P^4} \ll \frac{HN}{K^2 P^2 \log^2 P},$$

so

$$S_{12}(H, P, K) \ll \frac{H^2 N}{K^3 P^2 \log^2 P},$$

and by (3.1) again

$$S_{12}(H, P) \ll \sum_K K \, S_{12}(H, P, K) \ll \sum_K \min\left(\frac{KP^3}{\log^3 P}, \frac{H^2 N}{K^2 P^2 \log^2 P}\right)$$

$$\ll \sum_K \left(\frac{KP^3}{\log^3 P}\right)^{\frac{2}{5}} \left(\frac{H^2 N}{K^2 P^2 \log^2 P}\right)^{\frac{3}{5}} \ll \left(\frac{H^2 N}{\log^4 P}\right)^{\frac{3}{5}}.$$

By Proposition 2, this bound satisfies (1.9) for $\gamma < \frac{19}{5}$.

There are many ways of proceeding in Case 1(a) and Case 2. We write
(2.10) and (2.16) in the common form

$$q^2 t - p^2 u = g, \tag{3.3}$$

and use the Dirichlet interchange.

Lemma 7. *Let G, P, T, U be real numbers greater than or equal to one,
with $2G \leq P^2 U$. The number of quadruplets of positive integers p, q, t, u
in the ranges $P \leq p < 2P$, $t \leq T$, $tU \leq u \leq 2tU$ with $(p, q) = 1$ and
$|q^2 t - p^2 u| \leq G$ is $O(T^2 U + GT\sqrt{U})$.*

Proof. We have

$$\frac{q^2}{p^2} = \frac{u}{t} + O\left(\frac{G}{P^2 t}\right) = \frac{u}{t}\left(1 + O\left(\frac{G}{P^2 t U}\right)\right),$$

$$\frac{q}{p} = \sqrt{\frac{u}{t}}\left(1 + O\left(\frac{G}{P^2 t U}\right)\right) = \sqrt{\frac{u}{t}} + O\left(\frac{G}{P^2 t \sqrt{U}}\right).$$

There are at most $1 + O(G/t\sqrt{U})$ rational numbers with denominator less
than 2P in an interval of length $O(G/P^2 t\sqrt{U})$. The number of solutions is

$$O\left(\sum_{t \leq T} \sum_{u \leq 2tU} \left(1 + \frac{G}{t\sqrt{U}}\right)\right). \qquad \square$$

For Case 1(a) we put $p = p_2$, $q = p_1$ if $p_1 > p_2$, or $p = p_1$, $q = p_2$ if $p_1 < p_2$, and $G = H/KL$, $T = \sqrt{2}N/KLP^2$, $U = 1$ in Lemma 7. Hence

$$S_1'(H,P,K,L) \ll T^2 U + GT\sqrt{U} \ll \frac{N^2}{K^2 L^2 P^4} + \frac{HN}{K^2 L^2 P^2}.$$

Now by the construction of $S(H,P)$ we have

$$K \le \max(q_1, q_2, q_3) \le D \ll (\log H)^{\frac{3}{2}},$$
$$L \le \max(q_1, q_2) \le D' \ll \log^3 H.$$

From (3.2) we have

$$S_{11}(H,P,K) \ll \frac{H}{K} S_1'(H,P,K) \ll \frac{H}{K} \sum_L L\, S_1'(H,P,K,L)$$

$$\ll \frac{H}{K} \sum_L \min\left(\frac{LP^2}{\log^2 P}, \frac{N^2}{K^2 L P^4} + \frac{HN}{K^2 L P^2}\right)$$

$$\ll \frac{H}{K} \min\left(P^2 \log P, \frac{N^2}{K^2 P^4} + \frac{HN}{K^2 P^2}\right).$$

Hence

$$S_{11}(H,P) \ll \sum_K K\, S_{11}(H,P,K)$$

$$\ll \sum_K \min\left(HP^2 \log P, \frac{HN^2}{K^2 P^4} + \frac{H^2 N}{K^2 P^2}\right)$$

$$\ll \min\left(HP^2 \log^2 P, \frac{HN^2}{P^4} + \frac{H^2 N}{P^2}\right)$$

$$\ll HN^{\frac{2}{3}}(\log P)^{\frac{4}{3}} + H^{\frac{3}{2}} N^{\frac{1}{2}} \log P.$$

By Proposition 2, these terms satisfy (1.9) for $\gamma < \frac{11}{3}$.

For Case 2 we put $p = p_3$, $q = p_1 p_2$, $G = O(H^2/K)$, $T = O(HN/KP^4)$, $U = O(P^2)$ in Lemma 7. Each choice of q gives two possibilities for p_1 and p_2. Hence

$$S_2(H,P,K) \ll T^2 U + GT\sqrt{U} \ll \frac{H^2 N^2}{K^2 P^6} + \frac{H^3 N}{K^2 P^3},$$

and by (3.1)

$$S_2(H,P) \ll \sum_K K S_2(H,P,K) \ll \sum_K \min\left(\frac{KP^3}{\log^3 P}, \frac{H^2 N^2}{KP^6} + \frac{H^3 N}{KP^3}\right)$$

$$\ll \frac{HN}{(P\,\log P)^{\frac{3}{2}}} + \sqrt{\frac{H^3 N}{\log^3 P}}. \tag{3.4}$$

By Proposition 2, the second term in (3.4) satisfies (1.9) for $\gamma < 4$. By Lemma 2, the first term in (3.4) satisfies (1.9) for $\gamma < \frac{7}{2}$. More generally, the first term satisfies (1.10) with $\eta = \frac{3}{2}$ for

$$P \geq P_3(H) \asymp H^{(2\gamma-4)/3} (\log H)^{\frac{7}{3}}. \tag{3.5}$$

If P is small, then we argue as follows. We can choose p_1, p_2 in $O(P^2 / \log^2 P)$ ways, then t in (2.10) in $O(HN/KP^4)$ ways, g in (3.3) in $O(H^2/K)$ ways, and then, since $P \gg N^{1/6}$, there are boundedly many prime factors that could be p_3. This gives

$$S_2(H, P, K) \ll \frac{P^2}{\log^2 P} \cdot \frac{HN}{KP^4} \cdot \frac{H^2}{K} \ll \frac{H^3 N}{K^2 P^2 \log^2 P},$$

so by (3.1)

$$S_2(H, P) \ll \sum_K K S_2(H, P, K) \ll \sum_K \min\left(\frac{KP^3}{\log^3 P}, \frac{H^3 N}{KP^2 \log^2 P}\right)$$
$$\ll \sqrt{\frac{H^3 N P}{\log^5 P}}. \tag{3.6}$$

This bound satisfies (1.11) with $\eta = \frac{1}{2}$ for

$$P \leq P_4(H) \asymp \frac{N}{H^{2\gamma-3} \log^5 H}. \tag{3.7}$$

We extend the range for γ by excluding ranges of P for which (1.8) holds. If $n + 1 \leq p^2 q \leq n + H$ and $P \leq p < 2P$, then

$$\left\| \frac{n}{p^2} \right\| \leq \delta = \frac{H}{P^2}. \tag{3.8}$$

We use a general bound for integer points close to a curve from [9].

Proposition 3. (Huxley and Sargos) *Let B, P, Δ, δ be positive real numbers with $B \geq 1$, $P \geq 1$. Let $f(x)$ be a function with*

$$\frac{\Delta}{B} \leq |f^{(4)}(x)| \leq B\Delta \tag{3.9}$$

for $P \leq x < 2P$. There is a positive constant C constructed from B such that the number of integers m in $P \leq m < 2P$ with $\|f(m)\| \leq \delta$ is

$$\leq C\left(\Delta^{\frac{1}{10}} P + \delta^{\frac{1}{6}} P + \left(\frac{\delta}{\Delta}\right)^{\frac{1}{4}} + 1\right).$$

We can suppose that $n \geq N/2 - H \geq N/4$ in (3.8), so $f(x) = n/x^2$ satisfies the hypothesis (3.9) with $\Delta = N/P^6$ and some bounded B. Then

$$T(H,P) \ll N^{\frac{1}{10}} P^{\frac{2}{5}} + H^{\frac{1}{6}} P^{\frac{2}{3}} + (H/N)^{\frac{1}{4}} P,$$

and (1.8) holds if

$$P^4 \leq c_9 \min \left(\frac{H^5}{\log^6 N}, \frac{H^{10}}{N \log^{10} N}, \frac{H^3 N}{\log^4 N} \right). \tag{3.10}$$

By Proposition 2, (3.10) is satisfied when

$$P \leq P_5(H) = c_{10} \left(\frac{H^{10}}{N \log^{11} N} \right)^{\frac{1}{4}}. \tag{3.11}$$

The constants c_9 and c_{10} are chosen sufficiently small. If $\gamma < \frac{59}{16}$, then

$$P_3^5 \leq P_4 P_5^4,$$

with a suitable choice of constants in (3.5) and (3.7). Hence one of the conditions (3.5), (3.7) or (3.11) must hold.

We have thus established Conjecture 2 for $\gamma < \frac{11}{3}$. The critical ranges are

$$\frac{\log P}{\log N} \sim \frac{1}{3}, \qquad \frac{\log H}{\log N} \sim \frac{1}{5}$$

in Case 1(a). Swinnerton-Dyer's method used in [8] just fails to establish (1.8) in this range. The whole treatment of Case 1(a) is wasteful because two square roots have simultaneous rational approximation, but only one is used.

The proof of the asymptotic formula for the sum of the γth powers of gaps between k-free numbers, where $k \geq 3$ and

$$\gamma < 2k - 1 + \frac{2}{(k+1)}$$

follows the same course: three points are noteworthy.

The conclusion of Lemma 1 becomes

$$\log(\alpha(r+1)) \leq - \left(1 - \frac{3}{2^{k+1}} \right) r \left((k-2) \log r + k \log \log r + O(1) \right),$$

so the convergence appears to be more rapid for $k \geq 3$.

Secondly, the factor P inside the square root sign on the right of (3.6) is replaced by P^{5-2k}, which is a negative power for $k \geq 3$. This estimate gets better for large P, and renders the use of (3.4) and Proposition 3 unnecessary.

Thirdly, the critical ranges in Case 1(a) are

$$\frac{\log P}{\log N} \sim \frac{1}{2k+1}, \qquad \frac{\log H}{\log N} \sim \frac{1}{k+1},$$

out of reach at present even by Swinnerton-Dyer's method.

References

1 P. Erdős: Some problems and results in elementary number theory. *Publ. Math. Debrecen* **2** (1951), 103–109.

2 M. Filaseta: On the distribution of gaps between square-free numbers. *Mathematika* **40** (1993), 87–100.

3 M. Filaseta and O. Trifonov: On the gaps between square-free numbers II. *J. London Math. Soc.* (2) **45** (1992), 215–221.

4 M. Filaseta and O. Trifonov: The distribution of fractional parts with applications to gap results in number theory. *Proc. London Math. Soc.*, to appear.

5 S.W. Graham: Moments of gaps between k-free numbers: *J. Number Theory* **44** (1993), 105–117.

6 G.H. Hardy and E.M. Wright: *An Introduction to the Theory of Numbers*, 5th. edn. Oxford University Press (1979).

7 C. Hooley: On the distribution of square-free numbers. *Canadian J. Math.* 25 (1973), 1216–1223.

8 M.N. Huxley: The integer points close to a curve. *Mathematika* **36** (1989), 198–215.

9 M.N. Huxley and P. Sargos: Points entiers au voisinage d'une courbe plane de classe C^n. *Acta Arith.* 69 (1995), 359–366.

10 L. Mirsky: Arithmetical pattern problems related to divisibility by rth powers. *Proc. London Math. Soc.* **50** (1949), 497–508.

12. On the Ternary Additive Divisor Problem and the Sixth Moment of the Zeta-Function

Aleksandar Ivić

1. Bounding the $2k$th moment on the critical line

Let $k \geq 3$ be a fixed integer and $T^\epsilon \leq G \leq T^{1-\epsilon}$. To obtain upper bounds for the important integral

$$\int_0^T \left| \zeta\left(\tfrac{1}{2} + it\right) \right|^{2k} dt$$

it suffices to obtain upper bounds for

$$I_k = I_k(T, M, G) = \frac{1}{M} \int_{T-2G}^{T+2G} a(t) \left| \sum_{M < m \leq M'} \frac{d_k(m)}{m^{it}} \right|^2 dt. \tag{1.1}$$

Here, as usual, we shall denote by $d_k(m)$ the divisor function generated by $\zeta^k(s)$, and shall suppose $M < M' \leq 2M$, $1 \ll M \ll T^{k/2}$. This easily follows by partial summation and the approximate functional equation for $\zeta^k(s)$ given by Th. 4 of [5] (the so-called "reflection principle"). The function $a(t)$ is a smooth (C^∞) function such that

$$a(t) = \begin{cases} 1 & \text{if} \quad T - G \leq t \leq T + G \\ 0 & \text{if} \quad t \leq T - 2G \quad \text{or if} \quad t \geq T + 2G \end{cases}$$

and

$$a^{(R)}(t) \ll_R G^{-R} \quad \text{for} \quad R = 1, 2, \ldots. \tag{1.2}$$

Thus using R integrations by parts (with $R = R(\epsilon)$ sufficiently large) and (1.2) we obtain

$$I_k = 2 \operatorname{Re} \left\{ \frac{1}{M} \int_{T-2G}^{T+2G} a(t) \sum_{M < m \leq M'} \sum_{\substack{M < n \leq M' \\ 0 < n - m \leq M^{1+\epsilon}/G}} d_k(m) d_k(n) \left(\frac{n}{m} \right)^{it} dt \right\}$$

$$+ O(GT^\epsilon)$$

$$= 2 \operatorname{Re} I'_k + O(GT^\epsilon), \tag{1.3}$$

Research financed by the Mathematical Institute of Belgrade.

Sieve Methods, Exponential Sums, and their Applications in Number Theory
Greaves, G.R.H., Harman, G., Huxley, M.N., Eds. ©Cambridge University Press, 1996

say.

In (1.3) the diagonal terms $m = n$ were estimated trivially, while the contribution of the terms for which $|m - n| \geq M^{1+\epsilon}/G$ is negligible, by repeated integrations by parts and the use of (1.2). Setting $h = n - m > 0$ we obtain from (1.3)

$$I'_k = \frac{1}{M} \sum_{h \leq M^{1+\epsilon}/G} \sum_{M < n \leq M'} d_k(n)d_k(n+h)K(n,h) + O(GT^\epsilon)$$
$$= I''_k + O(GT^\epsilon), \tag{1.4}$$

say, where

$$K(x,h) = K(x,h;T,G) = \int_{T-2G}^{T+2G} a(t)\left(1 + \frac{h}{x}\right)^{it} dt. \tag{1.5}$$

We may restrict G to the range $T^\epsilon \leq G \leq \min(T^{1-\epsilon}, \frac{1}{2}M)$, since the contribution of M for which $M \leq 2G$ is $O(GT^\epsilon)$, by the mean value theorem for Dirichlet polynomials (see Th. 5.2 of [5]).

The expression for I''_k in (1.4) clearly shows that the problem of upper bounds for I_k is connected, after summation over h, to the quantity $\Delta(x,h)$ in the formula

$$\sum_{n \leq x} d_k(n)d_k(n+h) = x\, P_{2k-2}(\log x; h) + \Delta(x,h). \tag{1.6}$$

In (1.6) it is assumed that $P_{2k-2}(\log x; h)$ is a suitable polynomial of degree $2k - 2$ in $\log x$, whose coefficients depend on h, while $\Delta_k(x,h)$ is supposed to be the error term. This means that we should have

$$\Delta_k(x,h) = o(x) \quad \text{as} \quad x \to \infty, \tag{1.7}$$

but unfortunately (1.7) is not yet known to hold for any $k \geq 3$, even for fixed h. However, in estimating I''_k there is the extra summation over h, and we may reasonably hope that a certain cancellation will occur in the sum

$$\sum_{h \leq H} \Delta_k(x,h).$$

At the beginning we supposed that $k \geq 3$. Indeed, for $k = 1, 2$ the estimation of I_k is considerably less difficult (see [5] and [6] for an extensive account, as well as [7], [8], [9], [15] and [16]). Also for $k = 1$ the sum in (1.6) is trivial, while for $k = 2$ it was extensively studied by many authors, including Kuznetsov [11] and Motohashi [17]. The natural next step in (1.6) is the case $k = 3$, but the works of A.I. Vinogradov and Takhtadžjan [22], [23] and A.I. Vinogradov [19], [20], [21] show that the analytic problems connected with the Dirichlet series generated by $d_3(n)d_3(n + h)$ are overwhelmingly hard. The ensuing problems are connected with the group $SL(3, \mathbb{Z})$, and they are much more difficult than the corresponding problems connected with the group $SL(2, \mathbb{Z})$ which appear in the case $k = 2$. The latter involve the spectral theory of the non-Euclidean Laplacian, which was extensively developed in recent times by Kuznetsov (see [12]) and others. Thus at present in the case $k = 2$ we have sharp explicit formulas, while in the case $k > 2$ we have none.

If $E_k(T)$ denotes the error term in the asymptotic formula for

$$\int_0^T \left| \zeta(\tfrac{1}{2} + it) \right|^{2k} dt$$

(see Ch. 4 of [6]), then in the cases $k = 1$ and $k = 2$ there exist fairly explicit representations of $E_k(t)$. For a long time I have been trying to put this to advantage in the case $k = 3$ (i.e. the sixth moment), by using the principle "6 = 4 + 2". Namely the idea was to try to extract a bound for $E_3(t)$ from the expressions for $E_1(t)$ and $E_2(t)$. Unfortunately this approach has not so far proved to be successful. Therefore here I shall turn to the more natural approach "6 = 3 + 3", which entails the use of the asymptotic formula (1.6) for $k = 3$ in the sixth moment problem. It should be noted that until now even the structure of the polynomial $P_{2k-2}(y, h)$ remained mysterious, and there does not seem to exist an explicit form for it in the literature. In §5 of this work it will be shown how to obtain the coefficients of $P_4(y, h)$ explicitly, and the same procedure may be used in the case of the general $P_{2k-2}(y, h)$. Although with this expression for $P_4(y, h)$ it will not be shown that (1.7) holds (when $k = 3$) for any explicit h, it is reasonably clear that our definition of P_4 is the right one.

A.I. Vinogradov [21] conjectured that $\Delta_k(x, h) \ll x^{1-1/k}$, without stating for which range of h this sharp bound should hold. I feel that this bound is too strong, and that (even for fixed h) a power of a logarithm should be included on the right-hand side. More importantly, I hope that the bound

$$\sum_{h \leq H} \Delta_k(x, h) \ll_{k,\epsilon} H x^{1-1/k+\epsilon} \quad \text{for} \quad 1 \leq H \leq x^{(k-2)/k+\delta} \tag{1.8}$$

holds uniformly in H for fixed $k \geq 3$ and some $\delta = \delta_k > 0$. Note that Vinogradov's conjecture in the form $\Delta_k(x, h) \ll x^{1-1/k+\epsilon}$ trivially implies (1.8), but the important point is that there are no absolute value signs in the sum in (1.8). Hence one expects considerable cancellation among the summands of (1.8) to take place, and hopefully some version of (1.8) should be tractable by present-day methods. One can also assume (1.8) to hold in the case $k = 2$ for $1 \leq H \leq \sqrt{x}$, say. Then it would follow that the inequality

$$\int_{T-G}^{T+G} \left| \zeta(\tfrac{1}{2} + it) \right|^4 dt \ll GT^\epsilon$$

holds with $G = T^{5/6}$, whereas it is known (see e.g. [6]) that $G = T^{2/3}$ is unconditionally permissible. I also conjecture that for any $k \geq 2$ and $h \geq 1$ one has

$$\Delta_k(x, h) = \Omega(x^{1-1/k}).$$

For $k = 2$ and fixed h this conjecture has been recently proved by Motohashi [17].

Assuming (1.8) for $k = 3$ we shall prove

Theorem 1. *If* (1.8) *holds with* $k = 3$, *then*

$$\int_0^T \left| \zeta(\tfrac{1}{2} + it) \right|^6 dt \ll_\epsilon T^{1+\epsilon}. \tag{1.9}$$

Actually (1.8) yields in the general case

$$\int_0^T \left| \zeta(\tfrac{1}{2} + it) \right|^{2k} dt \ll_{\epsilon,k} T^{(k-1)/2+\epsilon} \quad \text{for} \quad k \geq 3, \tag{1.10}$$

which is weak for $k \geq 4$, since in that case sharper bounds than (1.10) follow from the fourth moment and the bound $\zeta(\tfrac{1}{2} + it) \ll |t|^{1/6}$. This shows that the use of (1.6) in the approach to the $2k$th moment seems promising in the case $k = 3$, while for $k \geq 4$ other methods of attack should be sought. Naturally it is difficult to ascertain how hard it would be to prove (1.8) with $k = 3$, assuming that it is true. On the other hand Theorem 1 follows also from a more general bound, namely (6.1) with $\alpha + \beta \leq 3$ (see Lemma 6), and this may be perhaps more tractable. It is interesting to note that it is unknown whether the classical conjecture $\Delta_k(x) \ll x^{(k-1)/(2k)+\epsilon}$ for fixed $k \geq 2$ can yield (1.9), where $\Delta_k(x)$ is the error term in the asymptotic formula for the summatory function of $d_k(n)$.

I wish to thank Professor M.N. Huxley for valuable remarks.

2. The Dirichlet series for $d_r(n)e(nh/k)$

To treat (1.6) when $k = 3$ we shall need a summation formula (see §3). Its derivation depends on the functional equation for the Dirichlet series generated by the function $d_3(n)e(nh/k)$, where h, k are natural numbers which satisfy the conditions $1 \leq h \leq k$, $(h, k) = 1$, and where $e(z) = e^{2\pi i z}$. To obtain this we define first, for a fixed integer $r \geq 1$,

$$E_r\left(s, \frac{h}{k}\right) = \sum_{n=1}^{\infty} \frac{d_r(n)e(nh/k)}{n^s} \quad \text{for} \quad \text{Re}\, s > 1. \tag{2.1}$$

Thus $E_r(s, 1/1) \equiv \zeta^r(s)$, and $E_2(s, h/k)$ is known in the literature as the Estermann zeta-function (see e.g. pp. 9–13 of Jutila [10]). For $\text{Re}\, s > 1$

$$E_r\left(s, \frac{h}{k}\right) = \sum_{n_1,\ldots,n_r=1}^{\infty} \frac{e(hn_1 \ldots n_r/k)}{(n_1 \ldots n_r)^s}$$

$$= \sum_{a_1=1}^{k} \cdots \sum_{a_r=1}^{k} e\left(\frac{ha_1 \ldots a_r}{k}\right) \sum_{m_1=0}^{\infty} \frac{1}{(a_1 + m_1 k)^s} \cdots \sum_{m_r=0}^{\infty} \frac{1}{(a_r + m_r k)^s}$$

$$= \frac{1}{k^{rs}} \sum_{a_1=1}^{k} \cdots \sum_{a_r=1}^{k} e\left(\frac{ha_1 \ldots a_r}{k}\right) \zeta\left(s, \frac{a_1}{k}\right) \ldots \zeta\left(s, \frac{a_r}{k}\right), \tag{2.2}$$

where

$$\zeta(s, \alpha) = \sum_{n=0}^{\infty} \frac{1}{(n + \alpha)^s} \quad \text{for} \quad 0 < \alpha \leq 1, \ \text{Re}\, s > 1$$

is the familiar Hurwitz zeta-function. The Hurwitz zeta-function may be analytically continued to the whole complex plane. Its only singularity is the simple pole at $s = 1$, where one has the Laurent expansion

$$\zeta(s, \alpha) = \frac{1}{s - 1} + \sum_{n=0}^{\infty} \gamma_n(\alpha)(s - 1)^n \tag{2.3}$$

with

$$\gamma_n(\alpha) = \frac{(-1)^n}{n!} \lim_{m \to \infty} \left(\sum_{k=0}^{m} \frac{\log^n(k + \alpha)}{k + \alpha} - \frac{\log^{n+1}(m + \alpha)}{n + 1} \right). \tag{2.4}$$

Therefore (2.2) provides an analytic continuation of $E_r(s, h/k)$ to the whole complex plane.

By using the functional equation for $\zeta(s,\alpha)$ and (2.2), one may obtain the functional equation for $E_r(s,h/k)$. This was worked out by Smith [18], and his result will be stated here as

Lemma 1. *The function $E_r(s,h/k)$, defined for $r \geq 2$ and $\operatorname{Re} s > 1$ by the series (2.1), may be analytically continued to the whole complex plane. It is everywhere holomorphic, except at $s = 1$, where it has a pole of order r. It satisfies the functional equation*

$$E_r\left(s, \frac{h}{k}\right) = \frac{\delta_r(1-s, h/k)}{\delta_r(s, h/k)} D_r^+\left(1-s, \frac{h}{k}\right)$$
$$+ i^{3r} \frac{\delta_r(2-s, h/k)}{\delta_r(1+s, h/k)} D_r^-\left(1-s, \frac{h}{k}\right), \qquad (2.5)$$

where

$$\delta_r\left(s, \frac{h}{k}\right) = \left(\frac{k}{\pi}\right)^{rs/2} \Gamma^r\left(\tfrac{1}{2}s\right),$$

$$D_r^\pm\left(s, \frac{h}{k}\right) = \sum_{\mathbf{m} \in \mathcal{P}_r} \frac{G_r^\pm(\mathbf{m}, h/k)}{N\mathbf{m}^s} \quad \text{if} \quad \operatorname{Re} s > 1,$$

$$G_r^\pm\left(\mathbf{a}, \frac{h}{k}\right) = \frac{1}{2k^{r/2}}\left\{ G_r\left(\mathbf{a}, \frac{h}{k}\right) \pm G_r\left(\mathbf{a}, -\frac{h}{k}\right)\right\},$$

$$\mathbf{a} = (a_1, \ldots, a_r), \quad \mathbf{x} = (x_1, \ldots, x_r), \quad \mathbf{a} \cdot \mathbf{x} = a_1 x_1 + \cdots + a_r x_r,$$

$$N\mathbf{x} = x_1 \ldots x_r, \quad \mathcal{P}_r = \left\{ \mathbf{x} \in \mathbb{Z}^r : x_i \geq 1 \quad \text{for} \quad i = 1, \ldots, r \right\},$$

$$G_r\left(\mathbf{a}, \frac{h}{k}\right) = \sum_{\mathbf{x} \pmod k} e\left(\frac{h N\mathbf{x} + \mathbf{a} \cdot \mathbf{x}}{k}\right).$$

We introduce the notation

$$A_r\left(n, \frac{h}{k}\right) = \frac{1}{2} \sum_{n_1 \ldots n_r = n} \sum_{x_1=1}^{k} \cdots \sum_{x_r=1}^{k} \left\{ e\left(\frac{n_1 x_1 + \cdots + n_r x_r + h x_1 \ldots x_r}{k}\right)\right.$$
$$\left. + e\left(\frac{n_1 x_1 + \cdots + n_r x_r - h x_1 \ldots x_r}{k}\right)\right\}, \qquad (2.6)$$

$$B_r\left(n, \frac{h}{k}\right) = \frac{1}{2} \sum_{n_1 \ldots n_r = n} \sum_{x_1=1}^{k} \cdots \sum_{x_r=1}^{k} \left\{ e\left(\frac{n_1 x_1 + \cdots + n_r x_r + h x_1 \ldots x_r}{k}\right)\right.$$
$$\left. - e\left(\frac{n_1 x_1 + \cdots + n_r x_r - h x_1 \ldots x_r}{k}\right)\right\}, \qquad (2.7)$$

and note that

$$\left| A_r\left(n, \frac{h}{k}\right)\right| \leq k^r d_r(n), \qquad \left| B_r\left(n, \frac{h}{k}\right)\right| \leq k^r d_r(n).$$

Then from Lemma 1 we obtain

Lemma 2. *For* $\operatorname{Re} s < 0$

$$E_r\left(s, \frac{h}{k}\right) = \frac{\pi^{r(2s-1)/2}}{k^{rs}} \left\{ \frac{\Gamma^r\left(\frac{1-s}{2}\right)}{\Gamma^r\left(\frac{s}{2}\right)} \sum_{n=1}^{\infty} A_r\left(n, \frac{h}{k}\right) n^{s-1} \right.$$
$$\left. + i^{3r} \frac{\Gamma^r\left(\frac{2-s}{2}\right)}{\Gamma^r\left(\frac{1+s}{2}\right)} \sum_{n=1}^{\infty} B_r\left(n, \frac{h}{k}\right) n^{s-1} \right\}, \quad (2.8)$$

where the two series on the right-hand side are absolutely convergent.

In the sequel we shall use the Laurent expansion of $E_3(s, h/k)$ at $s = 1$. We proceed now to obtain this, although of course we could also deal with the general case of $E_r(s, h/k)$. From (2.2) and (2.3) we obtain

$$E_3\left(s, \frac{h}{k}\right) = \frac{1}{k}\left(\frac{C}{(s-1)^3} + \frac{D}{(s-1)^2} + \frac{E}{s-1}\right) + \sum_{n=0}^{\infty} a_n(h,k)(s-1)^n \quad (2.9)$$

with

$$C = ka_{-3}(h,k) = \frac{1}{k^2} \sum_{a=1}^{k}\sum_{b=1}^{k}\sum_{c=1}^{k} e\left(\frac{abch}{k}\right), \quad (2.10)$$

$$D = ka_{-2}(h,k) = \frac{1}{k^2} \sum_{a=1}^{k}\sum_{b=1}^{k}\sum_{c=1}^{k} e\left(\frac{abch}{k}\right)\left(3\gamma_0\left(\frac{a}{k}\right) - 3\log k\right), \quad (2.11)$$

$$E = ka_{-1}(h,k)$$
$$= \frac{1}{k^2} \sum_{a=1}^{k}\sum_{b=1}^{k}\sum_{c=1}^{k} e\left(\frac{abch}{k}\right)\left(3\gamma_0\left(\frac{a}{k}\right)\gamma_0\left(\frac{b}{k}\right) - 9\gamma_0\left(\frac{a}{k}\right)\log k + \frac{9}{2}\log^2 k\right).$$
$$(2.12)$$

If in (2.10)–(2.12) we perform first the the summation on c, then it will be seen that the quantities C, D, E depend only on k and not on h, and that they are all $\ll_\epsilon k^\epsilon$. Namely

$$\sum_{a=1}^{k}\sum_{b=1}^{k}\sum_{c=1}^{k} e\left(\frac{abch}{k}\right) = k \sum_{a=1}^{k}\sum_{b=1}^{k} f(abh, k), \quad (2.13)$$

where we put

$$f(n,k) = \begin{cases} 1 & \text{if } k \,|\, n \\ 0 & \text{otherwise.} \end{cases}$$

But $k \mid abh$ implies $k \mid ab$, since $(h, k) = 1$. Thus the triple sum in (2.13) does not depend on h but only on k, and similarly for the right-hand sides of (2.11) and (2.12). Also we have

$$k \sum_{a=1}^{k} \sum_{b=1}^{k} f(abh, k) \le k \sum_{j=1}^{k} d(jk) \ll_\epsilon k^{2+\epsilon},$$

and we obtain, since (2.4) gives uniformly that $\gamma_n(a) \ll 1$,

$$a_{-j}(h, k) \ll_\epsilon k^{\epsilon-1} \quad \text{for} \quad j = 1, 2, 3. \tag{2.14}$$

3. A summation formula for $d_r(n)e(nh/k)$

When $0 < a < b, (h, k) = 1$ and $f(x) \in C^1[a, b]$, Jutila [10] established the summation formula

$$\sum_{a \le n \le b}{}' d(n)f(n)e\left(\frac{nh}{k}\right)$$

$$= \frac{1}{k} \int_a^b (\log x + 2\gamma - 2\log k) f(x) \, dx$$

$$+ \frac{1}{k} \sum_{n=1}^{\infty} d(n) \int_a^b \left\{ -2\pi e\left(\frac{-n\overline{h}}{k}\right) Y_0\left(4\pi \frac{\sqrt{nx}}{k}\right) \right.$$

$$\left. + 4e\left(\frac{n\overline{h}}{k}\right) K_0\left(4\pi \frac{\sqrt{nx}}{k}\right) \right\} f(x) \, dx, \tag{3.1}$$

where $\sum'$ means that if a or b is an integer then the first or last term in the sum in (3.1) is halved, γ is Euler's constant, $h\overline{h} \equiv 1 \pmod{k}$, and Y_0 and K_0 are the Bessel functions in standard notation. When $k = 1$, (3.1) reduces to the classical Voronoi summation formula (see [5] or [10]). What we need is an analogue of (3.1) when $d(n) \equiv d_2(n)$ is replaced by $d_3(n)$. However, despite the results of Berndt [1], [2], Hafner [4] and others on summation formulas involving coefficients of Dirichlet series, such a formula does not appear in the literature. In fact, it may be asked whether such a formula exists under the conditions for which (3.1) holds.

For the application that we have in mind, and this is the sixth moment of $|\zeta(\frac{1}{2} + it)|$, it is enough to require that $f(x)$ be a smooth function of compact support in $(0, \infty)$, which we henceforth assume throughout this section. Then, first of all, the Mellin transform

$$F(s) = \int_0^\infty f(x) x^{s-1} \, dx \tag{3.2}$$

is an entire function of fast decay. Namely, R integrations by parts yield

$$F(s) = \frac{(-1)^R}{s(s+1)\ldots(s+R-1)} \int_0^\infty f^{(R)}(x)x^{s+R-1}\,dx. \qquad (3.3)$$

Since $f^{(n)}(0) = f^{(n)}(\infty) = 0$ when $n \geq 0$, it follows that $F(s)$ has removable singularities at $s = 0, -1, -2, \ldots$, and (3.2) gives

$$F(s) \ll_{\sigma,R} \left(1 + |t|\right)^{-R} \quad \text{where} \quad s = \sigma + it, R = 1, 2, \ldots. \qquad (3.4)$$

Then by the inverse Mellin transform we have, for $r \geq 3$ and $c > 1$,

$$\sum_{n=1}^\infty f(n)d_r(n)e\left(\frac{nh}{k}\right) = \frac{1}{2\pi i}\int_{c-i\infty}^{c+i\infty} F(s)E_r\left(s,\frac{h}{k}\right)ds. \qquad (3.5)$$

In (3.5) we shift the line of integration to $\operatorname{Re} s = -c$ and apply the Residue Theorem. In view of Lemma 1 and Lemma 2 we obtain

$$\frac{1}{2\pi i}\int_{c-i\infty}^{c+i\infty} F(s)E_r\left(s,\frac{h}{k}\right)ds$$

$$= \operatorname*{Res}_{s=1} F(s)E_r\left(s,\frac{h}{k}\right)$$

$$+ \sum_{n=1}^\infty A_r\left(n,\frac{h}{k}\right)\left\{\frac{1}{2\pi i}\int_{-c-i\infty}^{-c+i\infty} \frac{\pi^{r(2s-1)/2}}{k^{rs}} \frac{\Gamma^r\left(\frac{1-s}{2}\right)}{\Gamma^r\left(\frac{s}{2}\right)} \frac{F(s)}{n^{1-s}}\,ds\right\}$$

$$+ i^{3r}\sum_{n=1}^\infty B_r\left(n,\frac{h}{k}\right)\left\{\frac{1}{2\pi i}\int_{-c-i\infty}^{-c+i\infty} \frac{\pi^{r(2s-1)/2}}{k^{rs}} \frac{\Gamma^r\left(\frac{2-s}{2}\right)}{\Gamma^r\left(\frac{1+s}{2}\right)} \frac{F(s)}{n^{1-s}}\,ds\right\},$$

$$\qquad (3.6)$$

where we could invert the order of summation and integration because of absolute convergence.

For $x > 0$ and $0 < \sigma < \frac{1}{2} - \frac{1}{r}$ let

$$U_r(x) = \frac{1}{2\pi i}\int_{\sigma-i\infty}^{\sigma+i\infty} \frac{\Gamma^r\left(\frac{s}{2}\right)}{\Gamma^r\left(\frac{1-s}{2}\right)} \frac{ds}{x^s}, \qquad V_r(x) = \frac{1}{2\pi i}\int_{\sigma-i\infty}^{\sigma+i\infty} \frac{\Gamma^r\left(\frac{1+s}{2}\right)}{\Gamma^r\left(\frac{2-s}{2}\right)} \frac{ds}{x^s}, \qquad (3.7)$$

and note that both integrals in (3.7) are absolutely convergent, which follows from Stirling's formula for the gamma-function. Further, by setting

$$G_r(s,n) = \frac{k^{rs}}{\pi^{rs}n^s} \frac{\Gamma^r\left(\frac{s}{2}\right)}{\Gamma^r\left(\frac{1-s}{2}\right)}, \qquad H_r(s,n) = \frac{k^{rs}}{\pi^{rs}n^s} \frac{\Gamma^r\left(\frac{1+s}{2}\right)}{\Gamma^r\left(\frac{2-s}{2}\right)},$$

we write (3.5) and (3.6) as

$$\sum_{n=1}^{\infty} f(n)d_r(n)e\left(\frac{nh}{k}\right) = \operatorname*{Res}_{s=1} F(s)E_r\left(s, \frac{h}{k}\right)$$

$$+ \frac{\pi^{r/2}}{k^r} \sum_{n=1}^{\infty} A_r\left(n, \frac{h}{k}\right)\left\{\frac{1}{2\pi i}\int_{-c-i\infty}^{-c+i\infty} F(s)G_r(1-s,n)\,ds\right\}$$

$$+ \frac{i^{3r}\pi^{r/2}}{k^r} \sum_{n=1}^{\infty} B_r\left(n, \frac{h}{k}\right)\left\{\frac{1}{2\pi i}\int_{-c-i\infty}^{-c+i\infty} F(s)H_r(1-s,n)\,ds\right\}.$$

$$(3.8)$$

But recall that (see, for example, (A.6) of [5])

$$\frac{1}{2\pi i}\int_{\sigma-i\infty}^{\sigma+i\infty} F(s)G(1-s)\,ds = \int_0^{\infty} f(x)g(x)\,dx \qquad (3.9)$$

if $F(s)$ and $G(s)$ are the Mellin transforms of $f(x)$ and $g(x)$, respectively. This relation is easily established if the integral on the left-hand side of (3.9) is absolutely convergent. Hence for $0 < \sigma < \frac{1}{2} - \frac{1}{r}$ we obtain, by applying Cauchy's Theorem and then (3.9),

$$\frac{1}{2\pi i}\int_{-c-i\infty}^{-c+i\infty} F(s)G_r(1-s)\,ds = \frac{1}{2\pi i}\int_{1-\sigma-i\infty}^{1-\sigma+i\infty} F(s)G_r(1-s)\,ds$$

$$= \int_0^{\infty} f(x)U_r\left(\frac{\pi^r nx}{k^r}\right)\,dx, \qquad (3.10)$$

and an analogous relation holds for the other integral in (3.8). Hence, and from (3.8) and (3.10), we obtain

Theorem 2. *With the above notation we have, for $r \geq 3$,*

$$\sum_{n=1}^{\infty} f(n)d_r(n)e\left(\frac{nh}{k}\right) = \operatorname*{Res}_{s=1} F(s)E_r\left(s, \frac{h}{k}\right)$$

$$+ \frac{\pi^{r/2}}{k^r} \sum_{n=1}^{\infty} A_r\left(n, \frac{h}{k}\right)\int_0^{\infty} f(x)U_r\left(\frac{\pi^r nx}{k^r}\right)\,dx$$

$$+ \frac{i^{3r}\pi^{r/2}}{k^r} \sum_{n=1}^{\infty} B_r\left(n, \frac{h}{k}\right)\int_0^{\infty} f(x)V_r\left(\frac{\pi^r nx}{k^r}\right)\,dx. \quad (3.11)$$

To apply Theorem 2 in practice one needs to know the asymptotic behaviour of $U_r(x)$ and $V_r(x)$ and their integrals. In the following lemma this will be provided in the case $r = 3$, but it is clear that the method of proof will work also in the general case.

Lemma 3. *If $U(x) \equiv U_3(x)$ and $V(x) \equiv V_3(x)$ are defined by (3.7), then for any fixed integer $K \geq 1$ and $x \geq x_0 > 0$*

$$U(x) = \sum_{j=1}^{K} \frac{c_j \cos\left(6x^{\frac{1}{3}}\right) + d_j \sin\left(6x^{\frac{1}{3}}\right)}{x^{j/3}} + O\left(\frac{1}{x^{(K+1)/3}}\right), \qquad (3.12)$$

$$V(x) = \sum_{j=1}^{K} \frac{e_j \cos\left(6x^{\frac{1}{3}}\right) + f_j \sin\left(6x^{\frac{1}{3}}\right)}{x^{j/3}} + O\left(\frac{1}{x^{(K+1)/3}}\right), \qquad (3.13)$$

$$\int U(x)\,dx = \sum_{j=0}^{K} \frac{g_j \cos\left(6x^{\frac{1}{3}}\right) + h_j \sin\left(6x^{\frac{1}{3}}\right)}{x^{(j-1)/3}} + O\left(\frac{1}{x^{K/3}}\right), \qquad (3.14)$$

$$\int V(x)\,dx = \sum_{j=0}^{K} \frac{k_j \cos\left(6x^{\frac{1}{3}}\right) + l_j \sin\left(6x^{\frac{1}{3}}\right)}{x^{(j-1)/3}} + O\left(\frac{1}{x^{K/3}}\right), \qquad (3.15)$$

with suitable constants $c_j, \ldots, l_j$, and in particular

$$c_1 = 0, \qquad d_1 = -\frac{2}{\sqrt{3\pi}}, \qquad e_1 = -\frac{2}{\sqrt{3\pi}}, \qquad f_1 = 0,$$

$$g_0 = \frac{1}{\sqrt{3\pi}}, \qquad h_0 = 0, \qquad k_0 = 0, \qquad l_0 = -\frac{1}{\sqrt{3\pi}}.$$

Proof. All these formulas are proved analogously, so only the details for (3.12) will be given. We note that $U(x)$ and $V(x)$ are real-valued functions of x, which follows from

$$U(x) = \frac{1}{2\pi i}\int_{(\sigma)} \frac{\Gamma^3\left(\frac{s}{2}\right)}{\Gamma^3\left(\frac{1-s}{2}\right)} \frac{ds}{x^s} = \frac{2}{2\pi i}\int_{(\sigma/2)} \frac{\Gamma^3(s)}{\Gamma^3(1-s)} \frac{ds}{x^{2s}}$$

and $\overline{\Gamma(s)} = \Gamma(\bar{s})$. Here we used the notation

$$\int_{(\sigma)} = \int_{\sigma-i\infty}^{\sigma+i\infty}.$$

The above integrals converge for $0 < \sigma < \frac{1}{2}$ and converge absolutely for $0 < \sigma < \frac{1}{6}$. This follows by use of the first derivative test and Stirling's formula in the form

$$\Gamma(\sigma + it) = \sqrt{2\pi}\,|t|^{\sigma-\frac{1}{2}}\exp\left\{-\frac{\pi}{2}|t| + i\left(t\log|t| - t + \frac{\pi t}{2|t|}\left(\sigma - \frac{1}{2}\right)\right)\right\}$$

$$\times\left(1 + O\left(\frac{1}{|t|}\right)\right), \qquad (3.16)$$

where $|t| \geq t_0 > 0$, $0 \leq \sigma \leq 1$. Thus by choosing $\sigma = \epsilon$ it follows by trivial estimation that, for any given $0 < \epsilon < \frac{1}{6}$ and $x > 0$,

$$U(x) = O(x^\epsilon), \qquad V(x) = O(x^\epsilon). \qquad (3.17)$$

Now let, for $\sigma > 0$,

$$H(s) := \frac{3^{6s - \frac{5}{2}} \Gamma^3(s) \Gamma\left(\frac{3}{2} - 3s\right)}{(-s)\Gamma^3(1 - s)\Gamma(3s - 1)} - 1 \qquad (3.18)$$

and use the full form of Stirling's formula, namely

$$\log \Gamma(s + b) = (s + b - \tfrac{1}{2}) \log s - s + \tfrac{1}{2} \log 2\pi + \sum_{j=1}^{K} \frac{a_j}{s^j} + O_\delta\left(\frac{1}{|s|^{K+1}}\right), \quad (3.19)$$

which is valid for b a constant, any fixed integer $K \geq 1$, $|\arg s| \leq \pi - \delta$ for $\delta > 0$, where the points $s = 0$ and the neighbourhoods of the poles of $\Gamma(s + b)$ are excluded, and the a_j are suitable constants. From (3.18) we infer that for $\sigma > 0$ the function $H(s)$ is regular except at the poles of $\Gamma\left(\frac{3}{2} - 3s\right)$, and (3.19) yields

$$H(s) = \sum_{j=1}^{K} \frac{b_j}{s^j} + O\left(\frac{1}{|s|^{K+1}}\right)$$

with suitable constants b_j.

Thus for $0 < \sigma < \frac{1}{4}$ we obtain

$$U(x) = \frac{2}{2\pi i} \int_{(\sigma)} \frac{(-s)\Gamma(3s - 1)}{\Gamma\left(\frac{3}{2} - 3s\right)} \frac{3^{5/2 - 6s}}{x^{2s}} \, ds$$
$$+ \frac{2}{2\pi i} \int_{(\sigma)} \frac{(-s)\Gamma(3s - 1)H(s)}{\Gamma\left(\frac{3}{2} - 3s\right)} \frac{3^{5/2 - 6s}}{x^{2s}} \, ds = I_1 + I_2,$$

say. We have, after the change of variable $3s - 1 = w$,

$$I_1 = \frac{-2}{6\pi i} \int_{(\sigma)} (1 + w) \frac{\Gamma(w)}{\Gamma\left(\frac{1}{2} - w\right)} \frac{3^{\frac{3}{2} - 2 - 2w}}{x^{2(1+w)/3}} \, dw = I_3 + I_4, \qquad (-1 < \sigma < \tfrac{1}{4})$$

say.

We have

$$I_4 = \frac{-6}{2\pi i \sqrt{3}} \int_{(\sigma)} \frac{w\Gamma(w)}{\Gamma(\frac{1}{2} - w)} \frac{dw}{\left(3x^{\frac{1}{3}}\right)^{2(1+w)}}.$$

In view of Stirling's formula the line of integration here can be moved to the left to $\mathrm{Re}\, s = -\infty$. There are simple poles of $\Gamma(w)$ at $w = -n$ for $n = 1, 2, \ldots$ with residues $(-1)^n/n!$. The Residue Theorem thus gives

$$I_4 = \frac{6}{\sqrt{3}} \sum_{n=1}^{\infty} \frac{(-1)^n n \left(3x^{\frac{1}{3}}\right)^{2n-2}}{n!\,\Gamma\left(n + \frac{1}{2}\right)} = \frac{6}{\sqrt{3\pi}} \sum_{n=1}^{\infty} \frac{(-1)^n \left(3x^{\frac{1}{3}}\right)^{2n-2} 2^n 2^{n-2} 2}{(n-1)!\, 2^{n-1} \cdot 1 \cdot 3 \cdots (2n-1)}$$

$$= \frac{12}{\sqrt{3\pi}} \sum_{n=1}^{\infty} \frac{(-1)^n \left(6x^{\frac{1}{3}}\right)^{2n-2}}{(2n-1)!} = \frac{-2}{\sqrt{3\pi}} \frac{\sin\left(6x^{\frac{1}{3}}\right)}{x^{\frac{1}{3}}}.$$

In the integral

$$I_3 = \frac{-6}{2\pi i \sqrt{3}} \int_{(\sigma)} \frac{\Gamma(w)}{\Gamma(\frac{1}{2} - w)} \frac{dw}{\left(3x^{\frac{1}{3}}\right)^{2(1+w)}}$$

the gamma-terms are of the order $|\mathrm{Im}\, w|^{2\sigma - \frac{1}{2}}$, so by the first derivative test and (3.16) the integral will converge for $\sigma < \frac{1}{4}$. Hence by the Residue Theorem we obtain

$$I_3 = \frac{6}{\sqrt{3\pi}\left(3x^{\frac{1}{3}}\right)^2} - \frac{6}{2\pi i \sqrt{3}} \int_{(\frac{1}{6})} \frac{\Gamma(w)}{\Gamma(-\frac{1}{2} + 1 - w)} \frac{dw}{\left(3x^{\frac{1}{3}}\right)^{2(1+w)}}$$

$$= \frac{6}{\sqrt{3\pi}\left(3x^{\frac{1}{3}}\right)^2} - \frac{6}{\sqrt{3}} \frac{J_{-\frac{1}{2}}\left(6x^{\frac{1}{3}}\right)}{\left(3x^{\frac{1}{3}}\right)^{\frac{3}{2}}}.$$

Here we used the well-known formula

$$\frac{1}{2\pi i} \int_{(\sigma)} \frac{\Gamma(s)}{\Gamma(\nu - s + 1)} \frac{ds}{x^{2s}} = \frac{J_\nu(2x)}{x^\nu} \quad \text{for} \quad 0 < \sigma < \frac{1}{2}\nu + \frac{3}{4} \qquad (3.20)$$

and the representation

$$J_{n+\frac{1}{2}}(x) = (-1)^n \sqrt{\frac{2}{\pi}}\, x^{n+\frac{1}{2}} \left(\frac{d}{x\, dx}\right)^n \frac{\sin x}{x},$$

$$J_{-(n+\frac{1}{2})}(x) = \sqrt{\frac{2}{\pi}}\, x^{n+\frac{1}{2}} \left(\frac{d}{x\, dx}\right)^n \frac{\cos x}{x},$$

$$(3.21)$$

valid for $n \geq 0$ an integer and $x > 0$, where $J_\nu(x)$ is the Bessel function of the first kind of order ν. Hence

$$I_3 = \frac{6}{\sqrt{3\pi}\,(3x^{\frac{1}{3}})^2} - \frac{2}{3\sqrt{3\pi}}\,\frac{\cos(6x^{\frac{1}{3}})}{x^{\frac{2}{3}}}.$$

Analogously we have

$$I_2 = \frac{\frac{2}{3}H(\frac{1}{3})}{x^{\frac{2}{3}}}\,\frac{3^{\frac{5}{2}-2}}{\Gamma(\frac{1}{2})} - \frac{2}{2\pi i}\int_{(\frac{1}{3}+\epsilon)}\frac{\Gamma(3s-1)H(s)}{\Gamma(\frac{3}{2}-3s)}\,\frac{3^{\frac{5}{2}-6s}}{x^{2s}}\,ds.$$

From (3.18) it follows that $H(\frac{1}{3}) = -1$, so that the first terms in the expressions for I_2 and I_3 cancel. In the integral in I_2 we make the change of variable $3s - 1 = w$ and use the asymptotic expansion of $H(s)$. We shall obtain terms of the form $A_j I_j(3x^{\frac{1}{3}})$, where the A_j are suitable constants and

$$I_j(x) := \frac{1}{2\pi i}\int_{(\frac{1}{3}+\epsilon)}\frac{x^{-2(1+w)}\Gamma(w)}{(w+1)^j\Gamma(\frac{1}{2}-w)}\,dw \quad \text{for} \quad j = 0,\,1,\dots;$$

note that our $I_j(x)$ has no connection with the modified Bessel functions of the first kind, for which this notation is usually used. Then by using $s\Gamma(s) = \Gamma(s+1)$, (3.20) and (3.21) we obtain

$$I_0(x) = \frac{\cos(2x)}{\sqrt{\pi}\,x^2},$$

$$I_1(x) = -\frac{1}{2\pi i}\int_{(\frac{1}{3}+\epsilon)}\frac{x^{-2(1+w)}\Gamma(w)}{(\frac{1}{2}-w)\Gamma(\frac{1}{2}-w)}\,dw$$

$$+\frac{3}{2}\cdot\frac{1}{2\pi i}\int_{(\frac{1}{3}+\epsilon)}\frac{x^{-2(1+w)}\Gamma(w)}{(w-\frac{1}{2})(w+1)\Gamma(\frac{1}{2}-w)}\,dw$$

$$= -\frac{1}{2\pi i}\int_{(\frac{1}{3}+\epsilon)}\frac{x^{-2(1+w)}\Gamma(w)}{\Gamma(\frac{3}{2}-w)}\,dw - \frac{3}{2}\cdot\frac{1}{2\pi i}\int_{(\frac{1}{3}+\epsilon)}\frac{x^{-2(1+w)}\Gamma(w)}{(w+1)\Gamma(\frac{3}{2}-w)}\,dw$$

$$= -\frac{1}{x^2}\frac{J_{\frac{1}{2}}(2x)}{x^{\frac{1}{2}}} - \frac{3}{2}I_{1,1}(x) = -\frac{\sin 2x}{\sqrt{\pi}\,x^3} - \frac{3}{2}I_{1,1}(x),$$

say. The same procedure can be applied K times to $I_{1,1}(x)$ and, in general, to all integrals $I_j(x)$. Bessel functions of the form (3.21) will emerge in all cases, and the last integrals will be estimated trivially by shifting the line of integration as far to the right as possible in the region of absolute convergence of the integral in question. This proves Lemma 3.

4. The δ-method

Before we proceed to the evaluation of the sum

$$D_3(x, h) := \sum_{n \leq x} d_3(n) d_3(n + h) \tag{4.1}$$

we need some technical preparation. There are several ways of tackling $D_3(x, h)$, most of which involve the use of Ramanujan sums

$$c_q(n) = \sum_{d \,(\mathrm{mod}\, q)}^{*} e\left(\frac{dn}{q}\right) = \sum_{d|(q,n)} d\mu\left(\frac{q}{d}\right), \tag{4.2}$$

where * denotes that d runs over a system of residues mod q which are coprime to q. It would be possible to separate the variables in $d_3(n + h)$ in (4.1) by using the Ramanujan expansion

$$d_3(n) = \sum_{q=1}^{\infty} f_3(q) c_q(n). \tag{4.3}$$

Explicit expressions for the coefficients $f_3(q)$ have recently been found by Lucht [13], [14], but the direct use of (4.3) would entail some delicate questions of convergence. To avoid them we shall employ the so-called δ-method, developed recently by Duke, Friedlander and Iwaniec [3], who used it to evaluate the sum

$$\sum_{am \mp bn = h} d(m) d(n) f(am, bn),$$

where $f(x, y)$ is a certain smooth function.

The δ-method involves the presence of an even, smooth function $w(n)$, supported in $[Q, 2Q] \cup [-2Q, -Q]$, such that

$$\sum_{q=1}^{\infty} w(q) = 1, \qquad w^{(R)}(u) \ll_R \frac{1}{Q^{R+1}} \quad \text{for} \quad R = 0, 1, \ldots.$$

Then if we define

$$\delta(n) = \sum_{m|n} \left(w(m) - w\left(\frac{n}{m}\right) \right), \tag{4.4}$$

we see that

$$\delta(0) = \sum_{q|0} w(q) = \sum_{q=1}^{\infty} w(q) = 1,$$

while for $n \geq 1$ we have

$$\delta(n) = \sum_{m|n} w(m) - \sum_{m|n} w(m) = 0,$$

and, since $w(u)$ is even, $\delta(-n) = 0$. Hence

$$\delta(n) = \begin{cases} 1 & \text{if} \quad n = 0 \\ 0 & \text{if} \quad n \neq 0, \ n \in \mathbf{Z}, \end{cases} \tag{4.5}$$

so that δ may be thought of as the discrete analogue of the Dirac δ-distribution.

Since

$$\sum_{qr=m} \frac{1}{qr} c_q(n) = \frac{1}{m} \sum_{r|m} c_{m/r}(n) = \frac{1}{m} \sum_{r|m} \sum_{\substack{1 \leq d \leq m/r \\ (d, m/r) = 1}} e\left(\frac{drn}{m}\right)$$

$$= \frac{1}{m} \sum_{1 \leq s \leq m} e\left(\frac{sn}{m}\right) = \begin{cases} 1 & \text{if} \quad m \mid n \\ 0 & \text{otherwise,} \end{cases}$$

it follows from (4.4) that

$$\delta(n) = \sum_{m=1}^{\infty} \left(w(m) - w\left(\frac{n}{m}\right) \right) \sum_{qr=m} \frac{c_q(n)}{qr} = \sum_{q=1}^{\infty} \sideset{}{^*}\sum_{d \,(\mathrm{mod}\, q)} e\left(\frac{dn}{q}\right) \Delta_q(n), \tag{4.6}$$

where

$$\Delta_q(u) := \sum_{r=1}^{\infty} \frac{1}{qr} \left(w(qr) - w\left(\frac{u}{qr}\right) \right), \tag{4.7}$$

and the summands in (4.7) vanish for

$$r \geq \frac{1}{q} \max\left(2Q, \frac{|u|}{Q} \right).$$

The nice properties of the smooth function $w(u)$ render the formulas (4.6) and (4.7), which are a sort of a Ramanujan expansion of $\delta(n)$, efficient in dealing with convergence problems. We have, if $f(u)$ is a smooth real-valued function of compact support, the following

Lemma 4. *For fixed $j \geq 1$ we have*

$$\int_{-\infty}^{\infty} f(u) \Delta_q(u) \, du = f(0) + O\left\{ \frac{q^j}{Q} \int_{-\infty}^{\infty} \left(Q^{-j} |f(u)| + Q^j |f^{(j)}(u)| \right) du \right\}. \tag{4.8}$$

Lemma 5.

$$\Delta_q(u) \ll \frac{1}{qQ + Q^2} + \frac{1}{qQ + |u|}, \tag{4.9}$$

and for $R \geq 1$

$$\Delta_q^{(R)} \ll_R \frac{1}{(qQ)^{R+1}}. \tag{4.10}$$

Lemma 4 is the corollary on p. 213 of [3], (4.9) is Lemma 2 of [3], and (4.10) follows from $w^{(R)}(u) \ll_R 1/Q^{R+1}$ and (4.7), which gives

$$\Delta_q^{(R)}(u) = -\frac{1}{q^{R+1}} \sum_{r=1}^{\infty} \frac{1}{r^{R+1}} w^{(R)}\left(\frac{u}{qr}\right).$$

5. The main term in the ternary additive divisor problem

Now we shall present an explicit expression for $D_3(x, h)$ in (4.1), by applying the δ-method of §4 in conjunction with Theorem 2 with $r = 3$. In particular in this section we shall discuss the main term in the ternary additive divisor problem, namely $xP_4(\log x; h)$ in (1.6). First of all we have, by using (2.9),

$$\operatorname*{Res}_{s=1} F(s)E_3\left(s, \frac{h}{k}\right) = \frac{1}{k}\left(E\,F(1) + D\,F'(1) + \tfrac{1}{2}C\,F''(1)\right)$$

$$= \frac{1}{k}\int_0^{\infty} P(\log x)f(x)\,dx,$$

where here and below we denote $P(z) = \frac{1}{2}Cz^2 + Dz + E$. Thus, for $r = 3$, (3.11) becomes

$$\sum_{n=1}^{\infty} f(n)d_3(n)e\left(\frac{hn}{k}\right) = \frac{1}{k}\int_0^{\infty} P(\log x)f(x)\,dx$$

$$+ \frac{\pi^{\frac{3}{2}}}{k^3}\sum_{n=1}^{\infty} A\left(n, \frac{h}{k}\right)\int_0^{\infty} f(x)U\left(\frac{\pi^3 n x}{k^3}\right)dx$$

$$+ \frac{\pi^{\frac{3}{2}}}{k^3}\sum_{n=1}^{\infty} B\left(n, \frac{h}{k}\right)\int_0^{\infty} f(x)V\left(\frac{\pi^3 n x}{k^3}\right)dx,$$

$$(5.1)$$

with

$$A_r\left(n, \frac{h}{k}\right) = \frac{1}{2}\sum_{n_1\ldots n_r=n}\sum_{x_1=1}^{k}\cdots\sum_{x_r=1}^{k}\left\{e\left(\frac{n_1 x_1 + \cdots + n_r x_r + h x_1 \ldots x_r}{k}\right)\right.$$

$$\left. + e\left(\frac{n_1 x_1 + \cdots + n_r x_r - h x_1 \ldots x_r}{k}\right)\right\}, \quad (5.2)$$

$$B_r\left(n, \frac{h}{k}\right) = \frac{1}{2}\sum_{n_1\ldots n_r=n}\sum_{x_1=1}^{k}\cdots\sum_{x_r=1}^{k}\left\{e\left(\frac{n_1 x_1 + \cdots + n_r x_r + h x_1 \ldots x_r}{k}\right)\right.$$

$$\left. - e\left(\frac{n_1 x_1 + \cdots + n_r x_r - h x_1 \ldots x_r}{k}\right)\right\} \quad (5.3)$$

and

$$U(x) = \frac{1}{2\pi i} \int_{\frac{1}{8}-i\infty}^{\frac{1}{8}+i\infty} \frac{\Gamma^3\left(\frac{s}{2}\right)}{\Gamma^3\left(\frac{1-s}{2}\right)} \frac{ds}{x^s}, \qquad V(x) = \frac{1}{2\pi i} \int_{\frac{1}{8}-i\infty}^{\frac{1}{8}+i\infty} \frac{\Gamma^3\left(\frac{1+s}{2}\right)}{\Gamma^3\left(\frac{2-s}{2}\right)} \frac{ds}{x^s}.$$

$$(5.4)$$

Now suppose that $f(x)$ is a smooth and piecewise monotonic function such that $f(x) = 1$ for $M \le x \le 2M$, $f(x) = 0$ for $x \le M - M/P$ or $x \ge 2M + M/P$, where

$$1 \le P \le \tfrac{1}{2}M, \qquad U = Q^2, \qquad Q = M^{\frac{1}{2}}/P^{\frac{1}{2}}, \qquad (5.5)$$

so that for $j \ge 1$

$$f^{(j)}(x) \ll_j \left(\frac{P}{M}\right)^j. \qquad (5.6)$$

For an explicit construction of $f(x)$ see, for example, Lemma 4.3 of [6]. Then we have, since $d_3(n) \ll_\epsilon n^\epsilon$ (see §7 for a discussion concerning a technical point in the definition of $\Delta_3(x,h)$),

$$\sum_{M<n\le2M} d_3(n)d_3(n+h) = D_3(2M,h) - D_3(M,h)$$

$$= \sum_{n=1}^{\infty} f(n)f(n+h)d_3(n)d_3(n+h) + O\left(M^{1+\epsilon}/P\right).$$

$$(5.7)$$

Applying (4.6), we may write the above series as

$$\sum_{n=1}^{\infty} f(n)f(n+h)d_3(n)d_3(n+h)$$

$$= \sum_{m=1}^{\infty}\sum_{n=1}^{\infty} f(m)d_3(m)f(n)d_3(n)\delta(m-n+h)\varphi(m-n+h)$$

$$= \sum_{q=1}^{\infty}\sum_{m=1}^{\infty}\sum_{n=1}^{\infty} f(m)d_3(m)f(n)d_3(n)$$

$$\times \sum_{d\,(\mathrm{mod}\,q)}^{*} e\left(\frac{dm-dn+dh}{q}\right)\Delta_q(m-n+h)\varphi(m-n+h), \quad (5.8)$$

where (as in [3]) φ is a smooth function satisfying $\varphi(0) = 1$, $\varphi(x) = 0$ for $|x| \ge U$ (see (5.5)) and $\varphi^{(j)}(x) \ll_j U^{-j}$ for $j \ge 1$. The seemingly trivial presence of φ has the effect of optimally truncating the series over q in (5.8). Namely we have $\varphi(m - n + h) = 0$ for $|m - n + h| \geq U$. Hence the triple sum in (5.8) is non-empty only if $|m - n + h| < U$. If $q \geq 2Q$, then

$$w(qr) = 0, \quad w\left(\frac{m - n + h}{qr}\right) = 0 \quad \text{for} \quad r \geq 1,$$

since by (5.5) we have

$$\frac{|m - n + h|}{qr} < \frac{U}{2Q} = \tfrac{1}{2}Q < Q.$$

Hence in view of (4.7) it follows from (5.7) and (5.8) that

$$\sum_{M < n \leq 2M} d_3(n)d_3(n + h)$$

$$= \sum_{q < 2Q} \sideset{}{^*}\sum_{d \,(\mathrm{mod}\, q)} e\left(\frac{dh}{q}\right) \sum_{m=1}^{\infty} d_3(m)e\left(\frac{dm}{q}\right) \sum_{n=1}^{\infty} d_3(n)e\left(\frac{-dn}{q}\right) E(m, n)$$

$$+ O(M^{1+\epsilon}/P), \tag{5.9}$$

where

$$E(x, y) := f(x)f(y)\Delta_q(x - y + h)\varphi(x - y + h). \tag{5.10}$$

Now we use (5.1) twice (with $k = q$) to transform the sums over m and n on the right-hand side of (5.9). This is the crucial step which yields

$$\sum_{M < n \leq 2M} d_3(n)d_3(n + h) = \sum_{j=1}^{9} S_j + O(M^{1+\epsilon}/P), \tag{5.11}$$

where

$$S_1 = \sum_{q < 2Q} \frac{1}{q^2} \sideset{}{^*}\sum_{d \,(\mathrm{mod}\, q)} e\left(\frac{dh}{q}\right) \int_0^{\infty} \int_0^{\infty} P(\log x)P(\log y) E(x, y) \, dx \, dy, \tag{5.12}$$

$$S_2 = \pi^{\frac{3}{2}} \sum_{q < 2Q} \frac{1}{q^4} \sideset{}{^*}\sum_{d \,(\mathrm{mod}\, q)} e\left(\frac{dh}{q}\right) \sum_{m=1}^{\infty} A\left(m, \frac{d}{q}\right)$$

$$\times \int_0^{\infty} \int_0^{\infty} U\left(\frac{\pi^3 xm}{q^3}\right) P(\log y) E(x, y) \, dx \, dy, \tag{5.13}$$

the sums S_3, S_4, S_5 are of similar type to S_2 and are estimated similarly,

$$S_6 = \pi^3 \sum_{q<2Q} \frac{1}{q^6} \sum_{d\,(\mathrm{mod}\,q)}^* e\left(\frac{dh}{q}\right) \sum_{m=1}^{\infty} A\left(m, \frac{d}{q}\right) \sum_{n=1}^{\infty} A\left(n, -\frac{d}{q}\right)$$

$$\times \int_0^{\infty} \int_0^{\infty} U\left(\frac{\pi^3 xm}{q^3}\right) U\left(\frac{\pi^3 yn}{q^3}\right) E(x,y)\, dx\, dy, \quad (5.14)$$

while the sums S_7, S_8, S_9 (which contain the functions B and V defined by (5.3) and Lemma 3 respectively) are of similar type to S_6 and are estimated similarly. The sum S_1 will provide the (heuristic) main term in the asymptotic formula for $D_3(x,h)$, while the sums S_j with $j \geq 2$ are to be thought of as error terms.

To deal with S_1, note first that the change of variable $x - y + h = u$ gives

$$I_q := \int_0^{\infty} \int_0^{\infty} P(\log x) P(\log y) E(x,y)\, dx\, dy$$

$$= \int_0^{\infty} P(\log x) f(x) \left\{ \int_{-\infty}^{\infty} P\left(\log(x-u+h)\right) f(x-u+h) \Delta_q(u) \varphi(u)\, du \right\} dx.$$

To the integral in curly braces we apply Lemma 4, with $j = R$ and $f(u)$ replaced by

$$F(u) = P\left(\log(x-u+h)\right) f(x-u+h) \varphi(u).$$

In view of (5.5) and

$$C = C(q) \ll_\epsilon q^\epsilon, \quad D = D(q) \ll_\epsilon q^\epsilon, \quad E = E(q) \ll_\epsilon q^\epsilon$$

(see (2.10)–(2.14)) we have

$$F^{(R)}(u) \ll_{R,\epsilon} \left(\frac{P}{M} + \frac{1}{U}\right)^R q^\epsilon \log^2 M \ll_{R,\epsilon} \frac{q^\epsilon \log^2 M}{U^R},$$

and consequently

$$I_q = \int_0^{\infty} P(\log x) f(x) \left\{ P\left(\log(x+h)\right) f(x+h) \right.$$

$$+ O_R\left(\frac{q^R}{Q} \int_{-\infty}^{\infty} \left(Q^{-R}|F(u)|\right.\right.$$

$$\left.\left.\left. + Q^R |F^{(R)}(u)|\right) du\right)\right\} dx$$

$$= \int_0^{\infty} f(x) f(x+h) P(\log x) P\left(\log(x+h)\right) dx$$

$$+ O_{R,\epsilon}\left\{ \frac{M^\epsilon q^\epsilon \log^2 M}{Q} \left(\left(\frac{q}{Q}\right)^R + \left(\frac{qQ}{U}\right)^R\right)\right\}.$$

But as $U = Q^2$ this means that the above error terms are $\ll_A M^{-A}$ for any fixed $A > 0$, provided that $q < Q^{1-\eta}$, $\eta > 0$, $Q \geq M^\delta$, and $R = R(A, \delta, \eta)$ is large enough. On the other hand by using (4.9) we have

$$I_q \ll q^{\epsilon/2} \int_0^\infty f(x) \log^4 M \, dx \int_{-U}^U |\Delta_q(u)| \, du$$

$$\ll q^{\epsilon/2} M \log^4 M \int_{-U}^U \left\{ \frac{1}{qQ + Q^2} + \frac{1}{qQ + |u|} \right\} du \ll M^{1+\epsilon}. \quad (5.15)$$

Therefore, for any fixed $\eta > 0$, we obtain

$$S_1 = \sum_{q < 2Q} \frac{c_q(h)}{q^2} I_q = O(1) + O\left(M^{1+\epsilon} \sum_{q > Q^{1-\eta}} q^{\epsilon-2} |c_q(h)| \right)$$

$$+ \sum_{q < Q^{1-\eta}} \frac{c_q(h)}{q^2} \int_M^{2M} P(\log x) P(\log(x+h)) \, dx,$$

$$(5.16)$$

where $P(z) = \frac{1}{2}Cz^2 + Dz + E$ as earlier in this section.

To estimate the error terms above note that, if $\sigma(n)$ denotes the sum of the divisors of n, then

$$|c_q(h)| = \left| \sum_{d|(q,h)} d\,\mu\left(\frac{q}{d}\right) \right| \leq \sigma((q,h)) \ll (q,h) \log \log((q,h) + 2). \quad (5.17)$$

This gives

$$\sum_{R < r \leq 2R} |c_r(h)| \ll \log \log R \sum_{R < r \leq 2R} (h, r)$$

$$\leq \log \log R \sum_{d|h, d \leq 2R} 1 \sum_{R < rd \leq 2R} 1 \ll R \log^2 R \quad (5.18)$$

uniformly in h. Consequently for $\epsilon < \eta$ it follows that

$$\sum_{q > Q^{1-\eta}} q^{\epsilon-2} |c_q(h)| = \sum_{j=1}^\infty \sum_{2^{j-1}Q^{1-\eta} < q \leq 2^j Q^{1-\eta}} q^{\epsilon-2} |c_q(h)|$$

$$\ll \sum_{j=1}^\infty 2^{j\epsilon-j} \log^2 (2^j Q) Q^{-1+\epsilon-\eta} \ll Q^{2\eta-1}. \quad (5.19)$$

By using (5.19) in (5.16) we obtain

$$S_1 = \sum_{q=1}^{\infty} \frac{c_q(h)}{q^2} \int_M^{2M} P(\log x) P\big(\log(x+h)\big)\, dx + O\big(M^{\frac{1}{2}+\epsilon} P^{\frac{1}{2}}\big). \quad (5.20)$$

If $1 \le h \le M/2$, then for $M \le x \le 2M$ we have

$$\log(x+h) = \log x + O\left(\frac{h}{M}\right),$$

and (5.20) gives

$$S_1 = \sum_{q=1}^{\infty} \frac{c_q(h)}{q^2} \int_M^{2M} P^2(\log x)\, dx + O\big(M^{\frac{1}{2}+\epsilon} P^{\frac{1}{2}} + M^\epsilon h\big). \quad (5.21)$$

Therefore it is clear that if we define

$$D_3(x;h) = \sum_{n\le x} d_3(n) d_3(n+h) = x\, P_4(\log x; h) + \Delta_3(x;h), \quad (5.22)$$

then we can define

$$P_4(\log x; h) = \frac{1}{x} \int_0^x \sum_{q=1}^{\infty} \frac{c_q(h)}{q^2} P^2(\log t)\, dt. \quad (5.23)$$

However in view of (5.20) perhaps it is more exact to define

$$\sum_{n\le x} d_3(n) d_3(n+h) = x\, p_4(x;h) + \Delta_3(x,h) \quad (5.24)$$

with

$$p_4(x;h) = \frac{1}{x} \int_0^x \sum_{q=1}^{\infty} \frac{c_q(h)}{q^2} P(\log t) P\big(\log(t+h)\big)\, dt. \quad (5.25)$$

In this way one avoids the error term $O\big(M^\epsilon h\big)$ present in (5.21), but for h relevant in applications to the sixth moment of $\big|\zeta(\frac{1}{2}+it)\big|$ it is irrelevant whether we define $\Delta_3(x;h)$ by (5.22) or (5.24). A similar analysis may be made concerning the general formula (1.6), and the (heuristic) main term may be defined analogously to (5.23).

6. Proof of Theorem 1

Having defined $\Delta_3(x, h)$ by (5.22) and (5.23) we may now prove Theorem 1 (the proof also works if $\Delta_3(x, h)$ is defined by (5.24) and (5.25)). Suppose that

$$\sum_{h \leq H} \Delta_3(x, h) \ll_\epsilon H^\alpha x^{\beta + \epsilon} \quad \text{for} \quad 1 \leq H \leq x^{\frac{1}{3} + \delta} \tag{6.1}$$

with some constants $\delta > 0$, $0 \leq \alpha \leq 1$, $0 \leq \beta \leq 1$, $\alpha + \beta \geq 1$. The assertion of Theorem 1 is that the sixth moment follows if (6.1) is true with $\alpha = 1$ and $\beta = \frac{2}{3}$.

We start from (1.4) (with $k = 3$) and (1.5), supposing that $G = T^{1-\epsilon}$. Then we have, in the notation of (1.4),

$$
\begin{aligned}
I_3'' &= \frac{1}{M} \sum_{h \leq M^{1+\epsilon}/G} \int_M^{M'} K(x, h)\, d\big(x\, P_4(\log x; h) + \Delta_3(x, h)\big) \\
&= \frac{1}{M} \sum_{h \leq M^{1+\epsilon}/G} \int_M^{M'} K(x, h)\big(P_4(\log x; h) + P_4'(\log x; h)\big)\, dx \\
&\quad + \frac{1}{M} \sum_{h \leq M^{1+\epsilon}/G} K(x, h)\Delta_3(x, h)\Big|_M^{M'} \\
&\quad - \frac{1}{M} \sum_{h \leq M^{1+\epsilon}/G} \int_M^{M'} \Delta_3(x, h)K'(x, h)\, dx \\
&= \Sigma_1 + \Sigma_2 - \Sigma_3,
\end{aligned}
$$

say.

We shall first treat Σ_1, observing that (5.19) and (5.23) give

$$
\begin{aligned}
P_4(\log x; h) + P_4'(\log x; h) &= \sum_{q=1}^\infty \frac{c_q(h)}{q^2} P^2(\log x) \\
&= \sum_{q \leq M/G} \frac{c_q(h)}{q^2} P^2(\log x) + O\big(M^{\epsilon-1}Gh^\epsilon \log^4 x\big),
\end{aligned}
$$

where $P(z) = \frac{1}{2}Cz^2 + Dz + E$ as before, and noting that ϵ denotes arbitrarily small constants which may not be the same at each occurrence. We obtain

$$
\begin{aligned}
\Sigma_1 = O\big(GT^\epsilon\big) + \frac{1}{M} \sum_{q \leq M/G} \frac{1}{q^2} \sum_{d \,(\text{mod } q)}^* \int_{T-2G}^{T+2G} a(t) \\
\times \int_M^{M'} P^2(\log x) \sum_{h \leq M^{1+\epsilon}/G} e\left(\frac{dh}{q} + \frac{t}{2\pi} \log\left(1 + \frac{h}{x}\right)\right) dx\, dt. \tag{6.2}
\end{aligned}
$$

Set

$$\Sigma_4 := \sum_{h \leq M^{1+\epsilon}/G} e\left(\frac{dh}{q}\right) e\left(\frac{t}{2\pi} \log\left(1 + \frac{h}{x}\right)\right).$$

Then by partial summation it follows for $M \ll x \ll M$, $T \ll t \ll T$ that

$$\Sigma_4 \ll \left| \sum_{h \leq M^{1+\epsilon}/G} e\left(\frac{dh}{q}\right) \right| + \int_1^{M^{1+\epsilon}/G} \frac{t}{u+x} \left| \sum_{h \leq u} e\left(\frac{dh}{q}\right) \right| du$$

$$\ll T^{2\epsilon} \sup_{1 \leq u \leq M^{1+\epsilon}/G} \left| \sum_{h \leq u} e\left(\frac{du}{q}\right) \right|.$$

If $q \geq 2$, then in $\sum_{d \,(\mathrm{mod}\, q)}^*$ we may suppose by periodicity that $|d| \leq \frac{1}{2}q$, $d \neq 0$ (since $(q,d) = 1$). Then applying Lemma 1.2 of [5] and the first derivative test we have

$$\sum_{h \leq u} e\left(\frac{d}{h}q\right) = \int_1^u e\left(\frac{d}{t}q\right) dt + O(1) \ll \frac{q}{|d|}. \tag{6.3}$$

Hence for $q \geq 2$ the contribution of Σ_4 to Σ_1 is

$$\ll \frac{T^{2\epsilon}}{M} \sum_{q \leq M/G} \frac{1}{q^2} \sum_{d(\mathrm{mod})\, q,\, 0<|d|\leq q/2} \frac{q^{1+\epsilon}}{|d|} GM \log^4 M \ll T^{1+\epsilon}, \tag{6.4}$$

since $G = T^{1-\epsilon}$.

In the case $q = 1$ the relevant expression is the real part of

$$\Sigma^* := \frac{1}{M} \int_{T-2G}^{T+2G} a(t) \int_M^{M'} P^2(\log x) \sum_{h \leq M^{1+\epsilon}/G} e\left(\frac{t}{2\pi} \log\left(1 + \frac{h}{x}\right)\right) dx\, dt.$$

Since $\log(1 + h/x) = h/x + O((h/M)^2)$, after summation over h we obtain

$$\mathrm{Re}\, \Sigma^* = \mathrm{Re}\left(\frac{1}{M} \int_M^{M'} P^2(\log x) \int_{T-2G}^{T+2G} \frac{e^{itM_0/x}}{1 - e^{-it/x}} a(t)\, dt\, dx\right) + O(GT^\epsilon),$$

where $M_0 = [M^{1+\epsilon}/G]-1$. In the inner integral we simplify the denominator by the first two terms of Taylor's formula, and apply integration by parts to

$$\int_{T-2G}^{T+2G} \frac{a(t)}{t} e^{itM_0/x}\, dt.$$

Each time the integrated terms will vanish, and the resulting integral will be similar to the original one, only the integrand will contain a new factor of order not exceeding $x/(M_0 G) \ll M^{-\epsilon}$, since $a'(t) \ll 1/G$. Thus performing integration by parts sufficiently many times we see that the contribution in the case $q = 1$ will be again $\ll GT^\epsilon$.

To deal with Σ_2 and Σ_3 note that from (1.5) we have

$$K(x, h) \ll G, \quad K'(x, h) = \frac{-ih}{x^2 + xh} \int_{T-2G}^{T+2G} a(t) \left(1 + \frac{h}{x}\right)^{it} dt \ll \frac{hTG}{M^2},$$
(6.5)

$$\frac{\partial K(x, h)}{\partial h} \ll \frac{TG}{M}, \quad \frac{\partial K'(x, h)}{\partial h} \ll \frac{TG}{M^2} + \frac{hT^2 G}{M^3}.$$

Now we use partial summation, (6.1) with $H = M^{1+\epsilon}/G$ and (6.4). If x denotes M or M', then

$$\sum_{h \leq H} K(x, h) \Delta_3(x, h)$$

$$= \sum_{h \leq H} \Delta_3(x, h) K(x, H) - \int_1^H \sum_{h \leq t} \Delta_3(x, t) \frac{\partial K(x, t)}{\partial t} dt$$

$$\ll GH^\alpha M^{\beta+\epsilon} + \int_1^H t^\alpha M^{\beta+\epsilon} \frac{TG}{M} dt \ll G^{1-\alpha} M^{\alpha+\beta+\epsilon}.$$

Since $\alpha + \beta \geq 1$, $G = T^{1-\epsilon}$ and $M \ll T^{\frac{3}{2}}$ this gives

$$\Sigma_2 = \frac{1}{M} \sum_{h \leq M^{1+\epsilon}/G} K(x, h) \Delta_3(x, h) \Big|_M^{M'} \ll_\epsilon T^{(\alpha+3\beta-1)/2+\epsilon}.$$
(6.6)

Similarly we obtain, for $M \leq x \leq M'$,

$$\sum_{h \leq H} \Delta_3(x, h) K'(x, h) \ll H^\alpha M^{\beta+\epsilon} \frac{HTG}{M^2} + \int_1^H \left(\frac{TG}{M^2} + \frac{tT^2 G}{M^3}\right) t^\alpha M^{\beta+\epsilon} dt$$

$$\ll M^{\alpha+\beta-1+\epsilon} G^{-\alpha} T + M^{\alpha+\beta-1+\epsilon} T^2 G^{-\alpha-1}.$$

Therefore

$$\Sigma_3 \ll_\epsilon T^{(\alpha+3\beta-1)/2+\epsilon},$$
(6.7)

and (6.4), (6.6) and (6.7) yield

Lemma 6. *Suppose that (6.1) holds for some constants $\delta > 0$, $0 \leq \alpha \leq 1$, $0 \leq \beta \leq 1$, $\alpha + \beta \geq 1$. Then for any given $\epsilon > 0$*

$$\int_0^T \left|\zeta(\tfrac{1}{2} + it)\right|^6 dt \ll_\epsilon T^{1+\epsilon} + T^{(\alpha+3\beta-1)/2+\epsilon}.$$
(6.8)

If (1.8) holds with $k = 3$, then (6.1) holds with $\alpha = 1$, $\beta = \frac{2}{3}$, and Theorem 1 follows directly from (6.8).

7. On the dyadic division in $\sum \Delta_3(x, h)$

In this section a technical point will be cleared. Namely, so far in treating $\Delta_3(x, h)$ in §5 we have been treating the error term not in the sum $\sum_{n \leq x} d_3(n)d_3(n + h)$, but the error term in the related sum

$$\sum_{M < n \leq 2M} d_3(n)d_3(n + h),$$

which will be denoted by $\Delta_3(M, 2M; h)$. Hence we shall obtain directly not a bound of the type (6.1), but a bound of the form

$$\sum_{h \leq M} \Delta_3(M, 2M; h) \ll_\epsilon \sum_{i=1}^{L} H^{a_i} M^{b_i + \epsilon} \quad \text{for} \quad 1 \leq H \leq M^\gamma \qquad (7.1)$$

with suitable constants for which $0 \leq a_i, b_i \leq 1$, $a_i + b_i \geq 1$, $\gamma > \frac{1}{3}$, the last condition being needed for applications to the sixth moment of $|\zeta(\frac{1}{2} + it)|$.

If (7.1) holds, then

$$\sum_{h \leq H} \Delta_3(x, h) = \sum_{j \geq 1} \sum_{h \leq H} \Delta_3\left(2^{-j}x, 2^{1-j}x; h\right) = \Sigma_1 + \Sigma_2,$$

say, where in Σ_1 summation is over j for which $j \geq (1 - \delta)\log x / \log 2$, and in Σ_2 summation is over j for which $j < (1 - \delta)\log x / \log 2$, where $0 < \delta < 1$ is a number to be determined. In Σ_1 one has $2^{-j}x \leq x^\delta$, so that trivially

$$\Sigma_1 \ll_\epsilon H x^{\delta + \epsilon} \qquad (7.2)$$

for any $H \geq 1$. In Σ_2 one has $2^{-j}x > x^\delta$, and to each sum over h one applies (7.1). For $1 \leq H \leq (2x^\delta)^\gamma$ one has $H \leq (2^{1-j}x)^\gamma$, so that we obtain in that case

$$\Sigma_2 \ll \sum_{j \geq 1} \sum_{i=1}^{L} H^{a_i}\left(\frac{x}{2^j}\right)^{b_i + \epsilon} \ll \sum_{i=1}^{L} H^{a_i} x^{b_i + \epsilon}. \qquad (7.3)$$

We need the range $H \leq x^{1/3 + \delta_1}$, so we take $\delta = 1/(3\gamma) + \delta_2$ for small $\delta_2 > 0$. Then (7.2) and (7.3) yield

$$\sum_{h \leq H} \Delta_3(x, h) \ll_\epsilon x^\epsilon \left(H x^{1/(3\gamma) + \delta_2} + \sum_{i=1}^{L} H^{a_i} x^{b_i}\right) \quad \text{for} \quad 1 \leq H \leq 2^\gamma x^{\frac{1}{3} + \gamma \delta_2},$$

$$(7.4)$$

where the $\ll$-constant depends only on ϵ. Thus if $\gamma \geq \frac{1}{2}$ holds in (7.1) then we have

$$H x^{1/(3\gamma)} \leq H x^{\frac{2}{3}},$$

and the last term is the one that appears in the conjectural bound (1.8) for $k = 3$.

8. Estimation of ternary exponential sums

In the sums $S_2, \ldots, S_9$ (see (5.13) and (5.14)) appearing in the error terms for $\Delta_3(x, h)$ (or $\Delta_3(M, 2M; h)$) we encounter the ternary sums $A(m, \pm d/q)$, $B(m, \pm d/q)$ defined by (5.2) and (5.3), respectively. By the Cauchy-Schwarz inequality the problem of the estimation of the sums in which these ternary sums appear will be reduced to the estimation of the sum

$$S^*(Q, N) := \sum_{Q < q \leq 2Q} \sum_{N < n \leq 2N} \sideset{}{^*}\sum_{d \,(\mathrm{mod}\, q),\, 0 < |d| \leq q/2} \frac{1}{|d|} A^2\left(n, \frac{d}{q}\right). \tag{8.1}$$

The bounds that we may obtain for $S^*(Q, N)$ will clearly also hold if in (8.1) we replace $A(n, d/q)$ by $B(n, d/q)$ and/or d by $-d$. By the Cauchy-Schwarz inequality we have

$$\sideset{}{^*}\sum_{d\,(\mathrm{mod}\,q)} A^2\left(n, \frac{d}{q}\right)$$

$$= \frac{1}{4} \sideset{}{^*}\sum_{d\,(\mathrm{mod}\,q)} \left| \sum_{n = n_1 n_2 n_3} \sum_{x_1, x_2, x_3 = 1}^{q} e\left(\frac{\pm d x_1 x_2 x_3 + n_1 x_1 + n_2 x_2 + n_3 x_3}{q}\right) \right|^2$$

$$\ll d_3(n) \sum_{n = n_1 n_2 n_3} \sum_{d\,(\mathrm{mod}\,q)} \left| \sum_{x_1, x_2, x_3 = 1}^{q} e\left(\frac{d x_1 x_2 x_3 + n_1 x_1 + n_2 x_2 + n_3 x_3}{q}\right) \right|^2.$$

The sum over d equals

$$\sum_{x_1, x_2, x_3,\,(\mathrm{mod}\,q)} \sum_{y_1, y_2, y_3\,(\mathrm{mod}\,q)} \left\{ \sum_{d\,(\mathrm{mod}\,q)} e\left(\frac{d(x_1 x_2 x_3 - y_1 y_2 y_3)}{q}\right) \right\}$$

$$\times e\left(\frac{n_1(x_1 - y_1) + n_2(x_2 - y_2) + n_3(x_3 - y_3)}{q}\right)$$

$$= q \sum_{\substack{x_1, x_2, x_3, y_1, y_2, y_3\,(\mathrm{mod}\,q) \\ x_1 x_2 x_3 \equiv y_1 y_2 y_3\,(\mathrm{mod}\,q)}} e\left(\frac{n_1(x_1 - y_1) + n_2(x_2 - y_2) + n_3(x_3 - y_3)}{q}\right).$$

Each of the variables y_1, y_2, y_3 may take the values from 1 to q, so that

$$x_1 x_2 x_3 = l_1 + a_1 q, \quad 1 \leq l_1 \leq q - 1, \quad 0 \leq a_1 \leq q^2 - 1.$$

Since there are $d_3(l_1 + a_1 q) \ll_\epsilon q^\epsilon$ solutions of $x_1 x_2 x_3 = l_1 + a_1 q$, we obtain

$$\sideset{}{^*}\sum_{d\,(\mathrm{mod}\,q)} A^2\left(n, \frac{d}{q}\right) \ll_\epsilon (qn)^\epsilon q^4. \tag{8.2}$$

This implies

$$A\left(n, \frac{d}{q}\right) \ll_\epsilon (qn)^\epsilon q^2 \qquad (8.3)$$

and gives trivially

$$S^*(Q, N) \ll_\epsilon (QN)^\epsilon Q^5 N. \qquad (8.4)$$

We shall now improve on (8.4) by proving

Lemma 7. *If $S^*(Q, N)$ is defined by (8.1), then*

$$S^*(Q, N) \ll_\epsilon (QN)^\epsilon Q^4 N. \qquad (8.5)$$

Proof. We have, for $a, b, c, \in \mathbb{N}, (d, q) = 1$,

$$A\left(n, \frac{d}{q}\right) = \tfrac{1}{2} \sum_{abc=n} \left(I(a, b, c, d; q) + I(a, b, c, -d; q)\right)$$

with

$$I = I(a, b, c, d; q) := \sum_{x,y,z \,(\mathrm{mod}\, q)} e\left(\frac{ax + by + cz + dxyz}{q}\right).$$

Then

$$I = \sum_{x,y \,(\mathrm{mod}\, q)} e\left(\frac{ax + by}{q}\right) \sum_{z \,(\mathrm{mod}\, q)} e\left(\frac{z(c + dxy)}{q}\right)$$

$$= q \sum_{\substack{x \,(\mathrm{mod}\, q)}} \sum_{\substack{y \,(\mathrm{mod}\, q) \\ c+dxy \equiv 0 \,(\mathrm{mod}\, q)}} e\left(\frac{ax + by}{q}\right).$$

In the last sum we have $xy \equiv -c\bar{d} \pmod{q}$, where $\bar{d}$ is the unique integer $\pmod{q}$ such that $d\bar{d} \equiv 1 \pmod{q}$. If $f = (x, q)$, $x = x_1 f$, then $f \mid c$ and

$$I = q \sum_{f \mid (c,q)} \sum_{\substack{1 \le x_1 \le q/f \\ (x_1, q/f)=1}} \sum_{x_1 f y \equiv -c\bar{d} \,(\mathrm{mod}\, q)} e\left(\frac{ax_1 f + by}{q}\right).$$

Thus $f \mid c\bar{d}$, $x_1 y \equiv -c\bar{d}/f \pmod{q/f}$, and if $x_1 \bar{x}_1 \equiv 1 \pmod{q/f}$ then we have

$$I = q \sum_{f \mid (c,q)} \sum_{\substack{1 \le x_1 \le q/f \\ (x_1, q/f)=1}} \sum_{\substack{y \,(\mathrm{mod}\, q) \\ y \equiv -\bar{x}_1 c\bar{d}/f \,(\mathrm{mod}\, q/f)}} e\left(\frac{ax_1 f + by}{q}\right).$$

The congruence $y \equiv -\bar{x}_1 c\bar{d}/f \pmod{q/f}$ has f solutions $\pmod{q}$, namely

$$y = -\frac{\bar{x}_1 c\bar{d}}{f} + \frac{rq}{f} \quad \text{with} \quad r = 0, 1, \ldots, f - 1.$$

Hence if $l = -c\overline{d}/f$ then

$$I = q \sum_{\substack{f|(c,q)}} \sum_{\substack{1 \le x_1 \le q/f \\ (x_1, q/f)=1}} e\left(\frac{ax_1 f + b\overline{x}_1 l}{q}\right) \sum_{r=0}^{f-1} e\left(\frac{br}{f}\right)$$

$$= q \sum_{f|(b,c,q)} f \sum_{\substack{1 \le x_1 \le q/f \\ (x_1, q/f)=1}} e\left(\frac{ax_1 + b\overline{x}_1 l/f}{q/f}\right)$$

$$= q \sum_{f|(b,c,q)} f\, S\left(a, \frac{bl}{f}; \frac{q}{f}\right), \tag{8.6}$$

where

$$S(m, n; c) = \sum_{\substack{d \,(\mathrm{mod}\ c),\, (d,c)=1 \\ d\overline{d} \equiv 1 \,(\mathrm{mod}\ c)}} e\left(\frac{md + n\overline{d}}{c}\right)$$

is the usual Kloosterman sum.

If we use the Selberg-Kuznetsov formula (see [12] for a proof)

$$S(m, n; c) = \sum_{d|(m,n,c)} d\, S\left(1, \frac{mn}{d^2}; \frac{c}{d}\right)$$

in (8.6) and set $f\delta = h$, then we obtain

$$I = q \sum_{f|(b,c,q)} f \sum_{\delta|(a, b\overline{c}/f^2, q/f)} \delta\, S\left(1, -\frac{abc\overline{d}}{f^2\delta^2}; \frac{q}{f\delta}\right)$$

$$= q \sum_{h|q} \Delta_1(a, b, c, h, q) h\, S\left(1, -\frac{abc\overline{d}}{h^2}; \frac{q}{h}\right)$$

$$= q \sum_{h|q} \Delta_1(a, b, c, h, q) h\, S\left(\frac{abc}{h^2}, -\overline{d}; \frac{q}{h}\right).$$

Here $\Delta_1(a, b, c, h, q)$ is a divisor function, that may be written down explicitly, for which one has $\Delta_1(a, b, c, h, q) \ll_\epsilon (qn)^\epsilon$, and we used the fact (since $(\overline{d}, q/h) = 1$) that for $(m, c) = 1$ one has the identity

$$S(1, mn; c) = S(m, n; c).$$

Thus we obtain

$$A\left(n, \frac{d}{q}\right) = q \sum_{h|q, h^2|n} \Delta(n, h, q) h\, S\left(\frac{n}{h^2}, -\overline{d}; \frac{q}{h}\right) \tag{8.7}$$

where $d\bar{d} \equiv 1 \pmod q$ and $\Delta(n, h, q)$ is a divisor function satisfying

$$\Delta(n, h, q) \ll_\epsilon (qn)^\epsilon. \tag{8.8}$$

Now if we use Weil's classical bound

$$\left| S(m, n; c) \right| \le (m, n, c)^{\frac{1}{2}} d(c) c^{\frac{1}{2}},$$

then it follows from (8.7) and (8.8) that

$$A\left(n, \frac{d}{q}\right) \ll_\epsilon (qn)^\epsilon q \sum_{h|q, h^2|n} h\left(\frac{q}{h}\right)^{\frac{1}{2}} \ll_\epsilon q^{\frac{3}{2}+\epsilon} n^{\frac{1}{4}+\epsilon}. \tag{8.9}$$

Note that (8.9) improves (8.3) for $n < q^2$. By using (8.9) we obtain

$$S^*(Q, N) \ll_\epsilon (QN)^\epsilon \sum_{Q<q\le 2Q} q^3 \sum_{\substack{N<n\le 2N}} \sum_{\substack{d\,(\mathrm{mod}\,q),\,(d,q)=1 \\ 0<|d|\le q/2}} \frac{1}{|d|} \sum_{h|q,\,h^2|n} h$$

$$\ll_\epsilon Q^{3+\epsilon} N^\epsilon \sum_{h=1}^{\infty} h \cdot \frac{Q}{h} \cdot \frac{N}{h^2} \ll_\epsilon (QN)^\epsilon Q^4 N, \tag{8.10}$$

which proves (8.5).

9. The error terms in the ternary additive divisor problem

The contribution to the error terms in the ternary additive divisor problem, namely (5.11), comes from the error term in (5.20) (or (5.21)) and the sums $S_2, \dots, S_9$ in (5.11). The sums S_2, S_3, S_4, S_5 will contain only one of the functions U, V, while the remaining sums S_6, S_7, S_8, S_9 will contain both of these functions, and consequently their estimation will be harder.

First note that, if $E(x, y)$ is defined by (5.10), then we have, for $q < 2Q$,

$$\frac{\partial^{i+j} E(x, y)}{\partial x^i \partial y^j} \ll_{i,j} \frac{1}{(qQ)^{i+j+1}}, \tag{9.1}$$

since

$$f^{(j)}(x) \ll_j (P/M)^j, \qquad \varphi^{(j)}(x) \ll_j 1/U^j = 1/Q^{2j},$$
$$\Delta_q^{(j)} \ll_j 1/(qQ)^{j+1}, \qquad Q = (M/P)^{\frac{1}{2}}.$$

In all the integrals in $S_2, \dots, S_9$ we perform in each variable a large number of integrations by parts. The integrated terms will vanish, and each integration of the functions $U(\cdot)$ and $V(\cdot)$ will bring in, by Lemma 3, a factor of order $qm^{-1/3}x^{2/3} \approx qm^{-1/2}M^{2/3}$ or $qn^{-1/3}M^{2/3}$, respectively, while each partial derivative of $E(\cdot)$ brings in a factor of order $(qQ)^{-1}$. This means that, after R integrations by parts in each variable, the factors

$$\left(\frac{M^{\frac{2}{3}}}{m^{\frac{1}{3}}Q}\right)^R, \quad \left(\frac{M^{\frac{2}{3}}}{n^{\frac{1}{3}}Q}\right)^R$$

will appear. Consequently if $R = R(\epsilon)$ is large, then the terms of these series in m and n with $m > M^{2+\epsilon}/Q^3$ and $n > M^{2+\epsilon}/Q^3$, respectively, will make a negligible contribution. Hence all the sums in $S_2, \ldots, S_9$ may be taken to be finite, and there will be no convergence problems, since the integrands are functions of bounded support.

We have

$$\sum_{h \leq H} S_2 = \sum_{q \leq 2Q} \frac{\pi^{\frac{3}{2}}}{q^4} \sum_{\substack{d \,(\mathrm{mod}\, q) \\ 0 < |d| \leq q/2}}^* \sum_{m \leq M^{2+\epsilon}/Q^3} A\left(m, \frac{d}{q}\right) \int_{-U}^{U} \Delta_q(u)\varphi(u) I \, du$$

(9.2)

with

$$I := \int_0^\infty U\left(\frac{\pi^3 x m}{q^3}\right) \sum_{h \leq H} e\left(\frac{dh}{q}\right) P\big(\log(x+h-u)\big) f(x) f(x+h-u) \, dx. \quad (9.3)$$

The sums containing S_3, S_4, S_5 are of similar type and their estimations will be analogous, so only (9.2) will be considered in detail. In I we perform first an integration by parts, then summation over h, and then finally we estimate the resulting expression trivially.

First we consider the values of m such that $m \gg q^3/M$. Since the support of I is $[M - M/P, 2M + M/P]$ we have from (3.14) of Lemma 3 that

$$\int U\left(\frac{\pi^3 x m}{q^3}\right) dx = \frac{K x^{\frac{1}{3}} q^2}{m^{\frac{2}{3}}} \cos\left(\frac{6\pi \sqrt[3]{xm}}{q}\right) + O\left(\frac{q^3}{m}\right) \quad (9.4)$$

with a suitable constant K, and we obtain

$$I = -\int_0^\infty \left\{ \frac{K x^{\frac{1}{3}} q^2}{m^{\frac{2}{3}}} \cos\left(\frac{6\pi \sqrt[3]{xm}}{q}\right) + O\left(\frac{q^3}{m}\right) \right\} \sum_{h \leq H} e\left(\frac{dh}{q}\right) \Phi_h'(x) \, dx,$$

(9.5)

where

$$\Phi_h(x) := P\big(\log(x+h-u)\big) f(x) f(x+h-u). \quad (9.6)$$

Note that the derivative of the log-factor is $\ll q^\epsilon M^{-1} \log^2 M$, since

$$|C| + |D| + |E| \ll q^\epsilon$$

in view of (2.10)–(2.14). Further we have $f'(x) \ll P/M$, the support of $f(x)$ is $\ll M/P$, and $\Phi_h'(x)$ is a piecewise monotonic function of h, so it can be removed by partial summation from the sum over h. Then by using (6.3) we obtain from (9.4)

$$I \ll M^{\frac{1}{3}+\epsilon} m^{-\frac{2}{3}} q^3 |d|^{-1}. \quad (9.7)$$

We insert (9.7) in (9.2), use the Cauchy-Schwarz inequality and (8.5) to obtain a contribution which is

$$\ll M^{\frac{1}{3}+\epsilon}\sum_{q<2Q}\frac{1}{q}\sum_{q^3/M\ll m\leq M^{2+\epsilon}/Q^3}\frac{1}{m^{\frac{2}{3}}}\sideset{}{^*}\sum_{\substack{d\,(\mathrm{mod}\,q)\\0<|d|\leq q/2}}\frac{1}{|d|}\left|A\left(m,\frac{d}{q}\right)\right|$$

$$\ll M^{\frac{1}{3}+\epsilon}\log^{\frac{1}{2}}Q\sum_{q<2Q}\frac{1}{q}\sum_{m\leq M^{2+\epsilon}/Q^3}\frac{1}{m^{\frac{2}{3}}}\left(\sideset{}{^*}\sum_{\substack{d\,(\mathrm{mod}\,q)\\0<|d|\leq q/2}}\frac{1}{|d|}\left|A^2\left(m,\frac{d}{q}\right)\right|\right)^{\frac{1}{2}}$$

$$\ll M^{\frac{1}{3}+\epsilon}\log^{\frac{5}{2}}M\max_{Q_1\leq Q}\max_{M_1\leq M^{2+\epsilon}/Q^3}\frac{1}{Q_1M_1^{\frac{2}{3}}}$$

$$\times\sum_{Q_1<q\leq 2Q_1}\sum_{M_1<m\leq 2M_1}\left(\sideset{}{^*}\sum_{\substack{d\,(\mathrm{mod}\,q)\\0<|d|\leq q/2}}\frac{1}{|d|}\left|A^2\left(m,\frac{d}{q}\right)\right|\right)^{\frac{1}{2}}$$

$$\ll M^{\frac{1}{3}+\epsilon}\log^{\frac{5}{2}}M\max_{Q_1\leq Q}\max_{M_1\leq M^{2+\epsilon}/Q^3}\frac{1}{Q_1M_1^{\frac{2}{3}}}(Q_1M_1)^{\frac{1}{2}}\left(S^*(Q_1,M_1)\right)^{\frac{1}{2}}$$

$$\ll M^{\frac{1}{3}+2\epsilon}\max_{Q_1\leq Q}\max_{M_1\leq M^{2+\epsilon}/Q^3}Q_1^{\frac{2}{3}}M_1^{\frac{1}{3}}\ll M^{1+3\epsilon}Q^{\frac{1}{2}}$$

$$=M^{\frac{5}{4}+3\epsilon}P^{-\frac{1}{4}}. \tag{9.8}$$

To estimate the contribution of m satisfying $m\ll q^3/M$ we use (3.17). Then we have

$$U\left(\frac{\pi^3 xm}{q^3}\right)\ll\left(\frac{Mm}{q^3}\right)^{\epsilon}\ll 1.$$

We perform first a summation over h, using (6.3), and then we estimate the resulting expression trivially to obtain a contribution which is

$$\ll\sum_{q<2Q}\frac{1}{q^4}\sum_{m\ll q^3/M}Mq\sideset{}{^*}\sum_{\substack{d\,(\mathrm{mod}\,q)\\0<|d|\leq q/2}}\frac{1}{|d|}\left|A\left(m,\frac{d}{q}\right)\right|$$

$$\ll M^{1+\epsilon}\max_{Q_1\leq Q}\max_{M_1\leq Q_1^3/M}\frac{1}{Q_1^3}\sum_{Q_1<q\leq 2Q_1}\sum_{M_1<m\leq 2M_1}\sum_{\substack{d\,(\mathrm{mod}\,q)\\0<|d|\leq q/2}}\frac{1}{|d|}\left|A\left(m,\frac{d}{q}\right)\right|$$

$$\ll Q^{\frac{5}{2}}M^{\epsilon}=M^{\frac{5}{4}+\epsilon}P^{-\frac{5}{4}}, \tag{9.9}$$

as was similarly done in (9.8). Thus from (9.8) and (9.9) it follows that

$$\sum_{h\leq H}S_2\ll M^{\frac{5}{4}+\epsilon}P^{-\frac{1}{4}}+M^{\frac{5}{4}+\epsilon}P^{-\frac{5}{4}}\ll M^{\frac{5}{4}+\epsilon}P^{-\frac{1}{4}}. \tag{9.10}$$

It remains to estimate the contributions of the sums $\sum_{h \leq H} S_j$ for $6 \leq j \leq 9$. These are all estimated analogously, so we shall consider in detail only

$$\sum_{h \leq H} S_6 = \pi^3 \sum_{q < 2Q} \frac{1}{q^6} \sum_{m \leq M^{2+\epsilon}/Q^3} \sum_{n \leq M^{2+\epsilon}/Q^3}$$

$$\times \sum_{\substack{d \,(\mathrm{mod}\, q) \\ 0 < |d| \leq q/2}}^{*} A\left(m, \frac{d}{q}\right) A\left(n, -\frac{d}{q}\right) \int_{-U}^{U} \Delta_q(u) \varphi(u) J \, du, \quad (9.11)$$

where

$$J := \int_0^\infty U\left(\frac{\pi^3 m x}{q^3}\right) \sum_{h \leq H} U\left(\frac{\pi^3 n(x+h-u)}{q^3}\right) e\left(\frac{dh}{q}\right) f(x) f(x+h-u) \, dx.$$

$$(9.12)$$

The integral J is similar to I, defined by (9.3), only it is more complicated because the expression (9.12) contains two functions $U(\cdot)$. We shall consider the two cases

$$m \ll q^3 M^{\epsilon-1}, \quad (9.13)$$
$$m \gg q^3 M^{\epsilon-1} \quad (9.14)$$

separately.

If (9.13) holds, and additionally $n \ll q^3 M^{\epsilon-1}$, then we perform the summation over h in (9.12) first, taking $\sigma = \epsilon$ in the definition (3.7) of $U(x)$. After partial summation we use (6.3) and the first derivative test, and then trivial estimation to obtain a contribution which is, by the use of (8.5),

$$\ll \sum_{q < 2Q} \frac{1}{q^6} M^{1+\epsilon} q \sum_{m \ll q^3 M^{\epsilon-1}} \sum_{n \ll q^3 M^{\epsilon-1}} \sum_{\substack{d \,(\mathrm{mod}\, q) \\ 0 < |d| \leq q/2}}^{*} \frac{1}{|d|} \left| A\left(m, \frac{d}{q}\right) A\left(n, -\frac{d}{q}\right) \right|$$

$$\leq \frac{1}{2} M^{1+\epsilon} \sum_{q < 2Q} \frac{1}{q^5} \sum_{m \ll q^3 M^{\epsilon-1}} \sum_{n \ll q^3 M^{\epsilon-1}}$$

$$\times \sum_{\substack{d \,(\mathrm{mod}\, q) \\ 0 < |d| \leq q/2}}^{*} \frac{1}{|d|} \left(A^2\left(m, \frac{d}{q}\right) + A^2\left(n, -\frac{d}{q}\right) \right)$$

$$\ll M^{1+\epsilon} \max_{Q_1 \leq Q} \max_{M_1 \leq Q^3 M^{\epsilon-1}} \frac{1}{Q_1^5} \sum_{Q_1 < q \leq 2Q_1} \sum_{M_1 < m \leq 2M_1}$$

$$\times \sum_{\substack{d \,(\mathrm{mod}\, q) \\ 0 < |d| \leq q/2}}^{*} \frac{1}{|d|} A^2\left(m, \frac{d}{q}\right) Q_1^3 M^{\epsilon-1}$$

$$\ll M^\epsilon \max_{Q_1 \leq Q} \max_{M_1 \leq Q^3 M^{\epsilon-1}} Q_1^{-2} Q_1^4 M_1 \ll Q^5 M^{\epsilon-1} \ll M^{\frac{3}{2}+\epsilon} P^{-\frac{5}{2}}. \quad (9.15)$$

Now we treat the case when (9.13) holds and $q^3 M^{\epsilon-1} \ll n \leq M^{2+\epsilon}/Q^3$ (this discussion is of course under the assumption that $q^3 \gg M^{1-\epsilon}$, otherwise the sums are empty). We use (3.12) of Lemma 3, noting that the condition $q^3 \gg M^{1-\epsilon}$ implies that the main contribution will come from the term $x^{-1/3} \sin 6x^{1/3}$, provided that K in Lemma 3 is taken to be sufficiently large. We perform summation over h first, removing by partial summation factors like $(x+h-u)^{1/3}$ and $f(x+h-u)$, which are piecewise monotonic. Then we shall encounter the exponential sum

$$\Sigma_{\pm} := \sum_{h \leq T} \exp\left(\frac{2\pi i}{q} \left(dh \pm 3n^{\frac{1}{3}}(x+h-ul)^{\frac{1}{3}} \right) \right) \quad \text{for} \quad T \geq 1 \qquad (9.16)$$

in which we have

$$D := \frac{1}{q} \frac{d}{dh} \left(dh \pm 3n^{\frac{1}{3}}(x+h-u)^{\frac{1}{3}} \right)$$
$$= \frac{1}{q} \left(d \mp n^{\frac{1}{3}}(x+h-u)^{-\frac{1}{3}} \right) = \frac{1}{q} \left(d + O(n^{\frac{1}{3}} M^{-\frac{2}{3}}) \right).$$

If we have, for some fixed constant $\delta > 0$,

$$P \leq M^{1-\delta}, \qquad (9.17)$$

then $n^{1/3} M^{-\frac{2}{3}} \ll M^{\epsilon}/Q = o(1)$ for sufficiently small ϵ, hence

$$\frac{|d|}{2q} \leq D \leq \frac{2|d|}{3q}.$$

Then by using (6.3) we obtain

$$\Sigma_{\pm} \ll \frac{q}{|d|}. \qquad (9.18)$$

Therefore from (9.18) we obtain, by using trivial estimation, that the contribution is

$$\ll \sum_{q < 2Q} \frac{1}{q^6} \sum_{\substack{m \ll q^3 M^{\epsilon-1} \\ q^3 M^{\epsilon-1} \ll n \leq M^{2+\epsilon}/Q^3}} {\sum_{d \,(\mathrm{mod}\, q)}}^* \frac{q}{|d|} \left| A\left(m, \frac{d}{q}\right) A\left(n, -\frac{d}{q}\right) \right| Mqm^{-\frac{1}{3}} M^{-\frac{1}{3}}$$

$$\ll M^{\frac{2}{3}+\epsilon} \sum_{q < 2Q} \frac{1}{q^4} \sum_{\substack{m \ll q^3 M^{\epsilon-1} \\ n \leq M^{2+\epsilon}/Q^3}} \frac{1}{n^{\frac{1}{3}}} {\sum_{d \,(\mathrm{mod}\, q)}}^* \frac{1}{|d|} \left(A^2\left(n, \frac{d}{q}\right) + A^2\left(n, -\frac{d}{q}\right) \right)$$

$$\ll M^{\frac{3}{2}+\epsilon} P^{-\frac{1}{2}}, \qquad (9.19)$$

as was seen similarly in (9.8). It follows therefore from (9.15) and (9.19) that the contribution of those m and n which satisfy $m \ll q^3 M^{\epsilon-1}$ or $n \ll q^3 M^{\epsilon-1}$ is, by symmetry,

$$\ll M^{\frac{3}{2}+\epsilon} P^{-\frac{1}{2}}. \tag{9.20}$$

It remains to consider m and n which satisfy

$$q^3 M^{\epsilon-1} \ll m, n \ll M^{2+\epsilon}/Q^3.$$

In J in (9.12) we use (3.12) of Lemma 3, noting that the main terms in

$$U\left(\frac{\pi^3 mx}{q^3}\right) U\left(\frac{\pi^3 n(x+h-u)}{q^3}\right)$$

will be

$$\frac{q^2}{(mn)^{\frac{1}{3}} x^{\frac{1}{3}} (x+h-u)^{\frac{1}{3}}} \exp\left(\frac{6\pi i}{q}\left((mx)^{\frac{1}{3}} \pm n^{\frac{1}{3}}(x+h-u)^{\frac{1}{3}}\right)\right), \tag{9.21}$$

provided that K in (3.12) is sufficiently large, so that the error terms will give, after trivial estimation, a negligible contribution. Let now

$$\Delta_{\pm} := m^{\frac{1}{3}} \pm n^{\frac{1}{3}}\left(\frac{x}{x+h-u}\right)^{\frac{2}{3}}. \tag{9.22}$$

In (9.21) we see, by using the Taylor series, that we may replace

$$(x+h-u)^{-\frac{1}{3}}$$

by $x^{-1/3}$, since $|u-h|$ is small when compared to x, which is of the order M. Then the main contribution to J will be

$$\frac{q^2}{(mn)^{\frac{1}{3}}} \sum_{h \leq H} e\left(\frac{dh}{q}\right) \psi(m,n,h), \tag{9.23}$$

where, denoting

$$\eta(x) = \exp\left(\frac{6\pi i}{q}\left((mx)^{\frac{1}{3}} \pm n^{\frac{1}{3}}(x+h-u)^{\frac{1}{3}}\right)\right),$$

we have

$$\psi(m,n,h)$$

$$= \int_0^\infty \eta(x) \frac{f(x)f(x+h-u)}{x^{\frac{1}{3}}} \, dx$$

$$= \frac{q}{2\pi i} \int_0^\infty \frac{f(x)f(x+h-u)}{x^{\frac{2}{3}}\left(m^{\frac{1}{3}}x^{-\frac{2}{3}} \pm n^{\frac{1}{3}}(x+h-u)^{-\frac{2}{3}}\right)} \, d\eta(x)$$

$$= -\frac{q}{2\pi i} \int_0^\infty \eta(x) \left(\frac{f(x)f(x+h-u)}{\Delta_\pm}\right)' \, dx$$

$$= -\frac{q}{2\pi i} \int_0^\infty \eta(x) \left\{ \frac{1}{\Delta_\pm}\left(f'(x)f(x+h-u) + f(x)f'(x+h-u)\right) \right.$$

$$\left. \pm \frac{1}{\Delta_\pm^2} \cdot \frac{\frac{2}{3}n^{\frac{1}{3}}f(x)f(x+h-u)}{\left(x/(x+h-u)\right)^{\frac{1}{3}}} \cdot \frac{u-h}{(x+h-u)^2} \right\} \, dx.$$

The contribution of Δ_+ will be

$$\ll \frac{q^4}{|d|(mn)^{\frac{1}{3}}}\left(\frac{1}{m^{\frac{1}{3}}} + \frac{Q^2+H}{n^{\frac{1}{3}}M}\right) \ll \frac{q^4}{m^{\frac{2}{3}}n^{\frac{1}{3}}|d|},$$

which gives then a total contribution which will be

$$\ll \sum_{q<2Q} \frac{\log M}{q^2} \sum_{m \leq M^{2+\epsilon}/Q^3} \frac{1}{m^{\frac{2}{3}}} \sum_{n \leq M^{2+\epsilon}/Q^3} \frac{1}{n^{\frac{1}{3}}} \sum_{d(\bmod q)}^* \frac{1}{|d|}\left|A\left(m,\frac{d}{q}\right)A\left(n,-\frac{d}{q}\right)\right|$$

$$\ll M^{\frac{3}{2}+\epsilon}P^{\frac{1}{2}}, \tag{9.24}$$

as was seen similarly in (9.8).

The contribution of Δ_- turns out to be the one that is largest. It is

$$\ll \frac{q^4}{(mn)^{\frac{1}{3}}|d|}\left(\frac{1}{|\Delta_-|} + \frac{(M^{2+\epsilon}Q^{-3})^{\frac{1}{3}}(H+Q^2)M^{-2}}{\Delta_-^2}\right)$$

$$\ll \frac{q^4}{(mn)^{\frac{1}{3}}|d|}\left(\frac{M^{\frac{4}{3}+\epsilon}Q^{-2}}{|m-nY|} + \frac{M^{\frac{8}{3}+\epsilon}Q^{-4}M^{\frac{2}{3}}Q^{-1}(H+Q^2)M^{-2}}{(m-nY)^2}\right)$$

$$= \frac{q^4}{(mn)^{\frac{1}{3}}|d|}\left(\frac{M^{\frac{1}{3}+\epsilon}P}{|m-nY|} + \frac{M^{\epsilon-\frac{7}{6}}P^{\frac{5}{2}}(H+Q^2)}{(m-nY)^2}\right),$$

where $Y = \left(x/(x+h-u)\right)^2$ at $x = M - m/P$ or $2M + M/P$.

If $|m - ny| \geq W$ (> 1), then the contribution of Δ_- will be

$$\sum_{q < 2Q} \frac{1}{q^2} \left(\frac{M^{\frac{1}{3}+\epsilon}P}{W} + \frac{M^{\epsilon - \frac{7}{6}}P^{\frac{5}{2}}(H + Q^2)}{W^2} \right)$$

$$\times \sum_{\substack{m \leq M^{2+\epsilon}/Q^3 \\ n \leq M^{2+\epsilon}/Q^3}} \frac{1}{m^{\frac{1}{3}}n^{\frac{1}{3}}} \sum_{d \,(\mathrm{mod}\, q)}^{*} \frac{1}{|d|} \left| A\left(m, \frac{d}{q}\right) A\left(n, -\frac{d}{q}\right) \right|$$

$$\ll \frac{M^{2+\epsilon}P^2}{W} + \frac{M^{\frac{1}{2}}P^{\frac{7}{2}}(H + Q^2)}{W^2}. \tag{9.25}$$

If $|m - nY| \leq W$ we fix n, and the contribution will be

$$\ll M^{\frac{1}{3}+\epsilon} \sum_{q < 2Q} \frac{1}{q^3} \sum_{n \leq M^{2+\epsilon}/Q^3} \frac{1}{n^{\frac{2}{3}}} \sum_{d \,(\mathrm{mod}\, q)}^{*} \frac{1}{|d|}$$

$$\times \sum_{nY - W \leq m \leq nY + W} \left| A\left(m, \frac{d}{q}\right) A\left(n, -\frac{d}{q}\right) \right|. \tag{9.26}$$

In (9.26) we shall apply the Cauchy-Schwarz inequality, but we cannot use Lemma 7 directly. However we observe that the argument leading to (8.10) actually gives

$$\ll \sum_{Q < q \leq 2Q} \sum_{N < n \leq N+M} \sum_{d \,(\mathrm{mod}\, q)}^{*} \frac{1}{|d|} A^2\left(n, \frac{d}{q}\right) \ll (QN)^{\epsilon} Q^4 (M + Q). \tag{9.27}$$

By using (9.27) we see that the expression in (9.26) is

$$\ll (W + Q) M^{\frac{1}{3}+\epsilon} Q \sum_{n \leq M^{2+\epsilon}/Q^3} \frac{1}{n^{\frac{2}{3}}} \ll (W + Q) M^{1+\epsilon}$$

$$= W M^{1+\epsilon} + M^{\frac{3}{2}+\epsilon} P^{-\frac{1}{2}}.$$

Thus from (9.10), (9.15), (9.20), (9.24) and (9.26) we obtain

$$\sum_{h \leq H} \Delta_3(M, 2M; h) \ll M^{\epsilon} \left(\frac{HM}{P} + HM^{\frac{1}{2}}P^{\frac{1}{2}} + \frac{M^{\frac{5}{4}}}{P^{\frac{1}{4}}} + M^{\frac{3}{2}}P^{\frac{1}{2}} \right.$$

$$\left. + \frac{M^2 P^2}{W} + \frac{M^{\frac{1}{2}}P^{\frac{7}{2}}}{W^2}(H + Q^2) + WM \right).$$

Now take

$$M^2 P^2 W^{-1} = WM, \quad W = M^{\frac{1}{2}}P.$$

242   A. Ivić

It follows that

$$\sum_{h \le H} \Delta_3(M, 2M; h) \ll M^\epsilon \left(HMP^{-1} + HM^{\frac{1}{2}}P^{\frac{1}{2}} \right.$$

$$\left. + M^{\frac{3}{2}}PM^{-\frac{1}{2}}P^{\frac{3}{2}}H + M^{\frac{1}{2}}P^{\frac{3}{2}}H + M^{\frac{1}{2}}P^{\frac{5}{2}} \right).$$

Since $P \ge 1$ this is worse than the trivial $HM^{1+\epsilon}$ (for $H \le M^{1/2}$). This means that we have not succeeded in obtaining an unconditional improvement of the known bound

$$\int_0^T \left| \zeta(\tfrac{1}{2} + it) \right|^6 \ll T^{\frac{5}{4}+\epsilon}.$$

To achieve this it seems necessary to be able to estimate double exponential sums (of the type (9.21)) weighted with factors $A(m, \pm d/q)$, $B(m, \pm d/q)$. This appears to be very hard and is a task for future investigations. Thus it seemed most appropriate to formulate our results concerning the sixth moment of $\zeta(s)$ conditionally in Theorem 1 and Lemma 6.

It is very likely that the ternary additive divisor problem may also be employed in dealing with the eighth moment of $\zeta(s)$, which reduces to the estimation of

$$\int_0^T \left| \zeta(\tfrac{1}{2} + it) \right|^6 \left| \sum_{N < n \le n' \le 2N} \frac{1}{n^{\frac{1}{2}+it}} \right|^2 dt \quad \text{for} \quad N \ll T^{\frac{1}{2}},$$

and the above integral may be treated similarly to I_k in (1.1).

References

1 B.C. Berndt: Identities involving the coefficients of a class of Dirichlet series I. *Trans. Amer. Math. Soc.* **137** (1969), 345–359 and VII, *ibid.* **201** (1975), 247–261.
2 B.C. Berndt: On the average order of a class of arithmetic functions I. *J. Number Theory* **3** (1971), 184–203.
3 W. Duke, J.B. Friedlander and H. Iwaniec: A quadratic divisor problem. *Invent. Math.* **115** (1994), 209–217.
4 J.L. Hafner: On the representation of the summatory function of a class of arithmetic functions. In *Analytic Number Theory* (K. Knopp., ed.), (Lecture Notes in Mathematics **899**), 148–165. Springer (1981).
5 A. Ivić: *The Riemann Zeta-function.* Wiley (1985).
6 A. Ivić: The Mean Values of the Riemann Zeta-Function. *Tata Institute Lecture Notes* **82**. Springer (1987).
7 A. Ivić: Power moments of the Riemann zeta-function over short intervals. *Archiv. Math.* **62** (1994), 418–424.
8 A. Ivić and Y. Motohashi: On the fourth power moment of the Riemann zeta-function. *J. Number Theory* **51** (1995), 16–45.

9 A. Ivić and Y. Motohashi: The mean square of the error term for the fourth moment of the Riemann zeta-function. *Proc. London Math. Soc.* (3) **69** (1994), 309–329.

10 M. Jutila: *A Method in the Theory of Exponential Sums* (Tata Institute Lecture Notes 80). Springer (1987).

11 N.V. Kuznetsov: Convolution of the Fourier coefficients of the Eisenstein-Maass series (in Russian). *Zap. Naučn. Sem. LOMI AN SSSR* **129** (1983), 43–84.

12 N.V. Kuznetsov: Petersson hypothesis for forms of weight zero and Linnik's conjecture. Sums of Kloosterman sums. *Math. USSR Sbornik* **39** (1981), 299–342.

13 L. Lucht: Weighted relationship theorems and Ramanujan expansions. *Acta Arith.* **70** (1995), 25–42.

14 L. Lucht: Ramanujan expansions revisited. *Archiv. Math.* **64** (1995), 121–128.

15 Y. Motohashi: The fourth power mean of the Riemann zeta-function. In *Proc. Amalfi Conf. Analytic Number Theory*, 325–344 (E. Bombieri, A. Perelli, S. Salerno and U. Zannier, eds.). Univ. di Salerno (1992).

16 Y. Motohashi: An explicit formula for the fourth power mean of the Riemann zeta-function. *Acta. Math.* **170** (1993), 181–220.

17 Y. Motohashi: The binary additive divisor problem. *Ann. Scient. École Norm. Sup*, 4e Sér. **27** (1994), 529–572.

18 R.A. Smith: The generalized divisor problem over arithmetic progressions. *Math. Ann.* **260** (1982), 255–268.

19 A.I. Vinogradov: Poincaré series on $SL(3,\mathbb{R})$ (in Russian). *Zap. Naučn. Sem. LOMI AN SSSR* **160** (1987), 37–40.

20 A.I. Vinogradov: Analytic continuation of $\zeta_3(s, k)$ to the critical strip. Arithmetic part (in Russian). *Zap. Naučn. Sem. LOMI AN SSSR* **162** (1987), 43–76.

21 A.I. Vinogradov: The SL_n-technique and the density hypothesis (in Russian). *Zap. Naučn. Sem. LOMI AN SSSR* **168** (1988), 5–10.

22 A.I. Vinogradov and L.A. Takhtadžjan: The theory of Eisenstein series for the group $SL(3,\mathbb{R})$ and its application to a binary problem (in Russian). *Zap. Naučn. Sem. LOMI AN SSSR* **76** (1978), 5–53.

23 A.I. Vinogradov and L.A. Takhtadžjan: The zeta-function of an additive divisor problem and the spectral decomposition of the automorphic Laplacian (in Russian). *Zap. Naučn. Sem. LOMI AN SSSR* **134** (1984), 84–117.

13. A Variant of the Circle Method

M. Jutila

1. The δ-function

As an analogue of the Dirac δ-function, define (for the integer variable m) the arithmetic function

$$\delta(m) = \begin{cases} 1 & \text{for} \quad m = 0, \\ 0 & \text{for} \quad m \neq 0. \end{cases}$$

If χ is the characteristic function of an arbitrary interval of length 1, then obviously

$$\delta(m) = \int_{-\infty}^{\infty} \chi(\alpha)\, e(m\alpha)\, d\alpha.$$

Any decomposition of the function χ into a sum of integrable functions χ_i gives rise to a decomposition of the δ-function into a sum of certain functions δ_i. In particular, splitting up the support of χ into subintervals gives an example of such a decomposition. A special case of this construction is the *Farey dissection*. Let $a'/q' < a/q < a''/q''$ be three consecutive rationals (in their lowest terms) having denominators in the interval $[1, Q]$, where Q is the order of the dissection. Then the *Farey interval* corresponding to the Farey fraction a/q is

$$\mathcal{M}(a/q) = \left[\frac{a+a'}{q+q'}, \frac{a+a''}{q+q''} \right] = \left[\frac{a}{q} - \frac{1}{q(q+q')}, \frac{a}{q} + \frac{1}{q(q+q'')} \right].$$

The basic identity of the circle method is the following decomposition of the δ-function:

$$\delta(m) = \sum_{q=1}^{Q} \sum_{a=1}^{q}{}^{*} \int_{\mathcal{M}(a/q)} e(m\alpha)\, d\alpha, \tag{1.1}$$

where the asterisk indicates the coprimality condition $(a, q) = 1$.

The irregularity of the lengths of the Farey intervals $\mathcal{M}(a/q)$ causes the *levelling problem*: how to invert the order of the summation and integration in (1.1) (after having introduced a local variable near a/q)? A classical device to overcome this difficulty is the *Kloosterman refinement* ([16], [17]) of the circle method. Its underlying idea is that after a suitable finer subdivision of the Farey intervals, the sum over a may be taken under the integral

Sieve Methods, Exponential Sums, and their Applications in Number Theory
Greaves, G.R.H., Harman, G., Huxley, M.N., Eds. ©Cambridge University Press, 1996

sign, where certain cancellations may occur. In applications to Fourier coefficients of modular forms, Kloosterman was led to the exponential sums

$$S(m, n; q) = \sum_{a=1}^{q}{}^{*} e\left(\frac{ma + n\bar{a}}{q}\right), \quad \text{where} \quad a\bar{a} \equiv 1 \pmod{q}$$

bearing now his name. Further refinements or variants of the Kloosterman
method were given in the eighties by Hooley [9], [10] and Iwaniec [11].
Another recent modification of the circle method is a discrete variant due
to Duke, Friedlander and Iwaniec [2]; its basic formula is a representation
of the δ-function as a suitable linear combination of Ramanujan sums.

The saving in Kloosterman's method depends on the nontriviality of the
estimates of individual Kloosterman sums. At this point, there is presently
no room for further improvement. The next step would be to sum nontriv-
ially even over the denominators q, and the potential importance of such a
possibility was indeed pointed out by Hooley [8] in his address to ICM-83,
following an earlier suggestion by Linnik [18]. Nowadays this idea can be
substantiated by spectral methods allowing nontrivial summations over the
denominators of Kloosterman sums.

Turning to our variant of the circle method, we imitate the Farey dissection on choosing a family of short intervals related to a certain set of
rationals, but the construction deviates from the traditional one in several
aspects:

(1) The rationals a/q do not run over any full system of Farey fractions,
for q will be restricted to a certain interval $[Q, 2Q]$.

(2) The intervals in our system are chosen to be of the same length.

(3) The interval related to a/q is counted with a certain weight $w(q) \in [0, 1]$.

(4) The intervals may overlap; in fact, overlapping of high multiplicity
is an essential point allowing us to introduce statistical ideas into the circle
method.

An obvious advantage of the property (2) is removing the levelling problem; on the other hand, the price to be paid for this convenience is that the
resulting decomposition of the characteristic function of the unit interval will
be only approximate. The approximation error depends on the dispersion of
the multiplicity of the overlapping, or equivalently on the well-distribution
of the rationals having denominators of given order.

The general scheme of method will be presented in §2. Next, in §3, we
illustrate it by an application to a certain sum involving Fourier coefficients
of cusp forms, reproving an estimate of Goldfeld [5]. Finally, in §4, we
comment briefly on the scope of our method more generally.

2. Approximate decomposition of the characteristic function of the unit interval

Let us consider the following practical construction. A wall is to be built over the unit interval $[0,1]$ by small bricks. These are centered at rationals: to each rational number $a/q \in [0,1]$ with $(a,q) = 1$ and $q > 0$ we attach a brick of thickness $w(q)$ and length 2δ lying over the interval $[a/q-\delta, a/q+\delta]$. Here $\delta > 0$ is a parameter at our disposal, and $0 \le w(q) \le 1$. We suppose that $w(q) = 0$ if $q \notin [Q, 2Q]$, so the interval $[Q, 2Q]$ is the relevant range for q. (We place the bricks in some order, and if any of them eventually fails to fit nicely to its position, we break it into shorter pieces in order to avoid holes). Do we really get something like a wall of uniform height ? The answer will be yes, at least if the construction is looked at from a distance.

The volume of the brick attached to the rational a/q may be measured by the number $2\delta w(q)$, so the total volume of the wall is $\lambda = 2\delta L$, where

$$L = \sum_q w(q)\varphi(q).$$

Another interpretation of λ is the *expected height* of the wall at a random point.

The *exact height* of the wall at x equals

$$\lambda(x) = \sum_{|a/q-x| \le \delta} w(q).$$

The *dispersion* of the "random variable" $\lambda(x)$ is estimated in the next lemma.

Lemma 1. *Let $1/Q^2 \ll \delta \ll 1/Q$. Then for any fixed $\varepsilon > 0$ we have*

$$\int_0^1 \left(\lambda(x) - \lambda\right)^2 dx \ll \delta Q^{2+\varepsilon}.$$

In particular, if $L \gg Q^2$ (for instance, if $w(q) = 1$ for all $q \in [Q, 2Q]$), then

$$\int_0^1 \left(\lambda(x) - \lambda\right)^2 dx \ll \lambda Q^\varepsilon.$$

This follows from the main theorem in [13] if $w(q)$ is allowed to take values 0 or 1 only, and the generalization is immediate.

Next we restate the preceding lemma in terms of characteristic functions of intervals. Let $\chi(x)$ and $\chi_{a/q}(x)$ stand for the characteristic functions of the intervals $[0,1]$ and $[a/q - \delta, a/q + \delta]$, respectively, and define

$$\tilde{\chi}(x) = \frac{1}{\lambda}\sum_q w(q) \sum_{a=1}^{q}{}^* \chi_{a/q}(x) = \frac{\lambda(x)}{\lambda}. \tag{2.1}$$

Since $\lambda(x) \approx \lambda$, we have $\tilde{\chi}(x) \approx \chi(x)$. In the next corollary, the error of the latter approximation is estimated in the mean square sense.

Corollary. *If* $1/Q^2 \ll \delta \ll 1/Q$ *and* $L \gg Q^2$, *then*

$$\int_0^1 \left(\chi(x) - \tilde{\chi}(x) \right)^2 dx \ll \frac{Q^\varepsilon}{\delta Q^2}.$$

This follows from Lemma 1, except that the integrals over the intervals $[-\delta, 0]$ and $[1, 1+\delta]$ should be taken into account. Their contribution is $\ll \delta \ll 1/(\delta Q^2)$, because $\delta \ll 1/Q$.

We now apply the corollary in the usual context of the circle method. That is, given a uniformly convergent exponential series

$$F(\alpha) = \sum_{n=-\infty}^{\infty} a_n \, e(n\alpha),$$

its mth coefficient may be picked out by the familiar formula

$$a_m = \int_{-\infty}^{\infty} \chi(\alpha) F(\alpha) \, e(-m\alpha) \, d\alpha.$$

We rewrite this as

$$a_m = \int_{-\infty}^{\infty} \tilde{\chi}(\alpha) F(\alpha) \, e(-m\alpha) \, d\alpha + \int_{-\infty}^{\infty} \left(\chi(\alpha) - \tilde{\chi}(\alpha) \right) F(\alpha) \, e(-m\alpha) \, d\alpha.$$

Here the latter integral may be viewed as an error term. Writing the former integral explicitly by use of (2.1), we end up with the following expression for a_m, which in fact is the basic formula of our variant of the circle method.

Lemma 2. *Let* $F(\alpha)$ *be a uniformly convergent exponential series. Then, under the assumptions of the corollary, the* mth *coefficient of* $F(\alpha)$ *is*

$$a_m = \frac{1}{\lambda} \int_{-\delta}^{\delta} e(-m\eta) \sum_q w(q) \sum_{a=1}^{q} {}^* e\left(-\frac{ma}{q} \right) F\left(\frac{a}{q} + \eta \right) d\eta + e_m,$$

where

$$e_m \ll \|F\|_2 \delta^{-\frac{1}{2}} Q^{-1+\varepsilon},$$

and also

$$e_m \ll \|F\|_\infty \delta^{-\frac{1}{2}} Q^{-1+\varepsilon}.$$

Remark. Suppose that $F(\alpha)$ is an exponential sum of length N, or an exponential series which "looks" like such a finite sum (the convergence being rapid for $|n| > N$). Suppose further that

$$F(\alpha) \ll AN^{\frac{1}{2}+\varepsilon}$$

uniformly for all α; if the average order of $|a_n|^2$ is $\gg A^2$ then this bound is essentially best possible. An example of such a series, related to holomorphic cusp forms, will be encountered in the next section. Choose now

$$\delta \asymp 1/Q, \quad N^{\frac{1}{2}} \ll Q \ll N$$

(the symbol $\asymp$ meaning that both sides are of the same order of magnitude). Then

$$e_m \ll AN^{\frac{1}{2}+\varepsilon}/Q^{\frac{1}{2}}.$$

In particular, $e_m \ll AN^{\varepsilon}$ if $Q \asymp N$. If A here is something like the expected order of the $|a_n|$, then the error term e_m will be negligible as far as the estimation of a_m is concerned.

3. An application to cusp forms

The classical additive divisor problems are concerned with the sums

$$D(N;m) = \sum_{n \leq N} d(n)d(n+m), \qquad D(N) = \sum_{n < N} d(n)d(N-n).$$

For recent work on these problems, see e.g. [19], [14].

Consider analogous sums for cusp forms. Let

$$f(z) = \sum_{n=1}^{\infty} a(n)\,e(nz) \quad \text{for} \quad \text{Im } z > 0$$

be a holomorphic cusp form of weight k for the full modular group. Then, applying the Möbius transformation with the matrix

$$\begin{pmatrix} \bar{a} & (1 - a\bar{a})/q \\ -q & a \end{pmatrix}$$

in the modular group, we see that

$$f\left(\frac{a}{q} + \xi\right) = \frac{1}{(q\xi)^k} f\left(-\frac{\bar{a}}{q} - \frac{1}{q^2\xi}\right) \quad \text{for} \quad \text{Im } \xi > 0. \qquad (3.1)$$

The analogues of $D(N;m)$ and $D(N)$ are

$$A(N;m) = \sum_{n \leq N} \overline{a(n)}\, a(n+m), \qquad A(N) = \sum_{n < N} a(n)\, a(N-n).$$

By Deligne's theorem,

$$a(n) \ll n^{(k-1)/2+\varepsilon}.$$

Hence

$$A(N) \ll N^{k-\frac{1}{2}+\varepsilon},$$

since f^2 is a cusp form of weight $2k$, and the $A(n)$ are its Fourier coefficients.

As an illustration of our method, we consider the following weighted version of $A(N;m)$:

$$\tilde{A}(N;m) = \sum_{n=1}^{\infty} \overline{a(n)}\, a(n+m) \exp\left(-\frac{2\pi(2n+m)}{N}\right).$$

Let α be a constant such that for the Hecke eigenvalues $t_j(n)$ for the jth Maass wave forms we have

$$t_j(n) \ll n^{\alpha+\varepsilon}$$

for any $\varepsilon > 0$. The *Ramanujan-Petersson conjecture* for Maass wave forms asserts that $\alpha = 0$ is admissible; presently it is known that $\alpha \leq \frac{5}{28}$ (see [1]).

Theorem. *For $N \geq 1$ and $m \geq 1$, we have*

$$\tilde{A}(N;m) \ll N^{k-\frac{1}{2}+\varepsilon} m^{\alpha}, \tag{3.2}$$

where the implied constant depends only on ε and the form f.

This estimate is not new, in fact it was mentioned by Goldfeld [5] in his lecture at the Journées Arithmétiques Marseille-Luminy 1978 (without specifying the dependence of the estimate on m).

The proof starts from the obvious formula

$$\tilde{A}(N;m) = \int_0^1 \left| f\left(\alpha + \frac{i}{N}\right) \right|^2 e(-m\alpha)\, d\alpha.$$

Clearly, we may assume that $m \ll N \log N$, for otherwise the assertion is trivial.

Choose

$$Q = N, \quad \delta = 1/N, \quad \sum_q w(q) \gg N, \quad w \text{ smooth}.$$

It is an important classical fact, due to G. H. Hardy, that

$$f\left(\alpha + \frac{i}{N}\right) \ll N^{\frac{1}{2}k}$$

(with a saving $N^{1/2}$ over the "trivial" estimate). This also follows from the estimate [12]

$$S(x;\alpha) = \sum_{n \leq x} a(n)\,e(n\alpha) \ll x^{\frac{1}{2}k} \qquad (3.3)$$

by partial summation. (The classical estimate of J. R. Wilton for $S(x;\alpha)$ is weaker by a factor $\log x$). Now, by Lemma 2 and the subsequent remark, we have

$$\tilde{A}(N;m)$$

$$= \frac{1}{\lambda}\sum_q w(q)\sum_{a=1}^{q}{}^{*} e\left(-\frac{ma}{q}\right)\int_{-1/N}^{1/N}\left|f\left(\frac{a}{q}+\eta+\frac{i}{N}\right)\right|^2 e(-m\eta)\,d\eta$$

$$+ O(N^{k-\frac{1}{2}+\varepsilon}),$$

where $\lambda \asymp N$.

We apply here the transformation property (3.1), which gives

$$f\left(\frac{a}{q}+\eta+\frac{i}{N}\right) = \frac{1}{(q(\eta+i/N))^k}\,f\left(-\frac{\bar{a}}{q}-\frac{1}{q^2(\eta+i/N)}\right).$$

Thus

$$\tilde{A}(N;m)$$

$$= \frac{1}{\lambda}\sum_q w(q)\sum_{r,s=1}^{\infty}\overline{a(r)}\,a(s)S(-m,r-s;q)$$

$$\times \int_{-1/N}^{1/N}\frac{e(-m\eta)}{(q(\eta+\frac{i}{N}))^{2k}}\,e\left(\frac{(r-s)\eta}{q^2(\eta^2+\frac{1}{N^2})}\right)\exp\left(-\frac{2\pi N(r+s)}{q^2(1+\eta^2 N^2)}\right)d\eta$$

$$+ O(N^{k-\frac{1}{2}+\varepsilon}).$$

Putting $r - s = p$, there are now three cases to consider: $p = 0$, $p > 0$, and $p < 0$.

The contribution of $p = 0$ is $\ll N^{k-1}$.

Turning to positive values of p, fix η for a moment and apply summation by parts to the sum over s for given p and q. Then, summing also over p

and q, we see that the contribution of these terms can be reduced to sums of the type

$$\sum_{p \asymp P} \sum_q W(q,p) S(-m,p;q) a_p, \qquad (3.4)$$

where $1 \le P \ll N \log N$, W is a bounded smooth function and

$$a_p = \sum_{1 \le s \le x} a(s) \overline{a(s+p)} \quad \text{with} \quad x \ll N \log N.$$

The desired bound for the sum (3.4) is the estimate (3.2) multiplied by N^2, thus it is of the order $\ll N^{k+3/2+\epsilon} m^\alpha$. The case $p < 0$ is analogous.

The proof will be completed by appealing to the following two lemmas.

Lemma 3. *For $x, y \ge 1$, we have*

$$\sum_{0 \le p \le x} \left| \sum_{s \le y} a(s) \overline{a(s+p)} \right|^2 \ll (x+y)^k y^k.$$

Proof. Letting $P = [2x]$ and $\nu(p) = \max(0, 1 - |p|/P)$, consider the sum

$$\sum_p \nu(p) \left| \sum_{s \le y} a(s) \overline{a(s+p)} \right|^2$$
$$= \int_0^1 \left| \int_0^1 \overline{S(x+y;\beta)} S(y;\beta) \sum_p \sqrt{\nu(p)} \, e(p(\alpha+\beta)) \, d\beta \right|^2 d\alpha$$

with the convention that $a(s) = 0$ for $s \le 0$. Obviously it suffices to estimate this sum instead of the original one. Squaring out and integrating over α, we obtain

$$\int_0^1 \int_0^1 \overline{S(x+y;\beta_1)} S(y;\beta_1) S(x+y;\beta_2) \overline{S(y;\beta_2)} \Delta_P(\beta_1 - \beta_2) \, d\beta_1 \, d\beta_2,$$

where

$$\Delta_P(\alpha) = \sum_{p=-P}^P \nu(p) e(p\alpha) = \frac{1}{P} \left(\frac{\sin \pi P \alpha}{\sin \pi \alpha} \right)^2$$

is the Fejér kernel. It is non-negative and its average over the unit interval equals 1. Therefore, by (3.3), the preceding expression is $\ll (x+y)^k y^k$.

Lemma 4. *Let $m \geq 1$, $P > 0$, $Q > 0$, and let $g(x,y)$ be a function of class C^4 with support in $[P, 2P] \times [Q, 2Q]$ satisfying*

$$\frac{\partial^{i+j}}{\partial x^i \partial y^j} g(x,y) \ll \frac{1}{P^i Q^j} \quad \text{for} \quad 0 \leq i, j \leq 2.$$

Then, for any complex numbers a_p, we have

$$\sum_{P \leq p \leq 2P} \sum_{Q \leq q \leq 2Q} a_p g(p,q) S(m, \pm p; q)$$

$$\ll (m^{\frac{1}{2}} + Q) P^{\frac{1}{2}} \left(\sup_j |t_j(m)| \right) \left(\sum_p |a_p|^2 \right)^{\frac{1}{2}} (mPQ)^\varepsilon.$$

This is essentially due to Deshouillers and Iwaniec [3] except that they summed over m as well whilst we are keeping m fixed.

Lemmas 3 and 4 now imply an estimate of the required order for the sum (3.4).

4. Concluding remarks

To deal with the sum $A(N; m)$, the Fourier series of $F(a/q + \eta + yi)$ with $y \asymp 1/N$ has to be truncated, which can be done by Fourier analysis. The choice of the basic parameters Q and δ should be adapted to the new situation. Then $A(N; m)$ may be reduced to *oscillating* sums of Kloosterman sums. In the work of Deshouillers and Iwaniec [4] on $D(N; m)$, the situation was analogous. The rest of the argument also follows Deshouillers and Iwaniec in principle, the main tools being Kuznetsov's trace formula and Iwaniec's spectral large sieve. It seems that the result will be

$$A(N; m) \ll N^{k - \frac{1}{3} + \varepsilon} \quad \text{for} \quad m \ll N^{\frac{2}{3}}$$

(thus the saving over the trivial estimate is about $N^{1/3}$). The actual novelty here is the method; the result as such is not new. In earlier work on this problem by Good [6], [7] and the author [14], [15], the *Dirichlet series method* was used. However, the latter argument requires a deep result on inner products involving cusp forms and Maass wave forms, which we may dispense with.

The method may be modified so as to apply to non-holomorphic cusp forms as well. Moreover, the analogue of the sum $A(N)$ (in other words, the analogue of the "dual" additive divisor problem) for non-holomorphic cusp forms now becomes tractable; no non-trivial result on this problem seems to be known so far.

References

1 D. Bump, W. Duke, J. Hoffstein and H. Iwaniec: An estimate for the Hecke eigenvalues of Maass forms. *Int. Math. Res. Notes* **4** (1994), 75–81.
2 W. Duke, J. Friedlander, and H. Iwaniec: Bounds for automorphic *L*-functions. *Invent. Math.* **112** (1993), 1–8.
3 J.-M. Deshouillers and H. Iwaniec: Kloosterman sums and Fourier coefficients of cusp forms. *Invent. Math.* **70** (1982), 219–288.
4 J.-M. Deshouillers and H. Iwaniec: An additive divisor problem. *J. London Math. Soc.* (2) **26** (1982), 1–14.
5 D. Goldfeld: Analytic and arithmetic theory of Poincaré series. In *Journées Arithmétiques de Luminy*, 1978 (Asterisque 61) (1979), 95–107.
6 A. Good: Cusp forms and eigenfunctions of the Laplacian. *Math. Ann.* **255** (1981), 523–548.
7 A. Good: On various means involving the Fourier coefficients of cusp forms. *Math. Zeitschr.* **183** (1983), 95–129.
8 C. Hooley: Some recent advances in analytic number theory. In *Proc. ICM Warsaw* 1983, vol. I, 85-97.
9 C. Hooley: On Waring's problem. *Acta Math.* **157** (1986), 49–97.
10 C. Hooley: On nonary cubic forms. *J. Reine Angew. Math.* **386** (1988), 32–98.
11 H. Iwaniec: The circle method and the Fourier coefficients of modular forms. In *Number Theory and Related Topics* (Ramanujan Coll. Bombay 1988, Stud. Math. **12**), 47-55. Tata Institute of Fundamental Research (1989).
12 M. Jutila: On exponential sums involving the Ramanujan function. *Proc. Indian Math. Soc. (Math. Sci.)* **97** (1987), 157–166.
13 M. Jutila: Transformations of exponential sums. In *Proc. Amalfi Conf. on Analytic Number Theory*, 263–270 (E. Bombieri, A. Perelli, S. Salerno amd U. Zannier, eds.). Univ. di Salerno (1992).
14 M. Jutila: The additive divisor problem and its analogs for Fourier coefficients of cusp forms. I *Math. Zeitschr.*, to appear.
15 M. Jutila: The additive divisor problem and its analogs for Fourier coefficients of cusp forms. II *Math. Zeitschr.*, to appear.
16 H.D. Kloosterman: On the representation of numbers in the form $ax^2 + by^2 + cz^2 + dt^2$. *Acta Math.* **49** (1926), 407–464.
17 H.D. Kloosterman: Asymptotische Formeln für die Fourierkoeffizienten ganzer Modulformen. *Abhandlungen Math. Sem. Hamburg* **V** No. 4 (1927), 337–352.
18 Yu.V. Linnik: Additive problems and eigenvalues of the modular operators. In *Proc. ICM Stockholm* 1962, 270-284. Inst. Mittag-Leffler (1963).
19 Y. Motohashi: The binary additive divisor problem. *Ann. Sci. l'Ecole Norm. Sup.* **27** (1994), 529–572.

14. The Resemblance of the Behaviour of the Remainder Terms $E_\sigma(t)$, $\Delta_{1-2\sigma}(x)$ and $R(\sigma+it)$

Isao Kiuchi and Kohji Matsumoto

In this article we first survey known mean value results on the remainder terms given in the title, and then prove some new results on the power moments of $R(\sigma + it)$. Throughout this article, ϵ denotes an arbitrarily small positive number, not necessarily the same on each occurrence, and $a_1, a_2, \ldots$ denote certain absolute constants.

1. $\Delta(x)$ and $E(t)$

We first recall some facts, well-known to specialists, on the classical remainder terms $\Delta(x)$ and $E(t)$. Let $d(n)$ be the number of positive divisors of n, and define $\Delta(x)$ by

$$\sum_{n\leq x}{}' d(n) = x \log x + (2\gamma - 1)x + \tfrac{1}{4} + \Delta(x)$$

for $x \geq 2$, where γ is Euler's constant, and $\sum'$ indicates that the last term is to be halved if x is an integer. Dirichlet's original estimate was $\Delta(x) = O(x^{1/2})$, and Voronoi [48] improved it to show

$$\Delta(x) = O\big(x^{\frac{1}{3}} \log x\big). \tag{1.1}$$

Also, Voronoi [49] proved the explicit formula

$$\begin{aligned}
\Delta(x) &= -\frac{2}{\pi} x^{\frac{1}{2}} \sum_{n=1}^{\infty} \frac{d(n)}{n^{\frac{1}{2}}} \left\{ K_1\big(4\pi\sqrt{nx}\big) + \frac{\pi}{2} Y_1\big(4\pi\sqrt{nx}\big) \right\} \\
&= \frac{x^{\frac{1}{4}}}{\pi\sqrt{2}} \sum_{n=1}^{\infty} \frac{d(n)\cos\big(4\pi\sqrt{nx} - \frac{1}{4}\pi\big)}{n^{\frac{3}{4}}} + O\big(x^{-\frac{1}{4}}\big),
\end{aligned} \tag{1.2}$$

where K_1 and Y_1 are Bessel functions. Sometimes a truncated form is useful, which can be written as

$$\Delta(x) = \frac{x^{\frac{1}{4}}}{\pi\sqrt{2}} \sum_{n\leq N} \frac{d(n)\cos\big(4\pi\sqrt{nx} - \frac{1}{4}\pi\big)}{n^{\frac{3}{4}}} + O\big(x^\epsilon + x^{\frac{1}{2}+\epsilon}N^{-\frac{1}{2}}\big) \tag{1.3}$$

Sieve Methods, Exponential Sums, and their Applications in Number Theory
Greaves, G.R.H., Harman, G., Huxley, M.N., Eds. ©Cambridge University Press, 1996

for $1 \leq N \leq x^A$ with (any fixed) $A > 0$. It is also classically known that $\Delta(x) = \Omega(x^{1/4})$, and it is conjectured that $\Delta(x) = O(x^{1/4+\epsilon})$, but this conjecture is believed to be extremely difficult. Therefore it is natural to consider the mean values of $\Delta(x)$. Cramér [2] proved

$$\int_2^X \Delta^2(x)\,dx = \frac{1}{6\pi^2}\frac{\zeta^4(\frac{3}{2})}{\zeta(3)}X^{\frac{3}{2}} + \delta(X) \tag{1.4}$$

with $\delta(X) = O(X^{\frac{5}{4}+\epsilon})$, where $\zeta(s)$ is the Riemann zeta-function, and Tong [44] improved this to $\delta(X) = O(X\log^5 X)$. A simple proof of Tong's result was given by Meurman [35], while Preissmann [41] saved one log-factor in Tong's result.

It is well-known that there is an analogy between $\Delta(x)$ and $E(t)$; the latter is defined, for $t \geq 2$, by

$$\int_0^t \left|\zeta(\tfrac{1}{2}+iu)\right|^2 du = t\log t + (2\gamma - 1 - \log 2\pi)t + E(t). \tag{1.5}$$

An analogue of Voronoi's explicit formula for $E(t)$ was proved by Atkinson [1]. By using Atkinson's formula, Heath-Brown [3] proved, as the analogue of (1.4), that

$$\int_2^T E^2(t)\,dt = \frac{2}{3\sqrt{2\pi}}\frac{\zeta^4(\frac{3}{2})}{\zeta(3)}T^{\frac{3}{2}} + F(T) \tag{1.6}$$

with $F(T) = O(T^{5/4}\log^2 T)$. Meurman [35] and Motohashi [38], [39] (independently) improved this to $F(T) = O(T\log^5 T)$, and again one log-factor can be saved, as was noticed by the following four mathematicians (independently of each other) around 1987–91. Preissmann [42] and Ivić (Eq. (2.100) in [7]) used the same idea of applying the inequality of Montgomery and Vaughan as in Preissmann [41]. The methods of Motohashi and Tsang are related to the additive divisor problem. Motohashi's idea, using Shiu's theorem [43], is mentioned in the Notes at the end of Chapter 2 of Ivić [7]. Tsang's idea is based on a theorem of Heath-Brown [4] (see Tsang [46]).

As for the real order of $\delta(X)$ and $F(T)$, we should mention the recent work [23] of Lau and Tsang, in which they proved

$$\int_2^X \delta(x)\,dx = -\frac{1}{8\pi^2}X^2\log^2 X + a_1 X^2 \log X + O(X^2), \tag{1.7}$$

which obviously implies

$$\delta(X) = \Omega_-(X\log^2 X). \tag{1.8}$$

Lau and Tsang proposed the conjecture that

$$\delta(X) = -\frac{1}{4\pi^2} X \log^2 X + a_2 X \log X + O(X) \qquad (1.9)$$

(see also Tsang [46] and his excellent survey [47]). We note that, before this work of Lau and Tsang, Matsumoto [29] suggested the conjectures

$$\delta(X) \sim a_3 X \log^B X \qquad (1.10)$$
$$F(T) \sim a_4 T \log^B T \qquad (1.11)$$

with a certain positive constant B, which are of course much weaker than (1.9) but in the same direction.

On the fourth power mean of $\Delta(x)$, Tsang [45] proved

$$\int_2^X \Delta^4(x)\,dx = a_5 X^2 + O\big(X^{\frac{45}{23}+\epsilon}\big) \qquad (1.12)$$

with $a_5 > 0$. He also obtained an asymptotic formula for the third power mean. More general power moments were discussed by Ivić [6], and then Heath-Brown [5] proved the existence of the limit

$$\lim_{X\to\infty} \frac{1}{X^{1+\frac{1}{4}k}} \int_2^X |\Delta(x)|^k dx \qquad (1.13)$$

for any real number $k \in \big(0, \frac{28}{3}\big)$. In these works of Tsang, Ivić and Heath-Brown, similar results for $E(t)$ were also obtained.

2. $\Delta_{1-2\sigma}(x)$ and $E_\sigma(t)$

The remainder term $E_\sigma(t)$ in the mean-square formula of $\zeta(\sigma + it)$ for $\frac{1}{2} < \sigma < 1$, defined by

$$\int_0^t |\zeta(\sigma + iu)|^2\,du = \zeta(2\sigma)t + \frac{(2\pi)^{2\sigma-1}}{2-2\sigma}\,\zeta(2-2\sigma)\,t^{2-2\sigma} + E_\sigma(t), \qquad (2.1)$$

was first introduced by Matsumoto [28], in which the analogue of Atkinson's formula for $\frac{1}{2} < \sigma < \frac{3}{4}$ was proved, and some "critical" situation on the line $\sigma = \frac{3}{4}$ was pointed out. Then the function $E_\sigma(t)$ was studied extensively in subsequent works of Motohashi [40], Ivić [7], [9], Matsumoto and Meurman [33], [34], Laurinčikas [24], [25], [26], [27], Ivić and Matsumoto [12], Kiuchi [19], and Ivić and Kiuchi [11]. The upper bound

$$E_\sigma(t) = O\big(t^{1/(4\sigma+1)} \log^2 t\big)$$

was first proved for $\frac{1}{2} < \sigma < \frac{3}{4}$ by Matsumoto [28] as an application
of his Atkinson-type formula, and Motohashi [40] extended this result to
$\frac{1}{2} < \sigma < 1$. When Motohashi wrote this article the Atkinson-type formula
for $\frac{3}{4} \leq \sigma < 1$ was not known, hence Motohashi adopted the alternative
way of using weighted integrals. The analogue of Atkinson's formula for
$\frac{3}{4} \leq \sigma < 1$ was later given by Matsumoto and Meurman [34]. Motohashi's
article [40] is unpublished, but the contents of [40] are included in §2.7 of
Ivić's lecture note [7] . In that lecture note, combining Motohashi's argument with the method of exponent pairs, Ivić gave better upper bounds
for $E_\sigma(t)$ (Theorem 2.11 of [7]). There was an error in the proof of this
theorem, but it has been corrected in Ivić and Matsumoto [12], in which
the hitherto best upper-bound estimates have been given.

As for the mean square of $E_\sigma(t)$, the following formulas are now known:

$$\int_2^T E_\sigma^2(t)\,dt = \begin{cases} A_1(\sigma)T^{\frac{5}{2}-2\sigma} + O(T) & \text{if } \frac{1}{2} < \sigma < \frac{3}{4} \\ A_0 T \log T + O(T) & \text{if } \sigma = \frac{3}{4} \\ O(T) & \text{if } \frac{3}{4} < \sigma < 1, \end{cases} \qquad (2.2)$$

where

$$A_1(\sigma) = \frac{2}{5-4\sigma} (2\pi)^{2\sigma-\frac{3}{2}} \frac{\zeta^2(\frac{3}{2})}{\zeta(3)} \zeta(\tfrac{5}{2} - 2\sigma)\zeta(\tfrac{1}{2} + 2\sigma) \quad \text{for } \tfrac{1}{2} < \sigma < \tfrac{3}{4},$$

$$A_0 = \frac{\zeta^2(\frac{3}{2})\zeta(2)}{\zeta(3)}.$$

Matsumoto [28] first proved the formula for $\frac{1}{2} < \sigma < \frac{3}{4}$ with the weaker
error $O(T^{7/4-\sigma}\log T)$, and it was improved to $O(T)$ by Matsumoto and
Meurman [33]. The estimate $O(T)$ for $\frac{3}{4} < \sigma < 1$ was given by Matsumoto
and Meurman [34], in which the formula for $\sigma = \frac{3}{4}$ was also proved with
the slightly weaker error estimate $O(T\sqrt{\log T})$. Professor K.-M. Tsang
kindly informed the authors that his student K.-Y. Lam [22] has recently
improved this error term to $O(T)$. Note that the "critical" property of the
line $\sigma = \frac{3}{4}$ can be clearly seen from (2.2). Also, it is a direct consequence
of (2.2) that $E_\sigma(t) = \Omega(t^{3/4-\sigma})$ for $\frac{1}{2} < \sigma < \frac{3}{4}$. Ivić (see (3.40) in [7])
improved this to $\Omega_\pm(t^{3/4-\sigma})$ and further sharper Ω_+ and Ω_--results are
given by Matsumoto and Meurman [34] and by Ivić and Matsumoto [12]
respectively. These Ω-results suggest the conjecture

$$E_\sigma(t) = \begin{cases} O(t^{\frac{3}{4}-\sigma+\epsilon}) & \text{if } \frac{1}{2} < \sigma \leq \frac{3}{4} \\ O(t^\epsilon) & \text{if } \frac{3}{4} < \sigma < 1. \end{cases} \qquad (2.3)$$

For further discussions in this direction, see Ivić [9], Ivić and Matsumoto
[12], and Matsumoto's survey article [31].

As for the error terms $O(T)$ in (2.2), there is the following

Conjecture 1. *The error terms $O(T)$ in (2.2) could be replaced by*

$$A_2(\sigma)T + o(T)$$

with a certain constant $A_2(\sigma)$.

For $\frac{1}{2} < \sigma < \frac{3}{4}$ this conjecture was proposed by Matsumoto and Meurman [33] with $A_2(\sigma) = 4\pi^2\zeta^2(2\sigma - 1)$. For $\frac{3}{4} \leq \sigma < 1$, Matsumoto [30], [32] first mentioned this conjecture as a simple extension of the conjecture proposed in [33]. At least in the case $\frac{1}{2} < \sigma < \frac{3}{4}$ there are several reasons; a heuristic argument can be found in [31]. Another heuristic argument is as follows. Letting $\sigma \to \frac{1}{2} + 0$ in (2.1), we find that singularities appear from both of the explicit terms on the right-hand side. They cancel, and the result is the explicit terms in (1.5). In this process it is important that the orders of the two explicit terms in (2.1) are going to coincide when $\sigma \to \frac{1}{2} + 0$. Similarly we may guess that the appearance of a log-factor when $\sigma = \frac{3}{4}$ in (2.2) would be the result of such a cancellation process. This requires the existence of the term of order T^β in (2.2) for $\frac{1}{2} < \sigma < \frac{3}{4}$, $\beta \to 1$ when $\sigma \to \frac{3}{4} - 0$. Usually in such explicit terms the exponent β is non-increasing as a function of σ. However (2.2) implies $T^\beta = O(T)$, therefore we should conclude $\beta = 1$ for $\frac{1}{2} < \sigma < \frac{3}{4}$.

It is quite plausible that the analogue of (1.7) for $F(T)$ holds, and a conjecture similar to (1.9) could be formulated. If so, it would also be a strong support of our conjecture. In this direction it is an interesting problem to calculate

$$\int_2^T F_\sigma(t)\, dt \quad \text{for} \quad \tfrac{1}{2} < \sigma < \tfrac{3}{4}$$

by the method of Lau and Tsang, where

$$F_\sigma(t) = \int_2^t E_\sigma^2(u)\, du - A_1(\sigma)t^{\frac{5}{2} - 2\sigma}.$$

It should be noted, however, that Conjecture 1 (and also Conjectures 2 and 3 below) might be too optimistic. It is quite plausible that $F_\sigma(t) = \Omega(t)$ for $\frac{1}{2} < \sigma < \frac{3}{4}$, but at present we cannot exclude the possibility that $F_\sigma(t)$ might be oscillating.

Previously, several conjectures similar to Conjecture 1 have been mentioned by various authors (see (1.9), (1.10), (1.11), (3.3), (3.9), and Conjectures 2 and 3) but the origin of this type of conjecture is probably two papers written in 1991. One of them is Matsumoto and Meurman [33] as above, which was published three years later. Another is Kiuchi [17], written also in 1991, and (3.3) below is proposed in that article.

In §1 the analogy between $E(t)$ and $\Delta(x)$ was mentioned. Similarly, there is an analogy between $E_\sigma(t)$ and $\Delta_{1-2\sigma}(x)$; the latter is defined by

$$\sideset{}{'}\sum_{n \leq x} \sigma_{1-2\sigma}(n) = \zeta(2\sigma)x + \frac{\zeta(2 - 2\sigma)}{2 - 2\sigma}\, x^{2-2\sigma} - \tfrac{1}{2}\zeta(2\sigma - 1) + \Delta_{1-2\sigma}(x),$$

where

$$\sigma_{1-2\sigma}(n) = \sum_{d|n} d^{1-2\sigma}.$$

Improving Kiuchi's former result [15] , Meurman [36] has recently shown

$$\int_2^X \Delta_{1-2\sigma}^2(x)\,dx = \begin{cases} B_1(\sigma)X^{\frac{5}{2}-2\sigma} + O(X) & \text{if } \tfrac{1}{2} < \sigma < \tfrac{3}{4} \\ B_0 X \log X + O(X) & \text{if } \sigma = \tfrac{3}{4} \\ O(X) & \text{if } \tfrac{3}{4} < \sigma < 1, \end{cases} \tag{2.4}$$

which gives the complete analogue of (2.2). (Here, the values of the constants B_0 and $B_1(\sigma)$ can be explicitly written.) Therefore, as has been proposed in [32], the following conjecture can be formulated.

Conjecture 2. *The error terms $O(X)$ in (2.4) could be replaced by*

$$B_2(\sigma)X + o(X)$$

with a certain constant $B_2(\sigma)$.

3. The Mean Square of $R\left(\tfrac{1}{2} + it\right)$

The error term $R(s)$ in the approximate functional equation for $\zeta^2(s)$ is defined by

$$\zeta^2(s) = \sideset{}{'}\sum_{n \leq t/2\pi} \frac{d(n)}{n^s} + \chi^2(s) \sideset{}{'}\sum_{n \leq t/2\pi} \frac{d(n)}{n^{1-s}} + R(s),$$

where $\chi(s) = 2^s \pi^{s-1} \sin\left(\tfrac{1}{2}\pi s\right)\Gamma(1 - s)$. This is a classical object from the days of Hardy and Littlewood, and Titchmarsh, but it was Motohashi [37], [39] who discovered the remarkable relationship between $R(s)$ and $\Delta(x)$, that is

$$\chi(1 - s)\,R(s) = -\sqrt{2}\left(\frac{t}{2\pi}\right)^{-\frac{1}{2}}\Delta\left(\frac{t}{2\pi}\right) + O\left(t^{-\frac{1}{4}}\right). \tag{3.1}$$

We note that Jutila [14] gave another proof of (3.1).

In view of this relation, it is natural to consider the mean values of $R(s)$ as an analogue of the mean values of $\Delta(x)$. This direction of research was first cultivated by Kiuchi and Matsumoto [21], who proved

$$\int_2^T \left| R(\tfrac{1}{2} + it) \right|^2 dt = \sqrt{2\pi} C_0 T^{\frac{1}{2}} + K(T) \tag{3.2}$$

with $K(T) = O(T^{1/4} \log T)$, where

$$C_0 = \sum_{n=1}^{\infty} \frac{d^2(n)\, h^2(n)}{n^{\frac{1}{2}}}, \qquad h(n) = \sqrt{\frac{2}{\pi}} \int_0^\infty \frac{\cos(y + \pi/4)}{(y + n\pi)^{\frac{1}{2}}}\, dy.$$

Kiuchi [17] improved the estimate of $K(T)$ to $O(\log^5 T)$, and suggested the conjecture

$$K(T) \sim a_6 \log^3 T. \tag{3.3}$$

This observation of Kiuchi was one of the reasons why Matsumoto raised the conjectures (1.10), (1.11) for $\delta(X)$ and $F(T)$.

To prove the above mentioned estimates of $K(T)$, the relation (3.1) is not suitable; the error term $O(t^{-1/4})$ is too large. But Motohashi's result [39] is actually much more precise. He proved

$$\chi(1-s)\,R(s)$$
$$= -\sqrt{2}\left(\frac{t}{2\pi}\right)^{-\frac{1}{2}} \Delta\left(\frac{t}{2\pi}\right) - \frac{1}{\sqrt{2}}\left(1 - 2(\sigma - \tfrac{1}{2})i\right)\left(\frac{t}{2\pi}\right)^{-\frac{3}{2}} \Delta_1\left(\frac{t}{2\pi}\right)$$
$$+ \frac{1}{\pi\sqrt{2}}\left(\frac{t}{2\pi}\right)^{-\frac{1}{2}} \left(\tfrac{1}{6}\left(\log\frac{t}{2\pi} + 2\gamma + 6\right) - (\sigma - \tfrac{1}{2})\left(\log\frac{t}{2\pi} + 2\gamma\right)i\right)$$
$$+ \frac{1}{\sqrt{2\pi}}\left(\frac{t}{2\pi}\right)^{-\frac{1}{4}} \sum_{n=1}^{\infty} \frac{d(n)\, h_1(n) \sin\left(2\sqrt{2\pi t n} + \tfrac{1}{4}\pi\right)}{n^{\frac{1}{4}}}$$
$$- \frac{1}{32\pi\sqrt{2\pi}}\left(\frac{t}{2\pi}\right)^{-\frac{3}{4}} \sum_{n=1}^{\infty} \frac{d(n)\, h_1(n) \cos\left(2\sqrt{2\pi t n} + \tfrac{1}{4}\pi\right)}{n^{\frac{1}{4}}}$$
$$+ 3i(\sigma - \tfrac{1}{2})\frac{1}{2\sqrt{2\pi}}\left(\frac{t}{2\pi}\right)^{-\frac{3}{4}} \sum_{n=1}^{\infty} d(n)\, h_2(n) n^{\frac{1}{4}} \cos\left(2\sqrt{2\pi t n} + \tfrac{1}{4}\pi\right)$$
$$- 5\left(\frac{\pi}{2^7}\right)^{\frac{1}{2}}\left(\frac{t}{2\pi}\right)^{-\frac{3}{4}} \sum_{n=1}^{\infty} d(n)\, h_3(n) n^{\frac{5}{4}} \cos\left(2\sqrt{2\pi t n} + \tfrac{1}{4}\pi\right)$$
$$+ O\left(\frac{\log t}{t}\right), \tag{3.4}$$

where

$$\Delta_1(x) = \sum_{n \leq x}(x-n)d(n) - \tfrac{1}{2}x^2\left(\log x + 2\gamma - \tfrac{3}{2}\right) - \tfrac{1}{4}x = O(x^{\frac{3}{4}}),$$

$$h_j(n) = \int_0^\infty \frac{\cos\left(y + \tfrac{1}{4}(-1)^j\pi\right)}{(y+n\pi)^{j+\frac{1}{2}}}\,dy = O\left(\frac{1}{n^{j+\frac{1}{2}}}\right)$$

for $j = 1, 2, 3$. Also we note

$$h(n) = -\frac{1}{\pi\sqrt{n}} + \frac{1}{\sqrt{2\pi}}\,h_1(n) \sim -\frac{1}{\pi\sqrt{n}}. \tag{3.5}$$

Therefore, putting $s = \tfrac{1}{2} + it$ in (3.4) and using Voronoi's formula (1.2), we obtain

$$\chi\left(\tfrac{1}{2} - it\right)R\left(\tfrac{1}{2} + it\right) = \left(\frac{t}{2\pi}\right)^{-\frac{1}{4}}\sum_{n=1}^\infty \frac{d(n)\,h(n)\sin\left(2\sqrt{2\pi t n} + \tfrac{1}{4}\pi\right)}{n^{\frac{1}{4}}}$$

$$+\frac{1}{6\pi\sqrt{2}}\left(\frac{t}{2\pi}\right)^{-\frac{1}{2}}\left(\log\frac{t}{2\pi} + 2\gamma + 6\right) + O(t^{-\frac{3}{4}}), \quad (3.6)$$

which is the starting point of Kiuchi's proof of $K(T) = O\left(\log^5 T\right)$.

Let

$$g(t) = t^{\frac{1}{2}}\chi\left(\tfrac{1}{2} - it\right)R\left(\tfrac{1}{2} + it\right), \qquad G(t) = g(t) - \frac{1}{6\sqrt{\pi}}\left(\log\frac{t}{2\pi} + 2\gamma + 6\right).$$

Then (3.6) can be written as

$$G(t) = (2\pi t)^{\frac{1}{4}}\sum_{n=1}^\infty \frac{d(n)\,h(n)\cos\left(2\sqrt{2\pi t n} - \tfrac{1}{4}\pi\right)}{n^{\frac{1}{4}}} + O\left(\frac{1}{t^{\frac{1}{4}}}\right). \tag{3.7}$$

On the other hand, combining this with the truncated Voronoi formula (1.3), we can show that

$$G(t) = (2\pi t)^{\frac{1}{4}}\sum_{n \leq N} \frac{d(n)\,h(n)\cos\left(2\sqrt{2\pi t n} - \tfrac{1}{4}\pi\right)}{n^{\frac{1}{4}}} + O\left(t^\epsilon + t^{\frac{1}{2}+\epsilon}N^{-\frac{1}{2}}\right). \tag{3.8}$$

This was given by Kiuchi [17] (the idea first appeared in Kiuchi and Matsumoto [21]), where the error estimate was further refined by using Meurman's lemma from [35]. Noting (3.5), we can see that the analogy between (1.2) (resp. (1.3)) and (3.7) (resp. (3.8)) is indeed striking. This is the viewpoint emphasized by Ivić [8], in which he indicated how to construct a simple proof of $K(T) = O(\log^5 T)$ from this viewpoint. In the same paper, Ivić strengthened Kiuchi's conjecture (3.3) to propose the conjecture

$$K(T) = a_7 \log^3 T + a_8 \log^2 T + a_9 \log T + a_{10} + O(T^{-\frac{1}{4}+\epsilon}). \qquad (3.9)$$

Recently, inspired by Lau and Tsang's aforementioned work [23] on $\delta(x)$, Ivić [10] proved

$$\int_0^T K(t)\, dt = a_{11} T \log^3 T + a_{12} T \log^2 T + O(T \log T) \qquad (3.10)$$

with $a_{11} < 0$, which supports (3.3) and a part of (3.9). (In view of Lau and Tsang [23], it is perhaps safe to replace $a_{10} + O(T^{-1/4+\epsilon})$ in Ivić's conjecture (3.9) by $O(1)$.) From (3.10) it immediately follows that

$$K(T) = \Omega_-\left(\log^3 T\right), \qquad (3.11)$$

while Kiuchi [20] recently improved his former result to show

$$K(T) = O\left(\log^4 T\right). \qquad (3.12)$$

Therefore the difference between the O-result and the Ω-result on $K(T)$ is now just one log-factor!

4. The Mean Square of $R(\sigma + it)$

In the case $0 \le \sigma < \frac{1}{2}$, the hitherto sharpest mean square result is

$$\int_2^T |R(\sigma + it)|^2 dt = C_1(\sigma) T^{\frac{3}{2}-2\sigma} + O(T^{1-2\sigma} \log^4 T) \quad \text{when} \quad 0 \le \sigma < \frac{1}{2} \qquad (4.1)$$

with $C_1(\sigma) = (2\pi)^{2\sigma-\frac{1}{2}} C_0/(3 - 4\sigma)$, due to Kiuchi [18]. In the same paper Kiuchi proved

$$\int_2^T |R(\sigma + it)|^2 dt = \begin{cases} C_1(\sigma) T^{\frac{3}{2}-2\sigma} + O(1) & \text{if } \frac{1}{2} < \sigma < \frac{3}{4} \\ \pi C_0 \log T + O(1) & \text{if } \sigma = \frac{3}{4} \\ O(1) & \text{if } \frac{3}{4} < \sigma \le 1, \end{cases} \qquad (4.2)$$

which precisely corresponds to the formulas (2.2) and (2.4). However, in this case we can obtain deeper results. Matsumoto [32] proved that the error terms $O(1)$ in (4.2) can be replaced by

$$\begin{cases} C_2(\sigma) + O(T^{1-2\sigma} \log^4 T) & \text{if } \frac{1}{2} < \sigma \le \frac{3}{4} \\ C_2(\sigma) + C_1(\sigma) T^{\frac{3}{2}-2\sigma} + O(T^{\frac{1}{4}-\sigma}) & \text{if } \frac{3}{4} < \sigma \le 1, \end{cases}$$

with a certain constant $C_2(\sigma)$. This result in particular implies that the statement corresponding to Conjectures 1 and 2 in §2 is true for $R(\sigma + it)$. Therefore, here we again find the fact which may be regarded as a support of those conjectures. Kiuchi [20] gave a simple proof of Matsumoto's result for $\frac{1}{2} < \sigma \leq \frac{3}{4}$, and at the same time improved the error estimate $O(T^{1/4-\sigma})$ to $O(T^{1-2\sigma}\log^4 T)$ for $\frac{3}{4} < \sigma \leq 1$.

In view of his result proved in [32], Matsumoto suggested that the error term $o(T)$ in Conjecture 1 might be sharpened to $O(T^{2-2\sigma}\log^c T)$ for $\frac{1}{2} < \sigma < \frac{3}{4}$, with a certain $c \geq 0$. This suggestion first appeared in private correspondence, and was mentioned in the paper [11] of Ivić and Kiuchi. Meurman proposed the corresponding conjecture for $\Delta_{1-2\sigma}(x)$ in a private letter to the second-named author. Now, in view of Kiuchi [20], one might be able to extend those conjectures to $\frac{3}{4} \leq \sigma < 1$. That is,

Conjecture 3. *The error terms $O(T)$ in (2.2) (resp. $O(X)$ in (2.4)) could be replaced by $A_2(\sigma)T + O(T^{2-2\sigma}\log^c T)$ (resp. $B_2(\sigma)X + O(X^{2-2\sigma}\log^c X)$) for $\frac{1}{2} < \sigma < 1$.*

5. Power Moments for $R(\sigma + it)$

The general power moments of $R(\frac{1}{2} + it)$ were first discussed by Kiuchi [16]. Then Ivić [8] improved Kiuchi's result to show

$$\int_2^T \left|R(\tfrac{1}{2} + it)\right|^k dt = \begin{cases} D_1(k)T^{1-\frac{1}{4}k}(1+o(1)) \text{ as } T \to \infty & \text{if } 0 < k < 4 \\ D_0 \log T + D_3 + O(T^{-\frac{1}{23}+\epsilon}) & \text{if } k = 4 \\ D_2(k) + O(T^{1-\frac{1}{4}k}) & \text{if } 4 < k < \frac{28}{3} \\ D_2(k) + O(T^{-\frac{1}{11}(2k-4)+\epsilon}) & \text{if } k \geq \frac{28}{3}, \end{cases}$$
$$(5.1)$$

with certain constants D_0, $D_1(k)$, $D_2(k)$, D_3. (Here k is not necessarily an integer.) To prove (5.1), noticing the analogy between $\Delta(x)$ and $G(t)$ discussed in §3, Ivić applied the methods of Heath-Brown [5] and Tsang [45] which were originally developed for the study of $\Delta(x)$. (Recall (1.12) and (1.13).)

Now we consider the even power moments of $R(\sigma + it)$ for $0 \leq \sigma \leq 1$, and prove some new formulas. First we show, as an analogue of Heath-Brown's (1.13), that the limit

$$L(k) = \lim_{T \to \infty} \frac{1}{T^{1+\frac{1}{4}k}} \int_2^T \left|g(t)\right|^k dt \qquad (5.2)$$

exists for $k = 2, 4, 6, 8$. Heath-Brown's proof of (1.13) is based on the truncated Voronoi formula (1.3) and the estimate

$$\int_0^X \left|\Delta(x)\right|^K dx = O(X^{1+\frac{1}{4}K+\epsilon}) \qquad (5.3)$$

with $K = \frac{28}{3}$. The estimate (5.3) is essentially due to Ivić [6]; his value of K is $\frac{35}{4}$, but this is because he used a weaker estimate of $\Delta(x)$. Combining Ivić's argument with the new sharp estimate $\Delta(x) = O(x^{7/22+\epsilon})$ of Iwaniec and Mozzochi [13] we obtain (5.3). Using (3.1) we have

$$G(t) = -2\sqrt{\pi}\,\Delta\left(\frac{t}{2\pi}\right) + O(t^{\frac{1}{4}}). \tag{5.4}$$

Hence, and from (5.3), it immediately follows that

$$\int_2^T |G(t)|^K dt = O(T^{1+\frac{1}{4}K+\epsilon}) \tag{5.5}$$

for $K = \frac{28}{3}$. Using (3.8) and (5.5), we can show the existence of the limit

$$\lim_{T\to\infty} \frac{1}{T^{1+\frac{1}{4}k}} \int_2^T |G(t)|^k dt \tag{5.6}$$

for any $k \in (0, \frac{28}{3})$, by a method quite similar to Heath-Brown's proof of Theorem 2 of [5] . Since $g(t)$ is real for real t, $G(t)$ is also real. Therefore, if k is an even integer less than $\frac{28}{3}$, then

$$\int_2^T |g(t)|^k dt = \int_2^T |G(t)|^k dt$$
$$+ \sum_{l=0}^{k-1} \binom{k}{l} \int_2^T G(t)^l \left(\frac{1}{6\sqrt{\pi}}\left(\log\frac{t}{2\pi} + 2\gamma + 6\right)\right)^{k-l} dt,$$

and by using (5.6) and Hölder's inequality it is easy to see that the second term on the right-hand side is $O(T^{(k+3)/4}\log T)$. This implies the existence of the limit $L(k)$ for $k = 2, 4, 6, 8$. Note that Ivić [8] used, in his proof of (5.1) in the case $0 < k < 4$, the fact that $L(k)$ exists for $0 < k < 4$.

Later we will prove that $L(k) > 0$. Hence

$$M_1(\sigma, k) = (2\pi)^{k(\sigma-\frac{1}{2})} L(k) \frac{k+4}{k+4-4k\sigma}$$

is positive for $\sigma < \frac{1}{4} + \frac{1}{k}$ and negative for $\sigma > \frac{1}{4} + \frac{1}{k}$. Further, we put

$$M_0(k) = (2\pi)^{1-\frac{1}{4}k} L(k)\left(1 + \tfrac{1}{4}k\right),$$

which is of course positive. Using these notations, we can now state our results. The first result is

Theorem 1. *We have*

$$\int_2^T \left| R(\sigma + it) \right|^4 dt$$

$$= \begin{cases} M_1(\sigma, 4)T^{2-4\sigma} + O(T^{\frac{45}{23}-4\sigma+\epsilon}) & \text{if } 0 \le \sigma \le \frac{45}{92} \\ M_1(\sigma, 4)T^{2-4\sigma} + M_2(\sigma, 4) + O(T^{\frac{45}{23}-4\sigma+\epsilon}) & \text{if } \frac{45}{92} < \sigma < \frac{1}{2} \\ M_0(4)\log T + M_2(\frac{1}{2}, 4) + O(T^{-\frac{1}{23}+\epsilon}) & \text{if } \sigma = \frac{1}{2} \\ M_2(\sigma, 4) + M_1(\sigma, 4)T^{2-4\sigma} + O(T^{\frac{45}{23}-4\sigma+\epsilon}) & \text{if } \frac{1}{2} < \sigma \le 1, \end{cases}$$

$$(5.7)$$

with a certain constant $M_2(\sigma, 4)$.

Remark. The case $\sigma = \frac{1}{2}$ in (5.7) is included in Ivić's result (5.1). In particular $D_0 = M_0(4)$, though Ivić's expression of D_0 is different and more explicit.

Our second result is

Theorem 2. *For* $k = 6, 8$, *we have, when* $T \to \infty$,

$$\int_2^T \left| R(\sigma + it) \right|^k dt$$

$$= \begin{cases} M_1(\sigma, k)T^{1+\frac{1}{4}k-k\sigma}(1 + o(1)) & \text{if } 0 \le \sigma < \frac{1}{4} + \frac{1}{k} \\ M_0(k)\log T(1 + o(1)) & \text{if } \sigma = \frac{1}{4} + \frac{1}{k} \\ M_2(\sigma, k) + M_1(\sigma, k)T^{1+\frac{1}{4}k-k\sigma}(1 + o(1)) & \text{if } \frac{1}{4} + \frac{1}{k} < \sigma \le 1 \end{cases}$$

$$(5.8)$$

with a certain constant $M_2(\sigma, k)$.

Remark. For $k = 2, 4$, (5.8) also holds, but in these cases more precise results are known as mentioned above.

It is obvious that

$$M_2(\sigma, k) = \int_2^\infty \left| R(\sigma + it) \right|^k dt \quad \text{if } k = 4, 6, 8$$

for $\frac{1}{4} + \frac{1}{k} < \sigma \le 1$, but we cannot write down the value of $M_2(\sigma, 4)$ for $\sigma \le \frac{1}{2}$ in such a simple way.

From (4.2) and Theorems 1 and 2 we observe that the log-factor appears in the asymptotic formula when $\sigma = \frac{1}{4} + \frac{1}{k}$ for $k = 2, 4, 6, 8$. Moreover, the proof of Theorem 2 (given in the next section) shows that (5.8) would hold for any larger even integer k if we assumed the plausible statement that $L(k)$ exists for this k. However, this does not suggest the "critical" property of these lines. In fact, putting $s = \frac{1}{2} + it$ in (3.4) and subtracting it from the original form of (3.4), we see that

$$\chi(1 - s) R(s) = \chi(\tfrac{1}{2} - it) R(\tfrac{1}{2} + it) + F(s)$$

with $F(s) = O(t^{-1/2} \log t)$, hence

$$R(s) = \chi(s)\left(\chi(\tfrac{1}{2} - it) R(\tfrac{1}{2} + it) + F(s)\right). \tag{5.9}$$

From (3.2) we have $R(\frac{1}{2} + it) = \Omega(t^{-1/4})$, so we may conjecture that $R(\frac{1}{2} + it) = O(t^{-1/4+\epsilon})$. From Stirling's formula we have

$$\left|\chi(s)\right| = \left(\frac{t}{2\pi}\right)^{\frac{1}{2}-\sigma}\left(1 + O\left(\frac{1}{t}\right)\right). \tag{5.10}$$

Therefore, the "real order" of $\left|R(s)\right|^k$ would be $t^{k(1/4-\sigma)}$ up to the ϵ-factor, whose exponent is equal to -1 (hence the log-factor appears after integration) when $\sigma = \frac{1}{4} + \frac{1}{k}$. This argument suggests that the appearance of the log-factor at $\sigma = \frac{1}{4} + \frac{1}{k}$ for $k = 2, 4, 6, \ldots$ is caused by the factor $\chi(s)$ on the right-hand side of (5.9), on which the σ-dependence is concentrated (except for the "error part" $F(s)$).

6. Proof of Theorems 1 and 2

First we prove Theorem 2 for $k = 2, 4, 6, 8$. Kiuchi [20] discovered a simple argument to deduce sharp results on the mean square of $R(\sigma + it)$, and his starting point is (5.9). We use the same idea as in Kiuchi [20] to prove our theorems, hence we also begin with (5.9). From (3.4) we have

$$\mathrm{Re}\left(\chi(\tfrac{1}{2} - it) R(\tfrac{1}{2} + it)\right) = -\sqrt{2}\left(\frac{t}{2\pi}\right)^{-\frac{1}{2}} \Delta\left(\frac{t}{2\pi}\right) + O(t^{-\frac{1}{4}})$$

$$\mathrm{Im}\left(\chi(\tfrac{1}{2} - it) R(\tfrac{1}{2} + it)\right) = O\left(\frac{\log t}{t}\right).$$

Hence, and from (5.9), we have

$$\left|R(s)\right|^2 = \left|\chi(s)\right|^2\left(\left|R(\tfrac{1}{2} + it)\right|^2 + \left|F(s)\right|^2 + D(s)\right), \tag{6.1}$$

where

$$D(s) = 2\operatorname{Re}\left(\chi(\tfrac{1}{2} - it)R(\tfrac{1}{2} + it)\right)\operatorname{Re} F(s)$$

$$+ 2\operatorname{Im}\left(\chi(\tfrac{1}{2} - it)R(\tfrac{1}{2} + it)\right)\operatorname{Im} F(s) \tag{6.2}$$

$$= -2\sqrt{2}\left(\frac{t}{2\pi}\right)^{-\frac{1}{2}}\Delta\left(\frac{t}{2\pi}\right)\operatorname{Re} F(s) + O\left(\frac{\log t}{t^{\frac{3}{4}}}\right).$$

Put $J(s) = |F(s)|^2 + D(s)$, and write $k = 2m$ when $m = 1, 2, 3, 4$. Then from (6.1) we have

$$\int_{T_1}^{T_2} |R(\sigma + it)|^k\, dt = \sum_{j=0}^{m} \binom{m}{j} I_{j,k}(T_1, T_2), \tag{6.3}$$

where $T_1 \leq T_2 \leq 2T_1$ and

$$I_{j,k}(T_1, T_2) = \int_{T_1}^{T_2} |\chi(s)|^{2m} |R(\tfrac{1}{2} + it)|^{2(m-j)} J(s)^j\, dt.$$

We first consider $I_{0,k}(T_1, T_2)$. From (5.10) it follows that

$$|\chi(s)|^{2m} = \left(\frac{t}{2\pi}\right)^{m(1-2\sigma)} + H_m(t) \tag{6.4}$$

with $H_m(t) = O\left(t^{m(1-2\sigma)-1}\right)$. Also we have $|R(\tfrac{1}{2} + it)| = t^{-1/2}|g(t)|$. Therefore

$$I_{0,k}(T_1, T_2) = (2\pi)^{-m(1-2\sigma)}\int_{T_1}^{T_2} \frac{|g(t)|^{2m}}{t^{2m\sigma}}\, dt + \int_{T_1}^{T_2} \frac{H_m(t)|g(t)|^{2m}}{t^m}\, dt. \tag{6.5}$$

By integration by parts and (5.2) it follows that

$$\int_{T_1}^{T_2} \frac{|g(t)|^{2m}}{t^{2m\sigma}}\, dt$$

$$= \begin{cases} L(2m)\dfrac{m+2}{m+2-4m\sigma}t^{1+\frac{1}{2}m-2m\sigma}(1+o(1))\Big|_{T_1}^{T_2} & \text{if } \sigma \neq \dfrac{m+2}{4m} \\[2ex] L(2m)(1+2m\sigma\log t)(1+o(1))\Big|_{T_1}^{T_2} & \text{if } \sigma = \dfrac{m+2}{4m} \end{cases}$$

as $T_1 \to \infty$, and the second term on the right-hand side of (6.5) is absorbed into the above $o(1)$ terms. Therefore

$$I_{0,k}(T_1, T_2) = \begin{cases} M_1(\sigma, k)t^{1+\frac{1}{4}k-k\sigma}(1+o(1))\Big|_{T_1}^{T_2} & \text{if } \sigma \neq \tfrac{1}{4} + \tfrac{1}{k} \\[2ex] M_0(k)\log t\,(1+o(1))\Big|_{T_1}^{T_2} & \text{if } \sigma = \tfrac{1}{4} + \tfrac{1}{k}. \end{cases} \tag{6.6}$$

Next consider $I_{j,k}(T_1, T_2)$ for $j \geq 1$. From (3.1) and (6.2) we have

$$\left|R(\tfrac{1}{2} + it)\right|^{2(m-j)} J(s)^j \ll \left(t^{-(m-j)}\left|\Delta\left(\frac{t}{2\pi}\right)\right|^{2(m-j)} + t^{-\frac{1}{2}(m-j)}\right)$$

$$\times \left(t^{-j}\left|\Delta\left(\frac{t}{2\pi}\right)\right|^j \log^j t + t^{-\frac{3}{4}j}\log^j t\right).$$

Substituting this into the definition of $I_{j,k}(T_1, T_2)$ and using

$$\int_{T_1}^{T_2}\left|\Delta\left(\frac{t}{2\pi}\right)\right|^k dt = O(T^{1+\frac{1}{4}k}) \quad \text{if} \quad 0 < k < \tfrac{28}{3}$$

(see (1.13)), we have

$$I_{j,k}(T_1, T_2) = O(T_1^{1+\frac{1}{2}m-\frac{1}{4}j-2m\sigma}\log^j T_1) \quad \text{if} \quad 1 \leq j \leq m,$$

which is also absorbed into the $o(1)$ terms in (6.6). This completes the proof of Theorem 2 (for $k = 2, 4, 6, 8$).

Comparing the case $k = 4$ in Theorem 2 with Ivić's (5.1), we see that $D_0 = M_0(4)$, and so

$$\int_2^T \left|R(\tfrac{1}{2} + it)\right|^4 dt = M_0(4)\log T + D_3 + V(T) \tag{6.7}$$

with $V(T) = O(T^{-1/23+\epsilon})$. Following the method of Tsang [45], Ivić [8] gives the explicit value of D_0. His expression is

$$D_0 = \frac{3\pi}{4} \sum_{\substack{k,l,m,n=1 \\ \sqrt{k}+\sqrt{l}=\sqrt{m}+\sqrt{n}}}^{\infty} \frac{h(k)h(l)h(m)h(n)d(k)d(l)d(m)d(n)}{(klmn)^{\frac{1}{4}}}.$$

Since $h(n) < 0$ for any positive integer n (as can be easily verified by integration by parts), obviously $D_0 = M_0(4) > 0$, also $L(4) > 0$. Also, from (3.2) it follows that $M_1(\tfrac{1}{2}, 2) = \sqrt{2\pi}C_0 > 0$, hence $L(2) > 0$. Next we show that $L(6)$, $L(8)$ are also positive. In fact, by Schwarz's inequality we have

$$\frac{1}{T}\int_2^T \left|R(\tfrac{1}{4} + it)\right|^4 dt \leq \left(\frac{1}{T}\int_2^T \left|R(\tfrac{1}{4} + it)\right|^6 dt\right)^{\frac{1}{2}} \left(\frac{1}{T}\int_2^T \left|R(\tfrac{1}{4} + it)\right|^2 dt\right)^{\frac{1}{2}}.$$

When $T \to \infty$ the left-hand side tends to $M_1(\tfrac{1}{4}, 4) > 0$ by Theorem 2 (with $k = 4$), while the second factor on the right-hand side tends to $\sqrt{C_1(\tfrac{1}{4})} > 0$ by (4.1). Hence

$$\int_2^T \left|R(\tfrac{1}{4} + it)\right|^6 dt \gg T,$$

which implies $M_1\left(\frac{1}{4},6\right) > 0$, so $L(6) > 0$. Next, using

$$\frac{1}{T}\int_2^T \left|R(\tfrac{1}{4}+it)\right|^6 dt \leq \left(\frac{1}{T}\int_2^T \left|R(\tfrac{1}{4}+it)\right|^8 dt\right)^{\frac{1}{2}} \left(\frac{1}{T}\int_2^T \left|R(\tfrac{1}{4}+it)\right|^4 dt\right)^{\frac{1}{2}}$$

and discussing similarly, we obtain $L(8) > 0$.

Finally we prove Theorem 1. We begin with (6.3) with $k = 4$ (hence $m = 2$). From (6.7) we have

$$\left|R(\tfrac{1}{2}+it)\right|^4 = \frac{M_0(4)}{t} + V'(t), \tag{6.8}$$

therefore

$$\begin{aligned}
I_{0,4}(T_1,T_2) &= \int_{T_1}^{T_2} \left(\left(\frac{t}{2\pi}\right)^{2-4\sigma} + H_2(t)\right)\left(\frac{M_0(4)}{t} + V'(t)\right) dt \\
&= (2\pi)^{4\sigma-2} M_0(4) \int_{T_1}^{T_2} t^{1-4\sigma} dt + (2\pi)^{4\sigma-2}\int_{T_1}^{T_2} t^{2-4\sigma} V'(t)\, dt \\
&\quad + M_0(4)\int_{T_1}^{T_2} \frac{H_2(t)}{t}\, dt + \int_{T_1}^{T_2} H_2(t) V'(t)\, dt \\
&= Z_1(T_1,T_2) + Z_2(T_1,T_2) + Z_3(T_1,T_2) + Z_4(T_1,T_2),
\end{aligned}$$

say. Since $H_2(t) = O(t^{1-4\sigma})$, we have $Z_3(T_1,T_2) = O(T_1^{1-4\sigma})$. By using (6.8) and (5.1) (with $k = 4$) we have

$$Z_4(T_1,T_2) \ll \max_{T_1 \leq t \leq T_2} |H_2(t)| \int_{T_1}^{T_2}\left(\left|R(\tfrac{1}{2}+it)\right|^4 + \frac{1}{t}\right) dt \ll T_1^{1-4\sigma} \log T_1.$$

By integration by parts and the estimate $V(t) = O\left(t^{-\frac{1}{23}+\epsilon}\right)$, we have

$$Z_2(T_1,T_2) = O\left(T_1^{\frac{45}{23}-4\sigma+\epsilon}\right).$$

Hence if $\sigma \neq \frac{1}{2}$, we have

$$I_{0,4}(T_1,T_2) = O\left(T_1^{2-4\sigma}\right). \tag{6.9}$$

Also, we obtain

$$\begin{aligned}
&Z_2(2,T) + Z_3(2,T) + Z_4(2,T) \\
&= \begin{cases} O\left(T^{\frac{45}{23}-4\sigma+\epsilon}\right) & \text{if } \sigma \leq \frac{45}{92} \\ Z_2(2,\infty) + Z_3(2,\infty) + Z_4(2,\infty) + O\left(T^{\frac{45}{23}-4\sigma+\epsilon}\right) & \text{if } \sigma > \frac{45}{92}, \end{cases}
\end{aligned}$$

hence

$$
I_{0,4}(2,T)
$$

$$
= \begin{cases} \dfrac{(2\pi)^{4\sigma-2}}{2-4\sigma} M_0(4) T^{2-4\sigma} + O\big(T^{\frac{45}{23}-4\sigma+\epsilon}\big) & \text{if } \sigma \le \frac{45}{92} \\[3mm] M(\sigma) + \dfrac{(2\pi)^{4\sigma-2}}{2-4\sigma} M_0(4) T^{2-4\sigma} + O\big(T^{\frac{45}{23}-4\sigma+\epsilon}\big) & \text{if } \sigma > \frac{45}{92},\ \sigma \ne \frac{1}{2} \end{cases}
$$

$$(6.10)$$

with a certain constant $M(\sigma)$. Next, using (1.4) we can easily show that

$$
I_{2,4}(T_1,T_2) = O\big(T_1^{\frac{3}{2}-4\sigma} \log^2 T_1\big).
$$

By using this estimate, (6.9) and Schwarz's inequality we have

$$
I_{1,4}(T_1,T_2) = O\big(T_1^{\frac{7}{4}-4\sigma} \log T_1\big).
$$

Therefore the contributions of $I_{1,4}(2,T)$ and $I_{2,4}(2,T)$ are constant terms and errors which are smaller than the error term in (6.10). Hence we now arrive at the assertion of Theorem 1 for $\sigma \ne \frac{1}{2}$.

References

1 F.V. Atkinson: The mean-value of the Riemann zeta function. *Acta Math.* **81** (1949), 353–376.

2 H. Cramér: Über zwei Sätze von Herrn G.H. Hardy. *Math. Zeitschr.* **15** (1922), 200–210.

3 D.R. Heath-Brown: The mean value theorem for the Riemann zeta-function. *Mathematika* **25** (1978), 177–184.

4 D.R. Heath-Brown: The fourth power moment of the Riemann zeta-function. *Proc. London Math Soc.* (3) **38** (1979), 385–422.

5 D.R. Heath-Brown: The distribution and moments of the error term in the Dirichlet divisor problem. *Acta Arith.* **60** (1992), 389–415.

6 A. Ivić: Large values of the error term in the divisor problem. *Invent. Math.* **71** (1983), 513–520.

7 A. Ivić: *Lectures on Mean Values of the Riemann Zeta Zunction* (Lectures on Math. and Phys. 82). Tata Institute/Springer (1991).

8 A. Ivić: Power moments of the error term in the approximate functional equation for $\zeta^2(s)$. *Acta Arith.* **65** (1993), 137–145.

9 A. Ivić: Some problems on mean values of the Riemann zeta-function. *J. de Théorie des Nombres de Bordeaux* **8** (1996), 101–123.

10 A. Ivić: On the mean square for the error term in the approximate functional equation for $\zeta^2(s)$ (preprint).

11 A. Ivić and I. Kiuchi: On some integrals involving the Riemann zeta-function in the critical strip. *Univ. Beograd. Publ. Elektrotehn. Fak. Ser. Mat.* **5** (1994), 19–28.

12 A. Ivić and K. Matsumoto: On the error term in the mean square formula for the Riemann zeta-function in the critical strip. *Monatsh. Math.* **121** (1996), 213–229.

13 H. Iwaniec and C.J. Mozzochi: On the divisor and circle problems. *J. Number Theory* **29** (1988), 60–93.

14 M. Jutila: On the approximate functional equation for $\zeta^2(s)$ and other Dirichlet series. *Quart. J. Math. Oxford* (2) **37** (1986), 193–209.

15 I. Kiuchi: On an exponential sum involving the arithmetic function $\sigma_a(n)$. *Math. J. Okayama Univ.* **29** (1987), 193–205.

16 I. Kiuchi: Power moments of the error term for the approximate functional equation of the Riemann zeta-function. *Publ. Inst. Math. Beograd* **52** (66) (1992), 10–12.

17 I. Kiuchi: An improvement on the mean value formula for the approximate functional equation of the square of the Riemann zeta-function. *J. Number Theory* **45** (1993), 312–319.

18 I. Kiuchi: The mean value formula for the approximate functional equation of $\zeta^2(s)$ in the critical strip. *Arch. Math.* **64** (1995), 316–322.

19 I. Kiuchi: An integral involving the error term of the mean square of the Riemann zeta-function in the critical strip. *Math. J. Okayama Univ.* **36** (1994), 45–49.

20 I. Kiuchi: The mean value formula for the approximate functional equation of $\zeta^2(s)$ in the critical strip II. *Arch. Math.*, to appear.

21 I. Kiuchi and K. Matsumoto: Mean value results for the approximate functional equation of the square of the Riemann zeta-function. *Acta Arith.* **61** (1992), 337–345.

22 K.-Y. Lam: (in preparation).

23 Y.-K. Lau and K.-M. Tsang: Mean square of the remainder term in the Dirichlet divisor problem. *J. de Théorie des Nombres de Bordeaux* **7** (1995), 75–92.

24 A. Laurinčikas: The Atkinson formula near the critical line. In *New Trends in Probability and Statistics, Vol.* 2: *Analytic and Probabilistic Methods in Number Theory* (E. Manstavičius and F. Schweiger, eds.), 335–354. VSP/TEV (1992).

25 A. Laurinčikas: The Atkinson formula near the critical line II. *Liet. Mat. Rinkinys* **33** (1993), 302–313 (in Russian) = *Lithuanian Math. J.* **33** (1993), 234–242.

26 A. Laurinčikas: On the moments of the Riemann zeta-function near the critical line. *Liet. Mat. Rinkinys* **35** (1995), 332–359 (in Russian) = *Lithuanian Math. J.* **35** (1995), 262–283.

27 A. Laurinčikas: A uniform estimate of the error term in the mean square of the Riemann zeta-function. *Liet. Mat. Rinkinys* **35** (1995), 508–517 (in Russian) = *Lithuanian Math. J.* **35** (1995), 403–410.

28 K. Matsumoto: The mean square of the Riemann zeta-function in the critical strip. *Japanese J. Math.* **15** (1989), 1–13.

29 K. Matsumoto: Mean values of error terms in the theory of the Riemann zeta-function. In *Proc. Sympos. Analytic Number Theory and Related Topics*, 101–110 (in Japanese). Gakushuin Univ., Tokyo (1992).

30 K. Matsumoto: The mean square of the Riemann zeta-function in the strip $1/2 < \sigma < 1$. *Sûrikaiseki Kenkyûsho Kôkyûroku* **837** (1993), 150–163 (in Japanese).

31 K. Matsumoto: On the function $E_\sigma(T)$. *Sûrikaiseki Kenkyûsho Kôkyûroku* **886** (1994), 10–28.

32 K. Matsumoto: On the bounded term in the mean square formula for the approximate functional equation of $\zeta^2(s)$. *Arch. Math.* **64** (1995), 323–332.

33 K. Matsumoto and T. Meurman: The mean square of the Riemann zeta-function in the critical strip II. *Acta Arith.* **68** (1994), 369–382.

34 K. Matsumoto and T. Meurman: The mean square of the Riemann zeta-function in the critical strip III. *Acta Arith.* **64** (1993), 357–382.

35 T. Meurman: On the mean square of the Riemann zeta-function. *Quart. J. Math. Oxford* (2) **38** (1987), 337–343.

36 T. Meurman: The mean square of the error term in a generalization of Dirichlet's divisor problem. *Acta Arith.* **74** (1996), 351–364.

37 Y. Motohashi: A note on the approximate functional equation for $\zeta^2(s)$. *Proc. Japan Acad.* **59A** (1983), 393–396; II *ibid.* 469–472.

38 Y. Motohashi: A note on the mean value of the zeta and *L*-functions IV. *Proc. Japan Acad.* **62A** (1986), 311–313.
39 Y. Motohashi: *Lectures on the Riemann-Siegel Formula.* Ulam Seminar. Dept. of Math., Univ. of Colorado, Boulder (1987) (see also the present volume, 293–324).
40 Y. Motohashi: The mean square of $\zeta(s)$ off the critical line (unpublished manuscript).
41 E. Preissmann: Sur la moyenne quadratique du terme de reste du problème du cercle. *C. R. Acad. Sci. Paris Sér. I. Math.* **306** (1988), 151–154.
42 E. Preissmann: Sur la moyenne de la fonction zêta. In *Analytic Number Theory and Related Topics* (K. Nagasaka, ed.), 119–125. World Scientific (1993).
43 P. Shiu: A Brun-Titchmarsh theorem for multiplicative functions. *J. Reine Angew. Math.* **313** (1980), 161–170.
44 K.-C. Tong: On divisor problems III. *Acta Math. Sinica* **6** (1956), 515–541 (in Chinese).
45 K.-M. Tsang: Higher power moments of $\Delta(x)$, $E(t)$ and $P(x)$. *Proc. London Math. Soc.* (3) **65** (1992), 65–84.
46 K.-M. Tsang: Mean square of the remainder term in the Dirichlet divisor problem II. *Acta Arith.* **71** (1995), 279–299.
47 K.-M. Tsang: The remainder term in the Dirichlet divisor problem. *Sûrikaiseki Kenkyûsho Kôkyûroku* (to appear).
48 G.F. Voronoi: Sur un problème du calcul des fonctions asymptotiques. *J. Reine Angew. Math.* **126** (1903), 241–282.
49 G.F. Voronoi: Sur une fonction transcendante et ses applications à la sommation de quelques séries. *Ann. Sci. École Norm. Sup.* (3) **21** (1904), 207–268; *ibid.* 459–534.

Note added in proof. The upper bound estimates of $E_\sigma(t)$ given by [12] (see §2) were further improved by A. Kačėnas: The asymptotic behaviour of the second power moment of the Riemann zeta-function in the critical strip. *Liet. Mat. Rinkinys* **35** (1995), 315–331 (in Russian) = Lithuanian Math. J. **35** (1995), 249–261.

15. A Note on the Number of Divisors of Quadratic Polynomials

James McKee

1. Introduction and statement of the theorem

Let a, b and c be integers with $b^2 - 4ac = \Delta$ not a square. Let $d(n)$ denote the number of positive divisors of n. Consider

$$\lambda = \lim_{x \to \infty} \left(\sum_{n \leq x} d(an^2 + bn + c) \middle/ \sum_{n \leq x} d(n^2 + bn + ac) \right).$$

Let $\rho(n)$ be the number of solutions to the quadratic congruence

$$ax^2 + bx + c \equiv 0 \ (\mathrm{mod}\, n) : 0 \leq x < n,$$

and let $\rho_a(n)$ be the number of solutions to the congruence

$$x^2 + bx + ac \equiv 0 \ (\mathrm{mod}\, n) : 0 \leq x < n.$$

Both ρ and ρ_a are multiplicative functions, but not totally multiplicative. If p does not divide a, then $\rho(p^r) = \rho_a(p^r)$ for all r. One can show (adapting arguments in [2]) that

$$\lambda = \lim_{x \to \infty} \left(\sum_{n \leq x} \frac{\rho(n)}{n} \middle/ \sum_{n \leq x} \frac{\rho_a(n)}{n} \right).$$

Using techniques from §§4,7 of [1] and §4 of [2] we can replace this ratio of divergent sums by the ratio of the corresponding divergent Euler products, giving

$$\lambda = \lim_{x \to \infty} \left(\prod_{p \leq x} \left(1 + \frac{\rho(p)}{p} + \cdots \right) \middle/ \prod_{p \leq x} \left(1 + \frac{\rho_a(p)}{p} + \cdots \right) \right)$$

$$= \prod_{p \mid a} \left(\left(1 + \frac{\rho(p)}{p} + \frac{\rho(p^2)}{p^2} + \cdots \right) \middle/ \left(1 + \frac{\rho_a(p)}{p} + \frac{\rho_a(p^2)}{p^2} + \cdots \right) \right).$$

Sieve Methods, Exponential Sums, and their Applications in Number Theory
Greaves, G.R.H., Harman, G., Huxley, M.N., Eds. ©Cambridge University Press, 1996

This gives a convenient way of deducing the asymptotic behaviour of

$$\sum_{n \leq x} d(an^2 + bn + c)$$

from the special case when $a = 1$. If $\gcd(a, \Delta) = 1$ (or equivalently if $\gcd(a, b) = 1$), then this takes a particularly simple form, namely

$$\lambda = \prod_{p \mid a} \left(1 - \frac{1}{p+1}\right),$$

since in this case $\rho(p^r) = 1$ and $\rho_a(p^r) = 2$ whenever p divides a.

In [3], a compact expression was derived for the main term in

$$\sum_{n \leq x} d(n^2 + bn + c),$$

for the case $b^2 - 4c < 0$. Together with the argument sketched above, this can be extended to give the main term in the following theorem, although more careful estimates would be required to justify the error term. The purpose of these remarks is to make the theorem seem plausible before embarking on a detailed proof by other means.

Theorem. *Let a, b and c be integers with*

$$\Delta = b^2 - 4ac < 0, \quad \gcd(a, \Delta) = 1.$$

Then

$$\sum_{n \leq x} d(an^2 + bn + c) = \frac{12\,H(\Delta)}{\pi\sqrt{-\Delta}} \prod_{p \mid a}\left(1 - \frac{1}{p+1}\right) x \log x + O(x),$$

where the implied constant in the $O(x)$ term depends on a, b, c. Here $H(\Delta)$ is the weighted class number which counts all reduced binary quadratic forms $Ax^2 + Bxy + Cy^2$ with $B^2 - 4AC = \Delta$ (both primitive and imprimitive forms), but counts forms proportional to $x^2 + y^2$ with weight $\frac{1}{2}$, and forms proportional to $x^2 + xy + y^2$ with weight $\frac{1}{3}$.

In the next section we take a different approach, extending the method in [3] to prove the above theorem, including a justification of the error term.

2. Proof of the theorem

The proof proceeds by a series of lemmata. Throughout we assume

$$b^2 - 4ac = \Delta < 0, \quad \gcd(a, \Delta) = 1,$$

although the full strength of these restrictions is needed only in Lemma 2. Implied constants in $O(\cdot)$ estimates may depend on a, b and c.

Let $\omega(n)$ be the number of distinct prime divisors of n. Let (A, B, C) denote the binary quadratic form $Ax^2 + Bxy + Cy^2$. Given $B^2 - 4AC < 0$, let $\varepsilon(A, B, C)$ be the reciprocal of the number of automorphs of (A, B, C). Thus $\varepsilon(A, B, C)$ is usually $\frac{1}{2}$, but is occasionally $\frac{1}{4}$ or $\frac{1}{6}$. We note that the weighted class number $H(\Delta)$ defined above can be expressed as

$$H(\Delta) = 2 \sum_{\substack{(A,B,C) \text{ reduced} \\ B^2 - 4AC = \Delta}} \varepsilon(A, B, C).$$

Lemma 1.

$$\sum_{n \leq x} d(an^2 + bn + c) = 2x \sum_{n \leq \sqrt{ax}} \frac{\rho(n)}{n} + O\left(\sum_{n \leq \sqrt{ax}} \rho(n) \right) + O(x).$$

Proof. This is standard, resting on the ancient observation

$$d(n) = 2 \sum_{\substack{d|n \\ d \leq \sqrt{n}}} 1 + O(1).$$

Lemma 2.

$$\rho(n) = \frac{1}{2^{\omega(a)}} \sum_{\substack{(A,B,C) \text{ reduced} \\ B^2 - 4AC = \Delta}} \varepsilon(A, B, C) \sum_{\substack{Ar^2 + Brs + Cs^2 = an \\ \gcd(r,s) = 1}} 1,$$

where, in the inner sum, r and s can be any integers (positive, negative or zero).

Proof. Multiplying the defining congruence for $\rho(n)$ by $4a$ gives (with $|\cdot|$ denoting the cardinality of a set)

$$\rho(n) = \left| \{ x : (2ax + b)^2 \equiv \Delta \pmod{4an}, \ 0 \leq x < n \} \right|$$

$$= \left| \{ z : z^2 \equiv \Delta \pmod{4an}, \ b \leq z < 2an + b, \ z \equiv b \pmod{2a} \} \right|$$

$$= \tfrac{1}{2} \left| \{ z : z^2 \equiv \Delta \pmod{4an}, \ 0 \leq z < 4an, \ z \equiv b \pmod{2a} \} \right|.$$

Now, since $\gcd(a, b) = 1$, each solution to $z^2 \equiv b^2 \pmod{2a}$ lifts to the same number of solutions to $z^2 \equiv \Delta \pmod{4an}$, unless $4 \mid a$, when exactly half the solutions to $z^2 \equiv b^2 \pmod{2a}$ lift, but of those which do, all lift to the same number of solutions to $z^2 \equiv \Delta \pmod{4an}$. Thus

$$\rho(n) = \frac{1}{2^{\omega(a)+1}} \left| \{z : z^2 \equiv \Delta \pmod{4an}, \ 0 \leq z < 4an\} \right|$$

$$= \frac{1}{2^{\omega(a)}} \left| \{z : z^2 \equiv \Delta \pmod{4an}, \ 0 \leq z < 2an\} \right|.$$

Now the total number of proper representations of an by all reduced forms (A, B, C) with $B^2 - 4AC = \Delta$, counting representations by (A, B, C) with weight $\varepsilon(A, B, C)$, is

$$\left| \{z : z^2 \equiv \Delta \pmod{4an}, \ 0 \leq z < 2an\} \right|$$

(see [3]). This completes Lemma 2.

Lemma 3.

$$\sum_{n \leq y} \rho(n)$$

$$= \frac{1}{2^{\omega(a)}} \sum_{\substack{(A,B,C) \text{ reduced} \\ B^2 - 4AC = \Delta}} \varepsilon(A, B, C) \sum_{f \leq \sqrt{ay/A}} \mu(f) \sum_{\substack{1 \leq At^2 + Btu + cu^2 \leq ay/f^2 \\ f^2(At^2 + Btu + Cu^2) \equiv 0 \pmod{a}}} 1,$$

where t and u can be any integers.

Proof. Use the expression for $\rho(n)$ given in Lemma 2, and a simple sieve to deal with the restriction $\gcd(r, s) = 1$.

Lemma 4. *Let $N(f, a, A, B, C)$ be the number of solutions to the congruence $f^2(At^2 + Btu + Cu^2) \equiv 0 \pmod{a} : 0 \leq t, u < a$. Then*

$$\sum_{\substack{1 \leq At^2 + Btu + cu^2 \leq ay/f^2 \\ f^2(At^2 + Btu + Cu^2) \equiv 0 \pmod{a}}} 1 = \frac{2\pi y}{a\sqrt{-\Delta}} \frac{N(f, a, A, B, C)}{f^2} + O\left(\frac{\sqrt{y}}{f}\right).$$

Proof. Allowing t and u to assume real values, the ellipse

$$At^2 + Btu + Cu^2 \leq \frac{ay}{f^2}$$

has area $2\pi ay/f^2 \sqrt{-\Delta}$ and circumference $O(\sqrt{y}/f)$. We wish to count the number of points within this ellipse with integer coordinates (t, u) such that

$$f^2(At^2 + Btu + Cu^2) \equiv 0 \pmod{a},$$

excluding the point $(0, 0)$. In any $a \times a$ block of points where t and u each run through a consecutive integers, exactly $N(f, a, A, B, C)$ points are counted.

Lemma 5. *With $N(f, a, A, B, C)$ as in Lemma 4, we have*

$$N(f, a, A, B, C) = \left(\gcd(f^2, a)\right)^2 \prod_{p^r \| a/\gcd(f^2, a)} p^r \left(r + 1 - \frac{r}{p}\right).$$

Proof. Plainly

$$N(f, a, A, B, C) = \left(\gcd(f^2, a)\right)^2 \cdot N\left(1, a/\gcd(f^2, a), A, B, C\right).$$

It remains to show that for e dividing a,

$$N(1, e, A, B, C) = \prod_{p^r \| e} p^r \left(r + 1 - \frac{r}{p}\right).$$

Since $\gcd(a, \Delta) = 1$, we have $\gcd(e, \Delta) = 1$, and an observation of Gauss is that (A, B, C) must then represent integers prime to e. Since the expression $N(1, e, A, B, C)$ is unchanged if we replace (A, B, C) by an equivalent form, we may suppose $\gcd(e, A) = 1$. Now $N(1, e, A, B, C)$ is the number of solutions to the congruence $(2Ax + By)^2 \equiv \Delta y^2 \pmod{4e}$, where the parity of B is determined by Δ and we have supposed $\gcd(A, e) = 1$. Hence $N(1, e, A, B, C)$ is independent of (A, B, C), subject to $B^2 - 4AC = \Delta$, and we may replace (A, B, C) by the principal form. It is then easy to check the stated formula for $N(1, e, A, B, C)$.

Lemma 6. *Let*

$$\Sigma_1 = \sum_{f \le \sqrt{ay/A}} \mu(f) \sum_{\substack{1 \le At^2 + Btu + Cu^2 \le ay/f^2 \\ f^2(At^2 + Btu + Cu^2) \equiv 0 \, (\text{mod } a)}} 1.$$

Then

$$\Sigma_1 = \frac{12 \cdot 2^{\omega(a)}}{\pi \sqrt{-\Delta}} \prod_{p | a} \left(1 - \frac{1}{p+1}\right) y + O(\sqrt{y} \log y).$$

Proof. From Lemma 4 and Lemma 5, we have

$$\Sigma_1 = \sum_{f \le \sqrt{ay/A}} \mu(f) \left\{ \frac{2\pi y}{a\sqrt{-\Delta}} \frac{\left(\gcd(f^2, a)\right)^2}{f^2} \right.$$

$$\left. \times \prod_{p^r \| a/\gcd(f^2, a)} p^r \left(r + 1 - \frac{r}{p}\right) + O\left(\frac{\sqrt{y}}{f}\right) \right\}$$

$$= \frac{2\pi y}{\sqrt{-\Delta}} \prod_{p^r \| a} \left(r + 1 - \frac{r}{p}\right) \sum_{f \le \sqrt{ay/A}} \mu(f) F(f) + O(\sqrt{y} \log y),$$

where

$$F(f) = \frac{(\gcd(f^2,a))^2}{f^2} \prod_{p^r \| a/\gcd(f^2,a)} p^r \left(r+1-\frac{r}{p}\right) \Big/ \prod_{p^r \| a} p^r \left(r+1-\frac{r}{p}\right)$$

is a multiplicative function of f. Also,

$$\left| \sum_{f>\sqrt{ay/A}} \mu(f)F(f) \right| \le \sum_{f>\sqrt{ay/A}} \frac{a^2}{f^2} = O\left(\frac{1}{\sqrt{y}}\right).$$

Hence, with $a = \prod_{p|a} p^{r_p}$,

$$\sum_{f \le \sqrt{ay/A}} \mu(f)F(f)$$

$$= \prod_p (1 - F(p)) + O\left(\frac{1}{\sqrt{y}}\right)$$

$$= \prod_p \left(1-\frac{1}{p^2}\right) \prod_{p|a} \left(1-\frac{1}{p^2}\right)^{-1} \prod_{p\|a} \left(1-\frac{1}{p(2-1/p)}\right)$$

$$\times \prod_{p^2|a} \left(1 - p^2\left(r_p - 1 - \frac{r_p-2}{p}\right) \Big/ p^2\left(r_p+1-\frac{r_p}{p}\right)\right) + O\left(\frac{1}{\sqrt{y}}\right)$$

$$= \frac{6}{\pi^2} \prod_{p|a} \frac{p^2}{p^2-1} \prod_{p\|a} \frac{2(p-1)}{2p-1} \prod_{p^2|a} \frac{2(p-1)}{(r_p+1)p - r_p} + O\left(\frac{1}{\sqrt{y}}\right)$$

$$= \frac{6}{\pi^2} \prod_{p|a} \frac{2p^2}{(p+1)((r_p+1)p - r_p)} + O\left(\frac{1}{\sqrt{y}}\right).$$

Feeding this equation into the above expression for Σ_1 completes the proof of Lemma 6.

Lemma 7.

$$\sum_{n \le y} \rho(n) = \frac{6H(\Delta)}{\pi\sqrt{-\Delta}} \prod_{p|a} \left(1 - \frac{1}{p+1}\right) y + O(\sqrt{y}\log y).$$

Proof. This is immediate from Lemma 3, Lemma 6 and the expression for $H(\Delta)$ given just before Lemma 1.

Lemma 8.

$$\sum_{n \le \sqrt{ax}} \frac{\rho(n)}{n} = \frac{6H(\Delta)}{\pi\sqrt{-\Delta}} \prod_{p|a} \left(1 - \frac{1}{p+1}\right) \log x + O(1).$$

Proof. This follows from Lemma 7 by partial summation.

Given the estimates of Lemma 7 and Lemma 8, the Theorem follows from Lemma 1.

References

1 C. Hooley: On the representation of a number as the sum of a square and a product. *Math. Zeitschr.* **69** (1958), 211–227.
2 C. Hooley: On the number of divisors of quadratic polynomials. *Acta Math.* **110** (1963), 97–114.
3 J.F. McKee: On the average number of divisors of quadratic polynomials. *Math. Proc. Camb. Phil. Soc.* **117** (1995), 389–392.

16. On the Distribution of Integer Points in the Real Locus of an Affine Toric Variety

B. Z. Moroz

1.

The purpose of this note is to explain the recently developed theory, [9], [10], which allows us to study integer points on an affine toric variety. This theory generalises the classical set-up of algebraic number theory (cf. Example 1). In a series of papers (see [8] and references therein) we studied the distribution of integer points on norm-form varieties and the related problem of the distribution of integral ideals having equal norms. The theory to be sketched in the next few pages allows, in particular, to put these early results in a proper context and to answer some of the questions posed in [8]. The method developed by Draxl [3] plays a crucial role in these investigations. [1]

Let $K \,|\, \mathbb{Q}$ be a finite normal extension, and let G be the Galois group of K over $\mathbb{Q}$ (as usual $\mathbb{Q}$, $\mathbb{Z}$, $\mathbb{R}$ stand for the field of rational numbers, the ring of integers of $\mathbb{Q}$, and the field of real numbers respectively). Given an integral representation $\rho : G \to \mathbf{GL}(d, \mathbb{Z})$ and a sequence $\mathcal{A} = (\mathcal{A}_1, \ldots, \mathcal{A}_d)$ of fractional ideals of K, we describe a system of Diophantine equations with integral rational coefficients. This system of equations defines an algebraic variety of dimension d over $\mathbb{Z}$, to be denoted by $X_{\mathcal{A}}$. When regarded as an algebraic variety over $\mathbb{Q}$, it turns out to be a toric variety with respect to the action of the algebraic torus T defined by the data $(K \,|\, \mathbb{Q}, \rho)$. As $(K, \rho, \mathcal{A})$ runs through all the admissible triples of data, one obtains a vast class of Diophantine equations containing, as a rather special case, the systems of norm-form equations (cf. Example 2) studied in [8].

Given an Abelian group H, we write $v = u^g$ for $u \in H^d$, $g \in G$ with

$$v_i = \prod_{j=1}^{d} u_j^{r_{ji}(g)} \quad \text{for} \quad 1 \leq i \leq d,$$

---

[1] This method appears also to be the cornerstone of a recent study [1] relating to another Diophantine problem on a toric variety.

Sieve Methods, Exponential Sums, and their Applications in Number Theory
Greaves, G.R.H., Harman, G., Huxley, M.N., Eds. ©Cambridge University Press, 1996

where $\rho(g) = \big(r_{ij}(g)\big)$ $(1 \leq i, j \leq d)$ denotes the matrix $\rho(g)$ in $\mathbf{GL}(d, \mathbb{Z})$. Let $I(K)$ be the group of (fractional) ideals of K, and let

$$I(T) = \{\mathcal{A} \mid \mathcal{A} \in I(K)^d, \ g\mathcal{A} = \mathcal{A}^g \text{ for } g \in G\},$$

where $g(\mathcal{A}_1, \ldots, \mathcal{A}_d) := (g\mathcal{A}_1, \ldots, g\mathcal{A}_d)$. Fix $\mathcal{A} = (\mathcal{A}_1, \ldots, \mathcal{A}_d)$ in $I(T)$ and, for each j, a $\mathbb{Z}$-basis $\{\omega_i{}^{(j)} \mid 1 \leq i \leq n\}$ of $\mathcal{A}_j$, where $n := [K : \mathbb{Q}]$. We introduce a set of independent variables $x = \{x_{ji} \mid 1 \leq i \leq n, \ 1 \leq j \leq d\}$, and let $t = (t_1, \ldots, t_d)$ with

$$t_j = \sum_{i=1}^{n} x_{ji}\omega_i^{(j)}$$

for $1 \leq j \leq d$. The variety $X_\mathcal{A}$ is defined by the following system of equations: $gt = t^g$ for $g \in G$, where

$$gt := (gt_1, \ldots, gt_d), \qquad gt_j := \sum_{i=1}^{n} x_{ji}g\omega_i^{(j)}$$

for each j. In the new variables

$$y = \left\{ y_{jg} \mid y_{jg} = \sum_{i=1}^{n} x_{ji}g\omega_i^{(j)}, \text{ for } g \in G, \ 1 \leq j \leq d \right\}$$

this system of equations may be rewritten as follows (by a slight abuse of notation, we use the symbol "1" to denote the unit element of any group):

$$y_{ig} = \prod_{j=1}^{d} y_{j1}^{r_{ji}(g)} \quad \text{for} \quad 1 \leq i \leq d, \ g \in G. \tag{1}$$

Equations (1) define the geometric structure of the variety $X_\mathcal{A}$ over the splitting field K of the algebraic torus T. According to the second exercise on p. 19 of [4], these equations define an affine toric variety whose dual G-invariant cone $\check{\sigma}$ is generated by the set of vectors

$$\left\{ u_{ig} = \sum_{j=1}^{d} r_{ji}(g)e_j \mid g \in G, \ 1 \leq i \leq d \right\}$$

in the $\mathbb{Q}$-vector space $\check{V}$ with a basis $\{e_j \mid 1 \leq j \leq d\}$, which affords the contragredient representation $\check{\rho}$, so that

$$\rho\big(g^{-1}\big)e_i = \sum_{j=1}^{d} r_{ji}(g)e_j \quad \text{for} \quad 1 \leq i \leq d, \ g \in G.$$

Over the field $\mathbb{Q}$ any two varieties $X_\mathcal{A}$ and $X_\mathcal{B}$ are isomorphic:

$$X_\mathcal{A} \cong_\mathbb{Q} X_{(1)} \quad \text{for} \quad \mathcal{A} \in I(T), \ (1) := (o, \ldots, o),$$

o being the ring of integers of K. It is a subtle (and difficult) question to determine when $X_\mathcal{A}$ is isomorphic to $X_\mathcal{B}$ over the ring $\mathbb{Z}$ of rational integers for given $\mathcal{A}, \mathcal{B}$ in $I(T)$.

Example 1. Let $\mathbb{Q} \subseteq k \subseteq K$, $[k : \mathbb{Q}] = l$. The affine space A^l may be regarded as a toric variety to the algebraic torus $\mathrm{Res}_{k/\mathbb{Q}}G_{m,k}$. Any $\mathbb{Z}$-submodule of the field k can be canonically embedded as a sublattice into $\mathbb{R}^l$, the real locus of A^l (see, for instance, [12], Ch. V).

Example 2. Let $\mathbb{Q} \subseteq k \subseteq k_i \subseteq K$ for $1 \leq i \leq \nu$, and let $\mathcal{B}_i$ be a fractional ideal in k_i. Choose a $\mathbb{Z}$-basis $\{\omega_j^{(i)} \mid 1 \leq j \leq m_i\}$ of $\mathcal{B}_i, m_i := [k_i : \mathbb{Q}]$, and let (here $x_i := (x_{i1}, \ldots, x_{im_i})$)

$$f_i(x_i) = N_{k_i(x_i)/k(x_i)}\left(\sum_{j=1}^{m_i} x_{ij}\omega_j^{(i)}\right)N_{k_i/k}\mathcal{B}_i^{-1},$$

where $k_i(x_i)$ (resp. $k(x_i)$) denotes the field of rational functions of x_i over k_i (resp. k). Equations $f_1(x_1) = \cdots = f_\nu(x_\nu)$ define a toric variety of the type X_A considered above (cf. [8], where the case $k = \mathbb{Q}$ has been studied in detail under the additional supposition that the fields $k_1, \ldots, k_\nu$ be linearly disjoint).

We are concerned with the distribution of integer points $Y_A(\mathbb{Z})$ in the real locus $Y_A(\mathbb{R})$ of the (Zariski-) open subset Y_A of X_A defined by the condition

$$t_1(x) \ldots t_d(x) \neq 0,$$

thereby deleting the d exceptional hypersurfaces $t_j = 0$ for $1 \leq j \leq d$. The torus T may be defined by the system of equations $gw = w^g$ for $g \in G$ in conjunction with the equation $w_0(x)w_1(x)\ldots w_d(x) = 1$, where

$$w = (w_1, \ldots, w_d), \quad w_j = \sum_{i=1}^{n} x_{ji}\omega_i \quad \text{for} \quad 0 \leq j \leq d,$$

$\{\omega_1, \ldots, \omega_n\}$ being an integral basis of $K|\mathbb{Q}$, so that over $\mathbb{Q}$ the open subset Y_A may be regarded as the principal orbit under the action of T.[2] By a generalisation [11] of Dirichlet's unit theorem, $T(\mathbb{Z}) \cong A \times \mathbb{Z}^r$, where A is a finite Abelian group. Another important parameter is the dimension μ of the vector space

$$V^G = \{v \mid \rho(g)v = v \text{ for } g \in G\}$$

of G-invariants; the torus T is *anisotropic* if $\mu = 0$, and it is said to be *isotropic* if $\mu \geq 1$. It turns out (see, for instance, [10]) that $Y_A(\mathbb{R})$ is homeomorphic to $T(\mathbb{R}) = (\mathbb{R}_+^*)^\mu \times T^1(\mathbb{R})$, where

$$T^1(\mathbb{R}) := (\mathbb{R}_+^*)^r \times (\mathbb{Z}/2\mathbb{Z})^\nu \times (S^1)^{d_1}.$$

---

[2] It is important, however, to distinguish between $Y_A(\mathbb{Z})$ and $T(\mathbb{Z})$. In the situation of Example 1, for instance, $T(\mathbb{Z})$ is isomorphic to the group of units of k, while $Y_A(\mathbb{Z}) = \mathbb{Z}^l\backslash\{0\}$.

Here $\mathbb{R}_+^*$ and S^1 denote respectively the multiplicative groups of the positive real numbers and of the complex numbers of absolute value one, $\nu = \mu + r$, and $\nu + d_1 = d$. Moreover, one can introduce new coordinates u by letting

$$u_i = \prod_{j=1}^{d} t_j^{c_{ji}} \text{ for } 1 \le i \le d,$$

for a suitable unimodular matrix $C = (c_{ij})$ in $SL(d, \mathbb{Z})$, in such a way that

$$u_i(a) \in \mathbb{R}^* \text{ if } i \le \nu, \qquad u_i(a) \in S^1 \text{ if } i > \nu$$

for any $a \in Y_A(\mathbb{R})$, so that $l_u : a \mapsto b(a)$ with

$$b_i(a) = |u_i(a)|, \qquad b_{i+\nu}(a) = \begin{cases} u_i(a)/|u_i(a)| & \text{for } i \le \nu \\ u_i(a) & \text{for } i > \nu \end{cases}$$

defines the mentioned homeomorphism. If the torus T is anisotropic then $Y_A(\mathbb{Z})$ essentially coincides with $T(\mathbb{Z})$, and its structure is completely determined by the unit theorem for algebraic tori we have just cited. Thus from now on it will be assumed that the torus T is isotropic, that is, $\mu \ge 1$. Let V be the vector space dual to $\check{V}$; it affords the representation ρ, so that

$$\rho(g)f_i = \sum_{j=1}^{d} f_j r_{ij}(g)$$

in the basis $\{f_j \mid 1 \le j \le d\}$ of V dual to $\{e_j \mid 1 \le j \le d\}$. The cone

$$\sigma = \{v \in V, \, \rho(g)v \ge 0 \text{ for } g \in G\}$$

is easily seen to be dual to the cone $\check{\sigma}$, that is,

$$\sigma = \{v \mid v \in V, \, (u, v) \ge 0 \text{ for } u \in \check{\sigma}\}$$

(here we write

$$\sum_{i=1}^{d} v_i f_i \ge 0$$

if $v_i \ge 0$ for each i). In what follows we shall suppose that $\sigma \ne \{0\}$. Let

$$\left| \sum_{i=1}^{d} v_i f_i \right| := \sum_{i=1}^{d} v_i,$$

and let

$$\|v\| := \sum_{g \in G} |\rho(g)v|$$

for $v \in V$. The third important parameter is defined as follows:

$$\kappa = \min\{m \mid m \in \mathbb{Z}, \ m \geq 1, \ \sigma(m) \neq \emptyset\},$$

where $\sigma(m) := \{v \mid v \in \sigma \cap M, \ \|v\| = m\}$, and

$$M = \left\{ \sum_{i=1}^{d} v_i f_i \mid v_i \in \mathbb{Z}, \ 1 \leq i \leq d \right\}$$

is the $\mathbb{Z}[G]$-module of the representation ρ. By construction, $\sigma(\kappa)$ is a finite G-invariant subset of σ; for further reference, let B be the number of different G-invariant orbits in $\sigma(\kappa)$. In this notation, let

$$U(y) = \left\{ a \mid a \in Y_{\mathcal{A}}(\mathbb{R}), \ |Nt(a)| \leq y^{\kappa}, \ 1/y \leq |u_j(a)| \leq y \text{ for } \mu < j \leq \nu \right\},$$

where

$$Nt := \prod_{j=1}^{d} \prod_{g \in G} g t_j.$$

The following theorem (see [10]) gives an asymptotic formula for the number of integral points in $U(y)$ satisfying an additional congruence condition.

Theorem. *Let* $\mathcal{F} = \{m, \mathcal{F}_0\}$, $m \in \mathbb{Z}$, $\mathcal{F}_0 \subseteq (\mathbb{Z}/2\mathbb{Z})^{\nu}$, *and let*

$$Z_{\mathcal{F}} = \{a \mid a \in Y_{\mathcal{A}}(\mathbb{Z}), \ a \equiv 1 \ (\mathrm{mod}\,\mathcal{F})\}.$$

Then

$$\mathrm{card}(Z_{\mathcal{F}} \cap U(y)) = cy \log^{\beta} y \left(1 + O\left(\frac{1}{\log y} \right) \right)$$

with $\beta \in \mathbb{Z}$, $\beta \geq 0$, $c \geq 0$, *and* $c > 0$ *if* $Z_{\mathcal{F}} \neq \emptyset$.

Here the congruence $a \equiv 1 \ (\mathrm{mod}\,\mathcal{F})$ is to be understood in the spirit of classical algebraic number theory. It means that $t_i(a) \equiv 1 \ (\mathrm{mod}\,m)$ for $1 \leq i \leq d$ in the ring o and $(b_{\nu+1}(a), \ldots, b_{2\nu}(a)) \in \mathcal{F}_0$. Clearly, $Z_{\mathcal{F}} = Y_{\mathcal{A}}(\mathbb{Z})$ for $\mathcal{F} = \{1, (\mathbb{Z}/2\mathbb{Z})^{\nu}\}$.

2.

Let us outline the idea of the proof of this theorem. As in the classical case (see, for instance Ch. I, §1 of [6]), one can define a grossencharacter χ of the torus T as a continuous homomorphism $\chi : T(A)/T(\mathbb{Q}) \to S^1$, where A denotes the adèle ring of $\mathbb{Q}$.

There is a subgroup $T^1(A)$ of $T(A)$ such that $T(A) \cong (\mathbb{R}_+^*)^\mu \times T^1(A)$ and the group $T^1(A)/T(\mathbb{Q})$ is compact; a grossencharacter χ is said to be *normalised* if χ is trivial on $(\mathbb{R}_+^*)^\mu$. A normalised grossencharacter χ gives rise to a homomorphism

$$\chi : I^{(\mathcal{F})}(T) \to S^1,$$

where $\mathcal{F} = \{m, \mathcal{F}_0\}$ is the uniquely determined conductor of χ, and $I^{(\mathcal{F})}(T)$ denotes the subgroup of ideals in $I(T)$ prime to m. To be precise, let $I_0(K)$ be the monoid of integral ideals of K, and let

$$I^{(\mathcal{F})}(T) = \{\mathcal{A} \mid \mathcal{A} \in I(T), \ (\mathcal{A}, m) = 1\}.$$

We write $(\mathcal{A}, m) = 1$ if $\mathcal{A} = (\mathcal{A}_1, \dots, \mathcal{A}_d)$ with

$$\mathcal{A}_i = \mathcal{B}_i \mathcal{C}_i^{-1}, \quad \mathcal{B}_i \in I_0(K), \quad \mathcal{C}_i \in I_0(K),$$
$$\gcd(N_{K/\mathbb{Q}}(\mathcal{B}_i \mathcal{C}_i), m) = 1 \quad \text{for} \quad 1 \le i \le d.$$

As one may anticipate, the character χ is trivial on the "ray"

$$I_{pr}^{(\mathcal{F})}(T) = \{(\alpha) \mid \alpha \in T(\mathbb{Q}), \ \alpha \equiv 1 \ (\mathrm{mod}\, \mathcal{F})\},$$

where the congruence $\alpha \equiv 1 \ (\mathrm{mod}\, \mathcal{F})$ has the same meaning as at the end of section 1. Let $I_0(T) = I(T) \cap I_0(K)^d$, and let

$$N\mathcal{A} = \prod_{i=1}^{d} N_{K/\mathbb{Q}} \mathcal{A}_i$$

for $\mathcal{A} = (\mathcal{A}_1, \dots, \mathcal{A}_d)$ in $I(T)$. One continues a character χ to a multiplicative function $\chi : I_0(T) \to S^1 \cup \{0\}$ by letting $\chi(\mathcal{A}) = 0$ if $\mathcal{A} \notin I^{(\mathcal{F})}(T)$. One defines (see [9], [10]) a Dirichlet series

$$L(T; s, \chi) = \sum_{\mathcal{A} \in I_0(T)} \frac{\chi(\mathcal{A})}{N\mathcal{A}^{s/\kappa}},$$

convergent absolutely for $\mathrm{Re}\, s > 1$, and notes (loc. cit.) that in this half-plane $L(T; s, \chi)$ may be represented by an absolutely convergent Euler product of the following form:

$$L(T; s, \chi) = \prod_{\mathcal{P} \in P} \left(1 - \frac{\chi(\mathcal{P})}{N\mathcal{P}^{s/\kappa}}\right)^{-1} \prod_p Q_p\left(p^{-s/\kappa e(p)}\right),$$

where p ranges over the rational primes, $e(p)$ being the ramification index of p in K, while P denotes a certain subset of $I_0(T)$ (in [9], [10] the elements of P are called the *strict primes*), $Q_p(t) \in \mathbb{C}[t]$, and $Q_p(0) = 1$ for each p.

It turns out (see [9], [10]) that up to a finite number of Euler factors the L-series $L(T; s, \chi)$ coincides with one of the Draxl L-functions; therefore the following proposition follows from the main theorem in [3].

Proposition. *There are subfields K_j of K such that*

$$L(T; s, \chi) = \prod_{j=1}^{B} L(\chi_j, s) L^{(1)}(T; s, \chi), \qquad (2)$$

where $L(\chi_j, s)$ is the Hecke L-function of K_j with a grossencharacter χ_j for $1 \leq j \leq B$, while the function $L^{(1)}(T; s, \chi)$ is holomorphic in the half-plane

$$\operatorname{Re} s > 1 - \frac{1}{\kappa + 1}.$$

As an immediate consequence of this proposition, one obtains an asymptotic formula:

$$\sum_{\substack{\mathcal{A} \in I_0(T) \\ N\mathcal{A} < y^\kappa}} \chi(\mathcal{A}) = y R_\chi(\log y) + O(y^{1-\gamma}), \qquad (3)$$

where $\gamma > 0$ and $R_\chi(t)$ is a polynomial of degree $q(\chi) - 1$ with

$$q(\chi) = \operatorname{card}\{j \mid 1 \leq j \leq B, \ \chi_j = 1\};$$

we set $R_\chi(t) = 0$ if $q(\chi) = 0$. The asymptotic formula in the theorem (with $\beta < B + r$) can be deduced from (3). See [9], [10] for the details of this argument.

3.

In the situation of Example 1, the L-series $L(T; s, \chi)$ coincides with the Hecke L-series $L(\chi, s)$, the character χ being one of the grossencharacters of the field k, and the theory outlined here reduces to the classical "multidimensional" arithmetic (cf., for instance, Ch. I, §1 of [6] and references therein). For the norm-form tori of Example 2, the L-series $L(T; s, \chi)$ coincides with the scalar product of Hecke L-functions of the fields $k_1, \ldots, k_\nu$ over k (as defined, for instance, in [3] or in Ch. II, §5 of [6]). In both cases (actually Example 1 is, of course, a special case of Example 2) $\mu = 1$ and the following condition is satisfied:

$$\forall_j(\chi_j = 1) \Rightarrow \chi \text{ is of finite order}, \qquad (4)$$

where χ and χ_j are related as in formula (2). Moreover, $B = 1$ if and only if the fields $k_1, \ldots, k_\nu$ are linearly disjoint. We do not know whether (4) holds true for a general torus (although it may well be the case); on the other hand, if $\chi = 1$ then $\chi_j = 1$ for $1 \le j \le d$ by construction. In any case, assuming (4) one deduces from (3) an estimate

$$\sum_{\substack{\mathcal{A} \in I_0(T) \\ N\mathcal{A} < y^\kappa}} \chi(\mathcal{A}) = O\big(y \log^{B-2} y\big) \tag{5}$$

for any grossencharacter χ of infinite order. Estimate (5) would allow us to prove an equidistribution theorem for integral ideals (in the spirit of Ch. III, §2 of [6], although with a weaker estimate of the remainder term). If $B = 1$ then (3) may be rewritten as follows:

$$\sum_{\substack{\mathcal{A} \in I_0(T) \\ N\mathcal{A} < y^\kappa}} \chi(\mathcal{A}) = g(\chi_1) L^{(1)}(T; 1, \chi) y + O\big(y^{1-\gamma}\big), \tag{6}$$

where $\gamma > 0$, with

$$g(\chi_1) = \begin{cases} w & \text{if} \quad \chi_1 = 1 \\ 0 & \text{if} \quad \chi_1 \ne 1, \end{cases}$$

and w equal to the residue of the zeta-function of K_1. Condition (4) gives now

$$\sum_{\substack{\mathcal{A} \in I_0(T) \\ N\mathcal{A} < y^\kappa}} \chi(\mathcal{A}) = O\big(y^{1-\gamma}\big), \tag{7}$$

with $\gamma > 0$, for any grossencharacter χ of infinite order. Estimates (6) and (7) would allow us to prove an equidistribution theorem for integer points in the domains of the following form:

$$U(y, \tau) = \Big\{ a \mid a \in Y_\mathcal{A}(R), \; |Nt(a)| < y^\kappa, \; f(a) \in \tau \Big\},$$

where τ is a "smooth" subset of $T^1(\mathbb{R})$ and $f : Y_\mathcal{A}(\mathbb{R}) \to T^1(\mathbb{R})$ is the natural surjective map (obtained, for instance, by means of the homeomorphism l_u defined above).

Returning to the norm-form tori of Example 2, one can now prove a general equidistribution theorem for integral ideals having equal norms generalising the results of Ch. III, §2 of [6]. Moreover, if the fields $k_1, \ldots, k_\nu$ are linearly disjoint over k, one obtains an equidistribution theorem for integer points on the corresponding norm-form variety, which is somewhat stronger than the main theorem of [8] because the congruence conditions introduced in the theorem of Section 1 allow one to study integer points on the norm-form variety itself without introducing an intermediate variety W, in the notation of [8] (cf. Remark 2 in [8]).

Having continued his L-function to the half-plane $\text{Re}\, s > 0$, Draxl raised a question (see p. 466 in [3]) concerning the possibility of analytic continuation of this function to the whole complex plane. In the mid-eighties this problem was completely solved for the norm-form tori of Example 2 (see [5], [7]): in this case the function $s \mapsto L(T; s, \chi)$ may be analytically continued to the whole complex plane (as a meromorphic function of s) if and only if either $\text{card}\{i|k_i \neq k\} \leq 1$ or $[k_i : k] \leq 2$ for each i and $\text{card}\{i|k_i \neq k\} \leq 2$; in the light of this solution, one may now try to describe the class of toric varieties whose Draxl L-functions are meromorphic in the whole plane.

The methods developed in [9], [10] (and summarised in this note) may most probably be used to study integer points of any affine toric variety. In a somewhat similar development, several authors study the distribution of integer points on affine homogeneous spaces of semisimple algebraic groups (see, for example, [2] and references therein).

As on many occasions, I feel indebted to and wish to thank Professor F. Hirzebruch for his generosity and support of my work. I am grateful to the organisers of the symposium for their hospitality in Cardiff, and to the referee for his detailed comments.

References

1 V.V. Batyrev and Yu. Tschinkel: Rational points of bounded height on compactifications of anisotropic tori. *International Mathematics Research Notices* **95** No. 12, 591–635.

2 M. Borovoi and Z. Rudnick: Hardy-Littlewood varieties and semisimple groups. *Invent. Math.* **119** (1995), 37–66.

3 P.K.J. Draxl: L-Funktionen algebraischer Tori. *J. Number Theory* **3** (1971), 444–467.

4 W. Fulton: *Introduction to toric varieties* (Annals of Mathematical Studies No. 131). Princeton University Press (1993).

5 N. Kurokawa: On the meromorphy of Euler products (I,II). *Proc. London Math. Soc.* **53** (1986), 1–47, 209–236.

6 B.Z. Moroz: *Analytic Arithmetic in Algebraic Number Fields* (Lecture Notes in Mathematics **1205**). Springer (1986).

7 B.Z. Moroz: On a class of Dirichlet series associated to the ring of representations of a Weil group. *Proc. London Math. Soc.* **56** (1988), 209–228.

8 B.Z. Moroz: On the distribution of integral points on an algebraic torus defined by a system of norm-form equations. *Quart. J. Math. Oxford* (2) **45** (1994), 243–253.

9 B.Z. Moroz: Exercises in analytic arithmetic on an algebraic torus. *The F. Hirzebruch Festband* (Israel Mathematical Conferences Proceedings **9**) (1995), 347–359.

10 B.Z. Moroz: On the integer points of some toric varieties. *Quart. J. Math. Oxford* (2) **48** (1997), to appear.

11 J.M. Shyr: A generalization of Dirichlet's unit theorem. *J. Number Theory* **9** (1977), 213–217.

12 A. Weil: *Basic Number Theory*. Springer (1973).

17. An Asymptotic Expansion of the Square of the Riemann Zeta-Function

Yoichi Motohashi

1. Introduction

In the present paper we shall prove a ζ^2-analogue of the Riemann-Siegel formula: we shall investigate the symmetric case of the approximate functional equation for $\zeta^2(s)$ and establish an asymptotic expansion for the error-term.

To begin with, we put

$$E(s,x) = \zeta(s) - \sum_{n \leq x} \frac{1}{n^s} - \chi(s) \sum_{n \leq t/2\pi x} \frac{1}{n^{1-s}}. \tag{1}$$

Here $s = \sigma + it$, $\chi(s) = 2^s \pi^{s-1} \sin(\frac{1}{2}s\pi)\Gamma(1-s)$ as usual; and we shall assume always that $t \geq 1$. Riemann's method, which has been carried out in detail by Siegel [1], yields a very precise asymptotic formula for $E(s,x)$. In particular we have, for fixed $0 \leq \sigma \leq 1$,

$$\chi(1-s)^{\frac{1}{2}} E\left(s, \left(\frac{t}{2\pi}\right)^{\frac{1}{2}}\right) = \left((-1)^{[(t/2\pi)^{\frac{1}{2}}]-1}\right) \frac{\cos\left(2\pi(\delta^2 - \delta - \frac{1}{16})\right)}{\cos 2\pi\delta} \left(\frac{t}{2\pi}\right)^{-\frac{1}{4}}$$
$$+ O\left(t^{-\frac{3}{4}}\right), \tag{2}$$

where $\delta = \theta((t/2\pi)^{\frac{1}{2}})$ with $\theta(x)$ being the fractional part of x.

On the other hand, the error term in the approximate functional equation for $\zeta^2(s)$ can be defined to be

$$R(s,x) = \zeta^2(s) - {\sum_{n \leq x}}' \frac{d(n)}{n^s} - \chi^2(s) {\sum_{n \leq y}}' \frac{d(n)}{n^{1-s}}$$

where d is the divisor function, $xy = (t/2\pi)^2$ and the primes indicate that the summands $d(x)$ and $d(y)$ are to be halved if they ever appear. But nothing like (2) has been obtained for $R(s,x)$, though it has often been expressed that asymptotic formulas for $R(s,t/2\pi)$ would greatly facilitate the analysis of $\zeta(s)$.

Sieve Methods, Exponential Sums, and their Applications in Number Theory
Greaves, G.R.H., Harman, G., Huxley, M.N., Eds. ©Cambridge University Press, 1996

Our main result is an answer to this classical problem, and is embodied in the

Theorem. *We have, for $t \geq 1$ and fixed $0 \leq \sigma \leq 1$,*

$$\chi(1-s)R\left(s, \frac{t}{2\pi}\right)$$

$$= -2\left(\frac{t}{\pi}\right)^{-\frac{1}{2}} \Delta\left(\frac{t}{2\pi}\right) - \frac{(1 - 2(\sigma - \frac{1}{2})i)}{\sqrt{2}}\left(\frac{t}{2\pi}\right)^{-\frac{3}{2}} \Delta_1\left(\frac{t}{2\pi}\right)$$

$$+ \frac{1}{\pi\sqrt{2}}\left(\frac{t}{2\pi}\right)^{-\frac{1}{2}}\left\{\frac{1}{6}\left(\log\frac{t}{2\pi} + 2\gamma + 6\right) - (\sigma - \frac{1}{2})\left(\log\frac{t}{2\pi} + 2\gamma\right)i\right\}$$

$$+ \frac{1}{\sqrt{2\pi}}\left(\frac{t}{2\pi}\right)^{-\frac{1}{4}} \sum_{n=1}^{\infty} n^{-\frac{1}{4}} d(n) h_1(n) \sin\left(2(2n\pi t)^{\frac{1}{2}} + \frac{1}{4}\pi\right)$$

$$- \left(\frac{t}{2\pi}\right)^{-\frac{3}{4}} \left\{ \begin{array}{l} \dfrac{1}{32\pi^{\frac{3}{2}}\sqrt{2}} \sum_{n=1}^{\infty} n^{-\frac{1}{4}} d(n) h_1(n) \cos\left(2(2n\pi t)^{\frac{1}{2}} + \frac{1}{4}\pi\right) \\[2ex] + \dfrac{3(\sigma - \frac{1}{2})}{2i\sqrt{2\pi}} \sum_{n=1}^{\infty} n^{\frac{1}{4}} d(n) h_2(n) \cos\left(2(2n\pi t)^{\frac{1}{2}} + \frac{1}{4}\pi\right) \\[2ex] + \frac{5}{8}\left(\frac{1}{2}\pi\right)^{\frac{1}{2}} \sum_{n=1}^{\infty} n^{\frac{5}{4}} d(n) h_3(n) \cos\left(2(2n\pi t)^{\frac{1}{2}} + \frac{1}{4}\pi\right) \end{array} \right\}$$

$$+ O\left(\frac{\log t}{t}\right).$$

Here γ is the Euler constant, and

$$h_\nu(n) = \int_0^\infty \frac{\cos\left(\xi + \frac{1}{4}(-1)^\nu \pi\right)}{(\xi + n\pi)^{\nu + \frac{1}{2}}} \, d\xi,$$

$$\Delta(x) = \sum_{n \leq x}' d(n) - x(\log x + 2\gamma - 1) - \frac{1}{4}, \qquad \Delta_1(x) = \int_0^x \Delta(y) \, dy.$$

Corollary. *We have, for $t \geq 1$ and fixed $0 \leq \sigma \leq 1$,*

$$\chi(1-s)R\left(s, \frac{t}{2\pi}\right) = -2\left(\frac{t}{\pi}\right)^{-\frac{1}{2}} \Delta\left(\frac{t}{2\pi}\right) + O(t^{-\frac{1}{4}}).$$

The corollary shows that the study of $R(s, t/2\pi)$ is essentially equivalent to that of the divisor problem. It should also be remarked that the error term in the theorem can well be replaced by further approximations, but the above seems to be sharp enough for most applications.

Notation. The functions χ, θ, d, Δ and Δ_1 are defined in the above; note that $d(x) = 0$ for non-integral x. The complex variable s has the imaginary part $t \geq 1$ and the real part $0 \leq \sigma \leq 1$ which is fixed, so estimates below may not be uniform in σ. To simplify the presentation we reserve the letter u for $(t/2\pi)^{1/2}$, and ε for $e^{i\pi/4}$. Also

$$\int_{(\alpha)} \ldots dw$$

indicates that the contour of the integral is the straight line $\mathrm{Re}(w) = \alpha$.

2. Initial transformation

We consider, instead of $R(s, u^2)$,

$$R^*(s, u^2) = \zeta^2(s) - I,$$

where

$$I = \sum_{n \leq u^2} \frac{d(n)}{n^s} + \chi^2(s) \sum_{n \leq u^2} \frac{d(n)}{n^{1-s}}.$$

Our idea lies in the following relation between the error terms E and R^*:

$$\tfrac{1}{2}\chi(1-s)R^*(s, u^2) = \sum_{n \leq u} \frac{1}{n} + \tfrac{1}{2}\chi(1-s)E^2(s, u) + K(s) + \overline{K(1 - \bar{s})}, \quad (3)$$

where

$$K(s) = \chi(1-s) \sum_{m \leq u} \frac{E(s, u^2/m)}{m^s}.$$

To show this we apply the hyperbola argument of Dirichlet to the sum I and get

$$I = 2 \sum_{m \leq u} \frac{1}{m^s} \sum_{n \leq u^2/m} \frac{1}{n^s} + \frac{2\chi^2(s)}{m^{1-s}} \sum_{n \leq u^2/m} \frac{1}{n^{1-s}}$$

$$- \left(\sum_{n \leq u} \frac{1}{n^s}\right)^2 - \left(\chi(s) \sum_{n \leq u} \frac{1}{n^{1-s}}\right)^2.$$

Then we have, by (1),

$$
I = 2 \sum_{m \le u} \frac{1}{m^s} \left\{ \zeta(s) - \chi(s) \sum_{n \le m} \frac{1}{n^{1-s}} - E\left(s, \frac{u^2}{m}\right) \right\}
$$

$$
+ 2\chi^2(s) \sum_{m \le u} \frac{1}{m^{1-s}} \left\{ \zeta(1-s) - \chi(1-s) \sum_{n \le m} \frac{1}{n^s} - E\left(1-s, \frac{u^2}{m}\right) \right\}
$$

$$
- \left(\zeta(s) - E(s, u) \right)^2 + 2\chi(s) \sum_{m \le u} \frac{1}{m^s} \sum_{n \le u} \frac{1}{n^{1-s}} .
$$

Invoking the functional equation $\zeta(s) = \chi(s)\zeta(1-s)$ and $\chi(s)\chi(1-s) = 1$, we see that the last identity becomes

$$
I = \zeta^2(s) - 2 \sum_{m \le u} \frac{E(s, u^2/m)}{m^s} - 2\chi^2(s) \sum_{m \le u} \frac{E(1-s, u^2/m)}{m^{1-s}} - E^2(s, u)
$$

$$
- 2\chi(s) \left\{ \sum_{m \le u} \frac{1}{m^s} \sum_{n \le m} \frac{1}{n^{1-s}} + \sum_{m \le u} \frac{1}{m^{1-s}} \sum_{n \le m} \frac{1}{n^s} - \sum_{m \le u} \frac{1}{m^s} \sum_{n \le u} \frac{1}{n^{1-s}} \right\}.
$$

The sum in the last braces is equal to $\sum_{n \le u} 1/n$, and we obtain (3).

Now our problem is to estimate $K(s)$ asymptotically. For this sake we appeal to the

Lemma. *We have, uniformly for* $1 \le x \le t^A$ *with an arbitrary fixed* $A \ge 0$,

$$
\chi(1-s)E(s, x)
$$

$$
= \left(\frac{u^2}{x} \right)^{s-1} \sum_{n=0}^{N-1} \frac{a_n}{2\pi i} \int_L \exp\left\{ \begin{array}{l} \frac{i}{4\pi} \left(\frac{x}{u} \right)^2 (w - 2\pi i\vartheta)^2 \\[4pt] + x(w - 2\pi i\vartheta) - [x]w \end{array} \right\} \frac{(w - 2\pi i\vartheta)^n}{e^w - 1} \, dw
$$

$$
+ O\left(t^{-\frac{1}{6}N} \left(t^{-\frac{1}{2}} x^{1-\sigma} + x^{-\sigma} \right) \right). \tag{4}
$$

Here $\vartheta = \theta(u^2/x)$ *with an arbitrary fixed positive integer* N; L *is a straight line in the direction* $\mathrm{Arg}(w) = \pi/4$ *passing between* 0 *and* $2\pi i$. *The coefficients* a_n *are functions of* s *and* x, *of which the first six are of the form*

$$
a_0 = 1, \quad a_1 = (\sigma - 1)\frac{x}{it}, \quad a_2 = c_1 \left(\frac{x}{t} \right)^2, \quad a_3 = c_2 \left(\frac{x}{t} \right)^3 - \frac{x^3}{3t^2},
$$

$$
a_4 = c_3 \left(\frac{x}{t} \right)^4 + c_4 \frac{x^4}{t^3}, \quad a_5 = c_5 \left(\frac{x}{t} \right)^5 + c_6 \frac{x^5}{t^4},
$$

where c_j $(1 \le j \le 6)$ *are polynomials in* σ.

We do not dwell on the proof, since it amounts to a careful adaptation of Riemann's method to our situation (see [2], §§4.14–4.15; note that there the condition $1 \le x \le u$ is imposed but it is not essential and can be replaced by $1 \le x \le t^A$).

We apply (4) with $N = 6$ and $x = u^2/m$ to $K(s)$, and have

$$K(s) = \frac{1}{2\pi i} \sum_{\nu=0}^{5} \sum_{m \leq u} \frac{a_\nu \Psi_\nu(m)}{m} + O\left(\frac{\log t}{t}\right),$$

where

$$\Psi_\nu(m) = \int_L \exp\left(\frac{i}{4\pi}\left(\frac{u}{m}\right)^2 w^2 + \theta\left(\frac{u^2}{m}\right)w\right) \frac{w^\nu \, dw}{e^w - 1}.$$

As is easily seen, we have, for $\nu \geq 1$,

$$\Psi_\nu(m) = O\left(\left(\frac{m}{u}\right)^\nu\right).$$

Thus we have more precisely

$$K(s) = \frac{1}{2\pi i} \sum_{m \leq u} \frac{\Psi_0(m)}{m} - \frac{1}{4\pi^2}(\sigma - 1) \sum_{m \leq u} \frac{\Psi_1(m)}{m^2}$$
$$+ \frac{ti}{48\pi^4} \sum_{m \leq u} \frac{\Psi_3(m)}{m^4} + O\left(\frac{\log t}{t}\right). \tag{5}$$

Next, in the integral $\Psi_0(m)$ we replace the contour by the curve consisting of two parts

$$\{w = \varepsilon\lambda : -\infty < \lambda \leq -\eta, \ \eta \leq \lambda \leq \infty\}, \qquad \{w = \eta\varepsilon e^{i\tau} : 0 \leq \tau \leq \pi\}$$

with a small $\eta > 0$, and let η tend to 0. We then have

$$\Psi_0(m) = -\pi i + \varepsilon F_0\left(\frac{u^2}{4\pi m^2}, \theta\left(\frac{u^2}{m}\right)\right); \tag{6}$$

and similarly, for $\nu = 1, 3$,

$$\Psi_\nu(m) = \varepsilon^{\nu-1} F_\nu\left(\frac{u^2}{4\pi m^2}, \theta\left(\frac{u^2}{m}\right)\right), \tag{7}$$

where

$$F_\nu(a, b) = \int_0^\infty \left(\lambda^\nu e^{-a\lambda^2}\right) \frac{\sin\left((b - \frac{1}{2})\lambda/\varepsilon + \frac{1}{4}\pi(1 - (-1)^\nu)\right)}{\sin(\lambda/2\varepsilon)} \, d\lambda$$

with $a > 0$.

We next transform $F_0(a, b)$. This is, of course, a continuous function of b, a real variable. By Mellin's integral formula we have

$$F_0(a, b) = \frac{1}{2\pi i} \int_0^\infty \frac{\sin((b - \frac{1}{2})\lambda/\varepsilon)}{\sin(\lambda/2\varepsilon)} \int_{(\frac{1}{4})} \frac{\Gamma(w)}{(a\lambda^2)^w} \, dw \, d\lambda.$$

This double integral converges absolutely, providing $0 < b < 1$. Then, assuming this for a while, we have

$$F_0(a, b) = \frac{1}{2\pi i} \int_{(\frac{1}{4})} \frac{\Gamma(w)}{a^w} \int_0^\infty \frac{\sin((b - \frac{1}{2})\lambda/\varepsilon)}{\lambda^{2w} \sin(\lambda/2\varepsilon)} \, d\lambda \, dw.$$

We replace the contour of the inner integral by the curve C that starts at infinity on the positive real axis, encircles the origin once in the positive direction, excluding the poles, and returns to infinity. We then have

$$F_0(a, b) = \frac{1}{2\pi i} \int_{(\frac{1}{4})} \frac{\Gamma(w)}{a^w(e^{-4\pi i w} - 1)} \int_C \frac{\sin((b - \frac{1}{2})\lambda/\varepsilon)}{\lambda^{2w} \sin(\lambda/2\varepsilon)} \, d\lambda \, dw,$$

where $\lambda^{2w} = \exp(2w \log \lambda)$ and Arg λ varies from 0 to 2π round C. Shifting the contour of the w-integral to $\mathrm{Re}(w) = \frac{5}{4}$ we encounter poles at $w = \frac{1}{2}$ and 1, and computing the residues we get

$$F_0(a, b) = \sqrt{\pi}(b - \tfrac{1}{2})a^{-\frac{1}{2}}$$
$$+ \frac{1}{2\pi i} \int_{(\frac{5}{4})} \frac{\Gamma(w)}{a^w(e^{-4\pi i w} - 1)} \int_C \frac{\sin((b - \frac{1}{2})\lambda/\varepsilon)}{\lambda^{2w} \sin(\lambda/2\varepsilon)} \, d\lambda \, dw.$$

Here we observe that the last double integral converges absolutely when $0 \le b \le 1$, and by continuity the last identity holds in this range of b; thus hereafter we may assume this. Then expanding C to infinity and computing the residues arising from poles at $\lambda = \pm 2\pi\varepsilon, \pm 4\pi\varepsilon, \dots$, we get

$$F_0(a, b) = \sqrt{\pi}(b - \tfrac{1}{2})a^{-\frac{1}{2}} + \varepsilon \int_{(\frac{5}{4})} \frac{e^{\frac{1}{2}\pi i w}}{\cos \pi w} \cdot \frac{\Gamma(w)}{(4\pi^2 a)^w} \sum_{n=1}^\infty \frac{\sin 2\pi n b}{n^{2w}} \, dw. \qquad (8)$$

We differentiate both sides with respect to b and shift the line of integration to $\mathrm{Re}(w) = \frac{7}{4}$; and we get

$$F_1(a, b) = \varepsilon\sqrt{\pi}a^{-\frac{1}{2}} + \frac{\sqrt{\pi}}{4\varepsilon}\left(\tfrac{1}{12} - (b - \tfrac{1}{2})^2\right)a^{-\frac{3}{2}}$$
$$+ 2\pi i \int_{(\frac{7}{4})} \frac{e^{\frac{1}{2}\pi i w}}{\cos \pi w} \cdot \frac{\Gamma(w)}{(4\pi^2 a)^w} \sum_{n=1}^\infty \frac{\cos 2\pi n b}{n^{2w-1}} \, dw. \qquad (9)$$

As for $F_3(a, b)$ we shift the line of integration in this identity to $\mathrm{Re}(w) = \frac{9}{4}$, and differentiate the result twice with respect to b, getting

$$F_3(a, b) = \frac{1}{2}\varepsilon\sqrt{\pi}a^{-\frac{3}{2}} - (2\pi)^3 \int_{(\frac{9}{4})} \frac{e^{\frac{1}{2}\pi i w}}{\cos\pi w} \cdot \frac{\Gamma(w)}{(4\pi^2 a)^w} \sum_{n=1}^{\infty} \frac{\cos 2\pi nb}{n^{2w-3}} \, dw.$$

Again shifting the line of integration to $\mathrm{Re}(w) = \frac{11}{4}$ we get

$$F_3(a, b) = \frac{1}{2}\varepsilon\sqrt{\pi}a^{-\frac{3}{2}} + \frac{3\sqrt{\pi}}{8\varepsilon}\left(\frac{1}{12} - \left(b - \frac{1}{2}\right)^2\right)a^{-\frac{5}{2}}$$
$$- (2\pi)^3 \int_{(\frac{11}{4})} \frac{e^{\frac{1}{2}\pi i w}}{\cos\pi w} \cdot \frac{\Gamma(w)}{(4\pi^2 a)^w} \sum_{n=1}^{\infty} \frac{\cos 2\pi nb}{n^{2w-3}} \, dw. \qquad (10)$$

We collect (5)–(10), and find that

$$K(s) = -\frac{1}{2}\sum_{m \le u} \frac{1}{m} + \frac{1}{\varepsilon u}\sum_{m \le u}\left(\theta\left(\frac{u^2}{m}\right) - \frac{1}{2}\right)$$
$$- \frac{(3\sigma - 2)\varepsilon}{6\pi u}\sum_{m \le u}\frac{1}{m} - \frac{\sigma}{2\varepsilon u^3}\sum_{m \le u}m\left(\frac{1}{12} - \left(\theta\left(\frac{u^2}{m}\right) - \frac{1}{2}\right)^2\right)$$
$$+ \frac{1}{2\pi}\int_{(\frac{5}{4})}\frac{e^{\frac{1}{2}\pi i w}}{\cos\pi w} \cdot \frac{\Gamma(w)}{(\pi u^2)^w}\sum_{n=1}^{\infty}\frac{S_0(n, w)}{n^{2w}}\, dw$$
$$+ \frac{\sigma - 1}{2\pi i}\int_{(\frac{7}{4})}\frac{e^{\frac{1}{2}\pi i w}}{\cos\pi w} \cdot \frac{\Gamma(w)}{(\pi u^2)^w}\sum_{n=1}^{\infty}\frac{S_1(n, w)}{n^{2w-1}}\, dw$$
$$+ \frac{t}{6\pi}\int_{(\frac{11}{4})}\frac{e^{\frac{1}{2}\pi i w}}{\cos\pi w} \cdot \frac{\Gamma(w)}{(\pi u^2)^w}\sum_{n=1}^{\infty}\frac{S_3(n, w)}{n^{2w-3}}\, dw$$
$$+ O\left(\frac{\log t}{t}\right), \qquad (11)$$

where

$$S_\nu(n, w) = \sum_{m \le u} m^{2w-1-\nu}\sin\left(2\pi\frac{n}{m}u^2 + \frac{1}{4}\pi(1 - (-1)^\nu)\right).$$

3. Application of the saddle point method

Now we estimate asymptotically the sum $S_0(n, w)$; we stress that we have

$$\operatorname{Re}(w) = \tfrac{5}{4} . \tag{12}$$

By Poisson's sum formula we have, for any $0 < \delta < 1$,

$$S_0(n, w) = \int_\delta^u x^{2w-1} \sin(2n\pi u^2/x) \, dx$$

$$+ 2 \sum_{j=1}^\infty \int_\delta^u x^{2w-1} \sin(2n\pi u^2/x) \cos 2\pi j x \, dx.$$

But we have, by partial integration,

$$\sum_{j=1}^\infty \int_0^\delta x^{2w-1} \sin\frac{tn}{x} \cos 2\pi j x \, dx$$

$$= \delta^{2w-1} \sin\frac{tn}{\delta} \sum_{j=1}^\infty \frac{\sin 2\pi j \delta}{2\pi j} - \sum_{j=1}^\infty \int_0^\delta \frac{\sin 2\pi j x}{2\pi j} \left\{ \begin{matrix} (2w-1)x^{2w-2} \sin\dfrac{tn}{x} \\[2mm] - x^{2w-3} tn \cos\dfrac{tn}{x} \end{matrix} \right\} dx,$$

which is $\ll \delta^{\frac{1}{2}}$, disregarding the dependence on other parameters, because of (12) and the bounded convergence of the series $\sum_{j=1}^\infty \sin 2\pi j x / 2\pi j$. Thus we have

$$S_0(n, w) = \int_0^u x^{2w-1} \sin(2n\pi u^2/x) \, dx$$

$$+ 2 \sum_{j=1}^\infty \int_0^u x^{2w-1} \sin(2n\pi u^2/x) \cos 2\pi j x \, dx. \tag{13}$$

For the first term on the right side we have, by integrating twice by parts,

$$\int_0^u x^{2w-1} \sin(2n\pi u^2/x) dx = \frac{1}{2\pi n} u^{2w-1} \cos 2\pi n u + O\left(\frac{|w|^2 t^{\frac{1}{4}}}{n^2}\right). \tag{14}$$

On the other hand we have

$$\int_0^u x^{2w-1} \sin(2n\pi u^2/x) \cos 2\pi j x \, dx$$

$$= \frac{1}{4i} u^{2w} \left\{ Y(w, j, n) - \overline{Y(\overline{w}, j, n)} + Y(w, -j, n) - \overline{Y(\overline{w}, -j, n)} \right\}, \tag{15}$$

where

$$Y(w, j, n) = \int_0^1 x^{2w-1} \exp\big(2\pi i u(j x + n/x)\big) \, dx.$$

To estimate this integral we appeal to the saddle point method. The procedure is quite standard; and the most typical situation occurs when we have $j > n > 0$. The other cases can be treated similarly, so we may proceed on this assumption. We note also that in what follows all estimations are uniform in w.

Now, let $x_0 = (n/j)^{\frac{1}{2}}$, and let us introduce four oriented segments:

$$L_1 = \left[0, (1 - \tfrac{1}{\sqrt{2}}\varepsilon)x_0\right], \qquad L_2 = \left[(1 - \tfrac{1}{\sqrt{2}}\varepsilon)x_0, (1 + \tfrac{1}{\sqrt{2}}\varepsilon)x_0\right],$$
$$L_3 = \left[(1 + \tfrac{1}{\sqrt{2}}\varepsilon)x_0, 1 + \tfrac{1}{\sqrt{2}}\varepsilon\right], \qquad L_4 = \left[1 + \tfrac{1}{\sqrt{2}}\varepsilon, 1\right].$$

Also we define x^{2w-1} to be $\exp\big((2w - 1)\log x\big)$, where $\log x$ is real on the positive real axis. Then we have

$$Y(w, j, n) = \sum_{\nu=1}^{4} \int_{L_\nu} x^{2w-1} \exp\big(2\pi i u(jx + n/x)\big)\, dx. \tag{16}$$

We have

$$\int_{L_1} = \left(\frac{n}{j}\right)^w e^{-\frac{1}{2}\pi i w} \int_0^{\frac{1}{\sqrt{2}}} r^{2w-1} \exp\left(2\pi i n(jn)^{\frac{1}{2}}\left(\frac{r}{\varepsilon} + \frac{\varepsilon}{r}\right)\right) dr$$
$$\ll \left|\left(\frac{n}{j}\right)^w\right| \exp\big(\tfrac{1}{2}\pi|w| - \pi u(jn)^{\frac{1}{2}}\big), \tag{17}$$

since $\mathrm{Im}(r/\varepsilon + \varepsilon/r) \geq \frac{1}{2}$. We have also, for α such that $\sin\alpha = \frac{1}{3}$ when $0 < \alpha < \frac{1}{4}\pi$,

$$\int_{L_3} = e^{2i\alpha w} \int_{\frac{1}{2}\sqrt{10}x_0}^{\frac{1}{2}\sqrt{10}} r^{2w-1} \exp\left(2\pi i u\left(jre^{i\alpha} + \frac{n}{r}e^{-i\alpha}\right)\right) dr;$$

thus

$$\int_{L_3} \ll \exp\big(\tfrac{1}{2}\pi|w| - \tfrac{2}{3}\pi u(jn)^{\frac{1}{2}}\big). \tag{18}$$

As for the integral along L_4 we have, putting $x = 1 + \varepsilon r$,

$$\int_{L_4} = -\varepsilon \int_0^{\frac{1}{\sqrt{2}}} (1 + \varepsilon r)^{2w-1} \exp\left(2\pi i u\left(j + n + (j - n)\varepsilon r + \frac{inr^2}{1 + \varepsilon r}\right)\right) dr$$
$$= i\frac{e^{2\pi i u(j+n)}}{2\pi u(j - n)} \int_0^{\frac{1}{\sqrt{2}}} (1 + \varepsilon r)^{2w-1} \exp\left(-\frac{2n\pi ur^2}{1 + \varepsilon r}\right) \frac{d}{dr}\big(e^{2\pi\varepsilon^3 u(j-n)r}\big) dr.$$

Integrating by parts, we see that

$$\int_{L_4} = \frac{e^{2\pi i u(j+n)}}{2\pi i u(j - n)} + O\left(\frac{j}{(j - n)^3 t}\, e^{\frac{1}{2}\pi|w|}\right). \tag{19}$$

We next treat the integral along L_2. Putting $x = (1 + \varepsilon r)x_0$ and estimating crudely the part corresponding to $\eta < |r| \leq 1/\sqrt{2}$, where $\eta = (4\pi^2 n^2 jn)^{-1/5}$, we have

$$\int_{L_2} = \left(\frac{n}{j}\right)^w \exp\left(\left(4\pi u(jn)^{\frac{1}{2}} + \tfrac{1}{4}\pi\right)i\right) \int_{-\eta}^{\eta} (1 + \varepsilon r)^{2w-1} \exp\left(\frac{-2\pi u(jn)^{\frac{1}{2}} r^2}{1 + \varepsilon r}\right) dr$$

$$+ O\left(e^{\frac{1}{2}\pi|w| - \frac{1}{6}(tjn)^{\frac{1}{10}}}\right).$$

Here we note that

$$(1 + \varepsilon r)^{2w-1} = 1 + (2w - 1)\varepsilon r + (2w - 1)(w - 1)\varepsilon^2 r^2$$

$$+ 2(2w - 1)(w - 1)(2w - 3)(\varepsilon r)^3 \int_0^1 (\xi - 1)^2 (1 + \varepsilon r\xi)^{2w-4}\, d\xi,$$

in which the last term is $O\left(|w|^3 e^{\frac{1}{2}\pi|w|} r^3\right)$. Further we have, for $|r| \leq \eta$,

$$\left\{1 + (2w - 1)\varepsilon r + (2w - 1)(w - 1)\varepsilon^2 r^2\right\} \exp\left(\frac{-2\pi u(jn)^{\frac{1}{2}} r^2}{1 + \varepsilon r}\right)$$

$$= \left\{\begin{array}{l} 1 + (2w - 1)\varepsilon r + (2w - 1)(w - 1)\varepsilon^2 r^2 \\[4pt] + 2\varepsilon\pi u(jn)^{\frac{1}{2}} r^3 + 4\pi i(w - 1)u(jn)^{\frac{1}{2}} r^4 + 2\pi^2 iu^2 r^6 \end{array}\right\} e^{-2\pi u(jn)^{\frac{1}{2}} r^2}$$

$$+ O\left((tjn)^{\frac{3}{2}}\left(1 + |w|r\right)^2 e^{-2\pi u(jn)^{\frac{1}{2}} r^2} r^9\right).$$

From these we obtain

$$\int_{L_2} = \left(\frac{n}{j}\right)^w \exp\left(\left(4\pi u(jn)^{\frac{1}{2}} + \tfrac{1}{4}\pi\right)i\right)$$

$$\times \left\{\frac{\sqrt{\pi}}{(4\pi^2 u^2 jn)^{\frac{1}{4}}} + i\left(w^2 - \tfrac{1}{16}\right)\frac{\sqrt{\pi}}{(4\pi^2 u^2 jn)^{\frac{3}{4}}}\right\}$$

$$+ O\left(\left|\left(\frac{n}{j}\right)^w\right|\frac{|w|^3 e^{\frac{1}{2}\pi|w|}}{tjn}\right). \tag{20}$$

Hence, by (16)–(20), we find that if $j > n > 0$

$$Y(w, j, n) = \left(\frac{n}{j}\right)^w \exp\left(\left(4\pi u(jn)^{\frac{1}{2}} + \tfrac{1}{4}\pi\right)i\right)$$

$$\times \left\{\frac{\sqrt{\pi}}{(4\pi^2 u^2 jn)^{\frac{1}{4}}} + i\left(w^2 - \tfrac{1}{16}\right)\frac{\sqrt{\pi}}{(4\pi^2 u^2 jn)^{\frac{3}{4}}}\right\} + \frac{e^{2\pi iu(j+n)}}{2\pi u(j - n)i}$$

$$+ O\left\{\left(\left|\left(\frac{n}{j}\right)^w\right|\frac{|w|^3}{jn} + \frac{j}{(j - n)^3}\right)\frac{e^{\frac{1}{2}\pi|w|}}{t}\right\}.$$

In much the same way it can be shown that if $j = n > 0$

$$Y(w, j, n) = \exp\left(i(4\pi un + \tfrac{1}{4}\pi)\right) \left\{ \frac{\sqrt{\pi}}{2(4\pi^2 u^2 n^2)^{\frac{1}{4}}} - \frac{\varepsilon w}{2\pi un} \right.$$

$$\left. + \frac{i\left(w^2 - \tfrac{1}{16}\right)\sqrt{\pi}}{2(4\pi^2 u^2 n^2)^{\frac{3}{4}}} \right\} + O\left(\frac{e^{\frac{1}{2}\pi|w|}}{tn^2} \right),$$

that if $n > j > 0$

$$Y(w, j, n) = \frac{e^{2\pi i u(j+n)}}{2\pi i u(j-n)} + O\left(\frac{j}{(j-n)^3 t} e^{\frac{1}{2}\pi|w|} \right),$$

and that if $j > 0$, $n > 0$,

$$Y(w, -j, n) = -\frac{e^{2\pi i u(n-j)}}{2\pi i u(j+n)} + O\left(\frac{1}{(j+n)^2 t} e^{\frac{1}{2}\pi|w|} \right).$$

Into (15) we insert these results on $Y(w, j, n)$, and then we have, by (13) and (14),

$$\frac{S_0(n, w)}{u^{2w}} = -\frac{1}{u}\left(\theta(u) - \tfrac{1}{2}\right) \sin 4\pi nu - \frac{w}{2\pi nu} \cos 4\pi nu$$

$$+ \frac{n^{w-\frac{1}{4}}}{\sqrt{2u}} \sum_{j=n}^{\infty}{}' j^{-w-\frac{1}{4}} \sin\left(4\pi u(jn)^{\frac{1}{2}} + \tfrac{1}{4}\pi\right)$$

$$+ \frac{n^{w-\frac{3}{4}}}{(2u)^{\frac{3}{2}}\pi} \sum_{j=n}^{\infty}{}' j^{-w-\frac{3}{4}} \cos\left(4\pi u(jn)^{\frac{1}{2}} + \tfrac{1}{4}\pi\right)$$

$$+ O\left(\frac{n|w|^3 e^{\frac{1}{2}\pi|w|}}{t} \right),$$

where the prime indicates that the terms with $j = n$ are halved. Hence we find that on (12)

$$\sum_{n=1}^{\infty} \frac{S_0(n, w)}{n^{2w}} = -u^{2w-1}\left(\theta(u) - \tfrac{1}{2}\right) \sum_{n=1}^{\infty} \frac{\sin 4\pi nu}{n^{2w}}$$

$$- 2\pi w u^{2w-1} \sum_{n=1}^{\infty} \frac{\cos 4\pi nu}{n^{2w+1}} + \frac{u^{2w-\frac{1}{2}}}{2\sqrt{2}} \sum_{n=1}^{\infty} \frac{d(n)\sin\left(4\pi u\sqrt{n} + \tfrac{1}{4}\pi\right)}{n^{w+\frac{1}{4}}}$$

$$+ \left(w^2 - \tfrac{1}{16}\right) \frac{u^{2w-\frac{3}{2}}}{2^{\frac{5}{2}}\pi} \sum_{n=1}^{\infty} \frac{d(n)\cos\left(4\pi u\sqrt{n} + \tfrac{1}{4}\pi\right)}{n^{w+\frac{3}{4}}} + O\left(t^{\frac{1}{4}}|w|^3 e^{\frac{1}{2}\pi|w|}\right). \quad (21)$$

In much the same way we may perform the asymptotic estimation of $S_\nu(n, w)$ for $\nu = 1, 3$, and the result can be expressed as

$$\sum_{n=1}^{\infty} \frac{S_\nu(n, w)}{n^{2w-\nu}} = \frac{u^{2w-\nu-\frac{1}{2}}}{2\sqrt{2}} \sum_{n=1}^{\infty} \frac{d(n)\cos\left(4\pi u\sqrt{n} + \frac{1}{4}\pi\right)}{n^{w+\frac{1}{4}-\frac{1}{2}\nu}} + O\left(t^{\frac{5}{4}}e^{\frac{1}{2}\pi|w|}\right), \quad (22)$$

where $\operatorname{Re}(w) = \frac{1}{4}(2\nu + 5)$.

We now turn to the two sums in (11) which involve $\theta(u^2/m)$. We have first

$$\sum_{m \le u}\left(\theta\left(\frac{u^2}{m}\right) - \frac{1}{2}\right) = \sum_{m \le u}\left\{\frac{u^2}{m} - \sum_{n \le u^2/m} 1 - \frac{1}{2}\right\}$$

$$= u^2 \sum_{m \le u} \frac{1}{m} - \frac{1}{2}\sum_{m \le u^2} d(m) - \frac{1}{2}[u]^2 - \frac{1}{2}[u].$$

We note that for $x > 1$

$$\sum_{m \le x} \frac{1}{m} = \log x + \gamma - \frac{\theta(x) - \frac{1}{2}}{x} - \frac{\left(\theta(x) - \frac{1}{2}\right)^2 - \frac{1}{12}}{2x^2} + O\left(\frac{1}{x^3}\right). \quad (23)$$

Hence we have

$$\sum_{m \le u}\left(\theta\left(\frac{u^2}{m}\right) - \frac{1}{2}\right) = -\frac{1}{2}\Delta(u^2) - \frac{1}{4}d(u^2) + \frac{1}{24} - \left(\theta(u) - \frac{1}{2}\right)^2 + O\left(t^{-\frac{1}{2}}\right). \quad (24)$$

On the other hand, on noting that for $x > 0$

$$\left(\theta(x) - \frac{1}{2}\right)^2 = 2\sum_{l \le x} l - x^2 + 2\left(\theta(x) - \frac{1}{2}\right)x + \frac{1}{4},$$

we have

$$\sum_{m \le u} m\left(\left(\theta\left(\frac{u^2}{m}\right) - \frac{1}{2}\right)^2 - \frac{1}{12}\right) = \sum_{m \le u^2} m\, d(m) + \left(\sum_{m \le u} m\right)^2 + \frac{1}{6}\sum_{m \le u} m$$

$$- u^4 \sum_{m \le u} \frac{1}{m} + 2u^2 \sum_{m \le u}\left(\theta\left(\frac{u^2}{m}\right) - \frac{1}{2}\right).$$

Thus we have, by (23) and (24),

$$\sum_{m \le u} m\left(\left(\theta\left(\frac{u^2}{m}\right) - \frac{1}{2}\right)^2 - \frac{1}{12}\right)$$

$$= -\sum_{m \le u^2} (u^2 - m)d(m) + u^4\left(\log u + \gamma - \frac{3}{4}\right) + \frac{1}{4}u^2 + O\left(t^{\frac{1}{2}}\right)$$

$$= -\Delta_1(u^2) + O\left(t^{\frac{1}{2}}\right). \quad (25)$$

4. Rearrangements

Now we insert (21), (22), (24) and (25) into (11), and obtain

$$K(s) = -\frac{1}{2}\sum_{m \leq u}\frac{1}{m} - \varepsilon\frac{3\sigma - 2}{6\pi u}(\log u + \gamma) + \frac{1}{u}Q(t)$$

$$-\frac{1}{2\varepsilon u}\Delta(u^2) - \frac{1}{4\varepsilon u}d(u^2) - \frac{\sigma}{2\varepsilon u^3}\Delta_1(u^2)$$

$$+\frac{1}{4\pi\sqrt{2u}}\sum_{n=1}^{\infty}n^{-\frac{1}{4}}d(n)\sin\left(4\pi u\sqrt{n} + \tfrac{1}{4}\pi\right)g_1(n\pi)$$

$$-\frac{1}{128\sqrt{2}\pi^2 u^{\frac{3}{2}}}\sum_{n=1}^{\infty}n^{-\frac{3}{4}}d(n)\cos\left(4\pi u\sqrt{n} + \tfrac{1}{4}\pi\right)g_1(n\pi)$$

$$-i\frac{\sigma - \frac{1}{2}}{4\sqrt{2\pi}u^{\frac{3}{2}}}\sum_{n=1}^{\infty}n^{\frac{1}{4}}d(n)\cos\left(4\pi u\sqrt{n} + \tfrac{1}{4}\pi\right)g_2(n\pi)$$

$$+\frac{1}{24\sqrt{2}u^{\frac{3}{2}}}\sum_{n=1}^{\infty}n^{\frac{5}{4}}d(n)\cos\left(4\pi u\sqrt{n} + \tfrac{1}{4}\pi\right)g_3(n\pi)$$

$$+O\left(\frac{\log t}{t}\right), \tag{26}$$

where

$$g_\nu(x) = \int_{(\nu+\frac{1}{4})}\frac{e^{\frac{1}{2}\pi i w}}{\cos \pi w} \cdot \frac{\Gamma(w)}{x^w}\,dw,$$

and

$$Q(t) = -\frac{1}{\varepsilon}\left(\left(\theta(u) - \tfrac{1}{2}\right)^2 - \tfrac{1}{24}\right) + \frac{1}{4\pi i}\sum_{n=1}^{\infty}n\cos(4n\pi u)g_2(n^2\pi)$$

$$-\frac{1}{2\pi}\left(\theta(u) - \tfrac{1}{2}\right)\sum_{n=1}^{\infty}\sin(4n\pi u)g_1(n^2\pi).$$

We note that for $\nu \geq 1$

$$g_\nu(x) = 2e^{\frac{1}{2}\pi i(\frac{1}{2}-\nu)}\Gamma\left(\nu - \tfrac{1}{2}\right)x^{\frac{1}{2}-\nu} + g_{\nu-1}(x),$$

while for $\nu \geq 0$

$$\frac{d}{dx}g_\nu(x) = -i\,g_{\nu+1}(x).$$

Solving the differential equation arising from the last two relations we find that for $\nu \geq 1$

$$g_\nu(x) = (-1)^\nu 2\varepsilon^{2\nu+1}\Gamma\left(\nu + \tfrac{1}{2}\right)g_\nu^*(x), \tag{27}$$

where

$$g_\nu^*(x) = \int_0^\infty \frac{e^{i\xi}}{(\xi + x)^{\nu+\frac{1}{2}}}\,d\xi.$$

Inserting (26) and (27) into (3) and comparing the result with the statement of the Theorem, we see that what remains for us to prove is the relation

$$\mathrm{Re}\{Q(t)\} = \frac{1}{4\sqrt{2\pi}} - \tfrac{1}{4}u\chi(1-s)E(s,u)^2 + O\big(t^{-\frac{1}{2}}\log t\big),$$

or, by (2),

$$\mathrm{Re}\{Q(t)\} = \frac{1}{4\sqrt{2\pi}} - \left(\frac{\cos\big(2\pi(\delta^2 - \delta - \frac{1}{16})\big)}{4\cos 2\pi\delta}\right)^2 + O\left(\frac{\log t}{t^{\frac{1}{2}}}\right).$$

We shall show that this holds without the error term. To this end we put

$$\Phi(a) = \int_L \exp\Big(\frac{i}{4\pi}w^2 + aw\Big)\frac{dw}{e^w - 1},$$

where L is the line used in (4). According to Siegel [1] (or [2], (2.10.5)) we have

$$\Phi(a) = 2\pi\frac{\cos\big(\pi(\frac{1}{2}a^2 - a - \frac{1}{8})\big)}{\cos \pi a}e^{i\pi(\frac{1}{2}a^2 - \frac{5}{8})}; \qquad (28)$$

and so we have also

$$\Phi'(a) = \pi\frac{\sin \pi a}{\cos \pi a}\Phi(a) + i\pi a\Phi(a) - 2\pi^2(a-1)\frac{\sin\big(\pi(\frac{1}{2}a^2 - a - \frac{1}{8})\big)}{\cos \pi a}e^{i\pi(\frac{1}{2}a^2 - \frac{5}{8})}. \qquad (29)$$

On the other hand, the argument which was used to derive (8)–(10) yields

$$\Phi(a) = -\pi i + 2\pi\varepsilon\big(a - \tfrac{1}{2}\big) + i\sum_{n=1}^{\infty} \sin(2\pi na)g_1(n^2\pi)$$

as well as

$$\Phi'(a) = 2\varepsilon\pi + 2\varepsilon^2\pi^2\big((a - \tfrac{1}{2})^2 - \tfrac{1}{12}\big) + 2\pi i\sum_{n=1}^{\infty} n\cos(2\pi na)g_2(n^2\pi),$$

provided $0 \le a \le 1$. We apply these with $a = \theta(2u)$, and find that

$$Q(t) = \frac{1}{2\pi i}\big(\tfrac{1}{2} - \theta(u)\big)\Big(\Phi\big(\theta(2u)\big) + \pi i\Big) - \frac{1}{8\pi^2}\Big(\Phi'\big(\theta(2u)\big) - 2\varepsilon\pi\Big).$$

We then consider two cases separately according to whether $0 \leq \delta < \frac{1}{2}$ or $\frac{1}{2} \leq \delta < 1$: In the first case we have $\theta(2u) = 2\delta$. Thus we have, by (28) and (29),

$$Q(t) = \frac{\varepsilon}{4\pi} + \tfrac{1}{2}(\delta - \tfrac{1}{2})\left\{ \frac{i}{2\pi}\Phi(2\delta) + \frac{\sin\left(2\pi(\delta^2 - \delta - \frac{1}{16})\right)}{\cos 2\pi\delta} e^{2\pi i(\delta^2 - \frac{5}{16})} - 1\right\}$$
$$\quad - \frac{1}{8\pi}(i + \tan 2\pi\delta)\Phi(2\delta)$$
$$= \frac{\varepsilon}{4\pi} + \frac{i}{2}(\delta - \tfrac{1}{2})\tan 2\pi\delta - \frac{\cos\left(2\pi(\delta^2 - \delta - \frac{1}{16})\right)}{4\left(\cos 2\pi\delta\right)^2} e^{2\pi i(\delta^2 - \frac{5}{16})},$$

which obviously proves our claim. The other case can be treated in just the same way; and we end the proof of the Theorem.

References

1 C.L. Siegel: Über Riemanns Nachlass zur analytischen Zahlentheorie. *Quellen und Studien zur Geschichte der Math. Astr. und Physik*, Abt. B, Studien 2 (1932), 45–80.
2 E.C. Titchmarsh: *The Theory of the Riemann Zeta-Function.* Oxford University Press (1951).

A historical note

The manuscript of the present paper was prepared late in 1982, and has remained unpublished because of an unhappy circumstance, although a short announcement [5] (cf. [3], p. 125) appeared later and an expanded version [6] was printed in a very limited number of copies (the third part of [5] as well as [6] contain the *non-symmetric* case, too). The manuscript was studied by Jutila and he could devise his own argument [4] to prove a result same as the one stated in the above corollary but not the asymptotic formula in the theorem. Our work was submitted twice to periodicals but was rejected, each time with referees' opinion to the effect that Taylor [7] had obtained the same result as our theorem, which incidentally coincides with Heath-Brown's note given in his edition (page 94 of its first printing) of the book [2]. However, we found that Taylor's argument was incorrect and the error appears to be fatal, of which we informed editorial boards of the periodicals but in vain. It should be stressed that our argument is completely different from Taylor's plan. Titchmarsh did not mention Taylor's work in his book in spite of his listing it in the bibliography; a fact indicating his awareness of the incorrectness of Taylor's claim. Taylor perished in an air battle in 1943; and his work was communicated by Titchmarsh to Oxford University at the request of the bereaved family. We deeply regret that the mathematical career of this young number-theorist was cut short.

Additional References

3 A. Ivić: *The Riemann Zeta-Function.* Wiley (1985).
4 M. Jutila: On the approximate functional equation for $\zeta^2(s)$ and other Dirichlet series. *Quarterly J. Math. Oxford* (2) 37 (1986), 193–209.
5 Y. Motohashi: A note on the approximate functional equation for $\zeta^2(s)$. *Proc. Japan Acad.* 59A (1983), 392–396; *II.* ibid, 469–472; *III.* ibid 62A (1986), 410–412.
6 Y. Motohashi: Riemann-Siegel Formula. *Ulam Seminar* (Colorado Univ., 1987).
7 P.R. Taylor: On the Riemann zeta-function. *Quarterly J. Math. Oxford*, 16 (1945), 1–21.

18. The Mean Square of Dedekind Zeta-Functions of Quadratic Number Fields

Yoichi Motohashi

The aim of the present paper is to show an intrinsic relation between the Dedekind zeta-functions of quadratic number fields and the Hecke L-functions attached to cusp-forms over congruence subgroups by means of investigating the mean square of those zeta-functions on the critical line. We shall here restrict ourselves to the structural aspect of such a relation; and the quantitative consequences of the present work, which becomes subtle and interesting when the uniformity with respect to the discriminants is taken into account, will be studied elsewhere. What we are developing below is in fact an extension of our works [4], [7] on the Riemann zeta-function. Thus the argument is naturally similar and we shall be mostly brief save for necessary explanations on some non-trivial ramifications related to the fact that the underlying groups have levels larger than 1; for instance, discussion about convergence is left for readers. Also, we shall not treat the problem in its full generality, since we assume (1.1) to avoid unnecessary complexity.

1. Reduction

Let $\zeta_D(s)$ be the Dedekind zeta-function of the quadratic number field with the discriminant D. We have

$$\zeta_D(s) = \zeta(s)L(s, \chi),$$

where $\zeta(s)$ is the Riemann zeta-function, and $L(s, \chi)$ the Dirichlet L-function with a real primitive character $\chi \bmod |D|$ such that $\chi(-1) = \mathrm{sgn}(D)$. We assume throughout the paper that

$$q = |D| \text{ is square-free.} \tag{1.1}$$

When q is not square-free, it is divisible by 4 with a square-free quotient, and it can be seen that our argument below extends to those cases, too, at the cost of minor technical complexity.

Then, we are going to give an explicit formula for

$$I(T, G; D) = \frac{1}{G\sqrt{\pi}} \int_{-\infty}^{\infty} \left|\zeta_D\left(\tfrac{1}{2} + i(T + t)\right)\right|^2 \exp\left(-\left(\frac{t}{G}\right)^2\right) dt,$$

Sieve Methods, Exponential Sums, and their Applications in Number Theory
Greaves, G.R.H., Harman, G., Huxley, M.N., Eds. ©Cambridge University Press, 1996

where T, $G > 0$ are arbitrary. To this end we consider

$$I(u, v, w, z; \chi)$$
$$= \frac{1}{iG\sqrt{\pi}} \int_{(0)} \zeta(u+s)L(v+s,\chi)\zeta(w-s)L(z-s,\chi)\exp\left(\left(\frac{s}{G}\right)^2\right)ds,$$

where u, v, w, z are complex numbers with real parts larger than 1 and (α) denotes the vertical line passing through α. We have omitted G on the left side for the sake of notational simplicity; we note that similar conventions will often be employed below. Shifting the contour appropriately we see that $I(u, v, w, z; \chi)$ is meromorphic over the whole of $\mathbb{C}^4$. In particular we have

$$I(P_T; \chi) = I(T, G; D) + i\frac{(1 - \chi(-1))\tau(\chi)L^2(1,\chi)}{\sqrt{\pi}G}\operatorname{Re}\left[\exp\left(\left(\frac{\frac{1}{2}+iT}{G}\right)^2\right)\right],$$

where $\tau(\chi)$ is the Gaussian sum attached to χ and

$$P_T = \left(\tfrac{1}{2}+iT, \tfrac{1}{2}+iT, \tfrac{1}{2}-iT, \tfrac{1}{2}-iT\right).$$

On the other hand we have, in the region of absolute convergence,

$$I(u, v, w, z; \chi) = \sum_{m,n=1}^{\infty} \frac{\sigma_{u-v}(m,\chi)\sigma_{w-z}(n,\chi)}{m^u n^w}\exp\left(-\left(\tfrac{1}{2}G\log\frac{n}{m}\right)^2\right),$$

where

$$\sigma_\xi(n,\chi) = \sum_{d|n}\chi(d)d^\xi.$$

As in [4], we apply the Atkinson dissection to the last double sum, and have

$$I(u, v, w, z; \chi) = Y_0(u, v, w, z; \chi) + Y_1(u, v, w, z; \chi) + Y_1(w, z, u, v; \chi), \quad (1.2)$$

where Y_0 corresponds to the part with $m = n$ and Y_1 to $m < n$. We are going to show that Y_0 and Y_1 exist as meromorphic functions over $\mathbb{C}^4$ so that the last identity holds for all (u, v, w, z). The continuation of Y_0 is easy, for we have

$$Y_0(u, v, w, z; \chi) = \frac{\zeta(u+w)L(v+z,\chi_0)L(u+z,\chi)L(v+w,\chi)}{L(u+v+w+z,\chi_0)},$$

where χ_0 denotes the principal character $\operatorname{mod} q$. But the continuation of Y_1 requires the full machinery of the spectral theory of automorphic forms over the Hecke congruence subgroup $\Gamma_0(q)$, a brief account of which is to be given in the next section. Here we make an initial preparation, namely the reduction of Y_1 to a sum of Kloosterman sums.

We have

$$Y_1(u, v, w, z; \chi) = \sum_{m,n=1}^{\infty} \frac{\sigma_{u-v}(m, \chi)\sigma_{w-z}(m+n, \chi)}{m^{u+z}(m+n)^{w-z}} W(n/m, z),$$

where

$$W(x, z) = (1 + x)^{-z} \exp\left(-\left(\tfrac{1}{2}G\log(1 + x)\right)^2\right).$$

We separate the variables m, n in the last sum. For this sake we invoke the following extension of an identity of Ramanujan: If $\mathrm{Re}(\xi) > 0$, then we have

$$\frac{\sigma_\xi(n, \chi)}{n^\xi} = \frac{L(\xi + 1, \chi)}{\tau(\chi)} \sum_{l=1}^{\infty} \frac{1}{l^{\xi+1}} \sum_{\substack{h=1 \\ (h,ql)=1}}^{ql} \chi(h)e\left(\frac{h}{ql}n\right),$$

where $e(x) = \exp 2\pi i x$ as usual. This holds for all characters $(\bmod\ q)$ (including the principal character) as can be seen by expressing the condition $(h, ql) = 1$ in terms of the Möbius function and then verifying that the coefficients of $l^{-\xi}$ are equal on both sides of the identity

$$\frac{\sigma_\xi(n, \chi)\tau(\chi)}{L(1 + \xi, \chi)n^\xi} = \sum_{l=1}^{\infty} \frac{1}{l^{\xi+1}} \sum_{d|l} \mu(d)\chi(d) \sum_{h=1}^{ql/d} \chi(h)e\left(\frac{hd}{ql}n\right).$$

We need also the Mellin transform

$$W^*(s, z) = \int_0^{\infty} W(x, z)x^{s-1}dx.$$

This exists as a function over $\mathbb{C}^2$, which is meromorphic with respect to s having simple poles at non-positive integers and entire with respect to z. Further we introduce the domain

$$\mathcal{D}_\alpha = \{(u, v, w, z) : \mathrm{Re}(u) > 1, \mathrm{Re}(v) > 1, \mathrm{Re}(w) > \mathrm{Re}(z) + 1, \mathrm{Re}(z) > \alpha\},$$

where $\alpha > 1$.

Then we have, in $\mathcal{D}_\alpha$,

$$Y_1(u, v, w, z; \chi)$$
$$= \frac{L(w - z + 1, \chi)}{2\pi i \tau(\chi)} \sum_{l=1}^{\infty} l^{z-w-1} \sum_{\substack{h=1 \\ (h,ql)=1}}^{ql} \chi(h)$$
$$\times \int_{(\alpha)} D_\chi(u + z - s, u - v; e(h/ql))\zeta(s, e(h/ql))W^*(s, z)\,ds. \quad (1.3)$$

Here we have

$$D_\chi\big(s,\xi;e(h/ql)\big) = \sum_{n=1}^\infty \frac{\sigma_\xi(n,\chi)\,e(hn/ql)}{n^s}, \quad \zeta\big(s,e(h/ql)\big) = \sum_{n=1}^\infty \frac{e(hn/ql)}{n^s}.$$

We are going to shift the contour in (1.3) to the right appropriately. But, before doing it we need to quote some properties of $D_\chi\big(s,\xi;e(h/ql)\big)$ as a function of s and ξ. By expressing it in terms of Hurwitz zeta-functions, we see readily that it is meromorphic over $\mathbb{C}^2$ and that if $\xi \neq 0$ and $(h,ql)=1$ then it has a simple pole at $s = 1 + \xi$ with the residue

$$\frac{\chi(h)\tau(\chi)L(\xi+1,\chi)}{ql^{\xi+1}},$$

and is regular for all other s. Moreover we have the functional equation

$$
\begin{aligned}
&D_\chi\big(s,\xi;e(h/ql)\big) \\
&= \chi(-\overline{h})(2\pi)^{2s-\xi-2}(ql)^{\xi-2s+1}\Gamma(1-s)\Gamma(1+\xi-s) \\
&\quad \times \left[\begin{array}{l} \big\{e(\tfrac14\xi)+\chi(-1)e(-\tfrac14\xi)\big\}D_\chi\big(1-s+\xi,\xi;e(\overline{h}/ql)\big) \\[4pt] -\big\{e(\tfrac14(2s-\xi))+\chi(-1)e(-\tfrac14(2s-\xi))\big\}D_\chi\big(1-s+\xi,\xi;e(-\overline{h}/ql)\big) \end{array} \right],
\end{aligned}
\tag{1.4}
$$

where $h\overline{h} \equiv 1 \bmod ql$.

We then introduce the new domain

$$\mathcal{E}_\beta = \big\{(u,v,w,z) : \operatorname{Re}(u+z) < \beta,\ \operatorname{Re}(v+z) < \beta,\ \operatorname{Re}(u+v+w+z) > 3\beta\big\},$$

where β is to be sufficiently large; and for a while we restrict (u,v,w,z) in $\mathcal{D}_\alpha \cap \mathcal{E}_\beta$, which is not empty if $\beta > \alpha + 1$. We then shift the contour (α) in (1.3) to (β). We may rearrange the resulting expression of $Y_1(u,v,w,z)$ by virtue of (1.4), and we find that

$$
\begin{aligned}
&Y_1(u,v,w,z;\chi) \\
&= \{Y_1^+ + Y_1^-\}(u,v,w,z;\chi) + q^{1-v-z}\prod_{p\mid q}(1-p^{v+z-2}) \\
&\quad \times \frac{\zeta(u+w)\zeta(v+z-1)L(w-z+1,\chi)L(u-v+1,\chi)}{L(u-v-z+w+2,\chi_0)}\,W^*(v+z-1,z). \tag{1.5}
\end{aligned}
$$

Here

$$Y_1^{\pm}(u,v,w,z;\chi) = \frac{\chi(-1)}{\tau(\chi)}\left(\frac{q}{2\pi}\right)^{w-z+1}L(w-z+1,\chi)$$

$$\times \sum_{m,n=1}^{\infty} m^{\frac{1}{2}(1-u-v-w-z)}n^{\frac{1}{2}(v+z-u-w-1)}\sigma_{u-v}(n,\chi)$$

$$\times \sum_{\substack{l=1 \\ l\equiv 0 \bmod q}}^{\infty} \frac{1}{l}S(m,\pm n\,;l)\psi_{\pm}\left(\frac{4\pi}{l}\sqrt{mn};u,v,w,z;\chi\right) \quad (1.6)$$

with the Kloosterman sum

$$S(m,n\,;l) = \sum_{\substack{h=1 \\ (h,l)=1}}^{l} e\left(\frac{mh+n\overline{h}}{l}\right), \quad \text{where} \quad h\overline{h}\equiv 1 \bmod l,$$

and

$$\psi_+(x;u,v,w,z;\chi)$$

$$= \frac{1}{2\pi i}\Big\{e(\tfrac{1}{4}(u-v)) + \chi(-1)e(-\tfrac{1}{4}(u-v))\Big\}$$

$$\times \int_{(\beta)} \left(\tfrac{1}{2}x\right)^{u+v+w+z-2s-1}\Gamma(s-u-z+1)\Gamma(s-v-z+1)W^*(s,z)\,ds,$$

$$\psi_-(x;u,v,w,z;\chi)$$

$$= -\frac{1}{2\pi i}\int_{(\beta)} \Big\{e(\tfrac{1}{4}(u+v+2z-2s)) + \chi(-1)e(-\tfrac{1}{4}(u+v+2z-2s))\Big\}$$

$$\times \left(\tfrac{1}{2}x\right)^{u+v+w+z-2s-1}\Gamma(s-u-z+1)\Gamma(s-v-z+1)W^*(s,z)\,ds.$$

Since the right side of (1.6) is absolutely convergent throughout $\mathcal{E}_\beta$, the expression (1.5) gives a meromorphic continuation of $Y_1(u,v,w,z;\chi)$ to $\mathcal{E}_\beta$. However, $\mathcal{E}_\beta$ does not contain the point P_T; so we need further continuation of $Y_1(u,v,w,z;\chi)$, which will be accomplished with a spectral argument.

2. Spectral theory

Our problem has now been reduced to a sum of Kloosterman sums whose moduli are divisible by q. Thus we may appeal to the spectral theory of automorphic forms over the group $\Gamma_0(q)$. Here we shall give its essentials as much as we need in dealing with our problem; for a general account see [1]. It should be stressed that we have the assumption (1.1).

By the general theory we have

$$\{\text{cusps}\} \equiv \left\{ \frac{1}{a} : a \mid q \right\} \bmod \Gamma_0(q).$$

We map each point $1/a$ to a point equivalent $\bmod \Gamma_0(q)$ in a special way. For this sake we write $q = ab$, and also $q = a_i b_i$ in the sequel. Then we consider the congruence (cf. p. 534 in [2])

$$V_a := \begin{pmatrix} k & l \\ q & g \end{pmatrix} \begin{pmatrix} 1 & \\ a & 1 \end{pmatrix} \begin{pmatrix} 1 & f \\ & 1 \end{pmatrix} \equiv \begin{cases} \begin{pmatrix} 1 & \\ & 1 \end{pmatrix} & \bmod a, \\ \begin{pmatrix} & -1 \\ 1 & \end{pmatrix} & \bmod b, \end{cases}$$

where

$$\begin{pmatrix} k & l \\ q & g \end{pmatrix} \in \Gamma_0(q).$$

This has a solution such that $k = 1 + a - a\bar{a}$ and

$$a\bar{a} \equiv 1 \bmod b, \quad l \equiv -1 \bmod b, \quad l \equiv -f \bmod a, \quad f \equiv -\bar{a} \bmod b.$$

Let V_a stand for such a solution. Then we write

$$[a] = V_a(\infty).$$

The points $[a]$ with $a \mid q$, constitute obviously a representative set of inequivalent cusps of $\Gamma_0(q)$. Further, we put

$$V_a^* = V_a \begin{pmatrix} \sqrt{b} & \\ & \frac{1}{\sqrt{b}} \end{pmatrix},$$

so that we have

$$V_a^*(\infty) = [a], \qquad (V_a^*)^{-1} \Gamma_{[a]} V_a^* = \left\{ \begin{pmatrix} 1 & n \\ & 1 \end{pmatrix} : n \in \mathbb{Z} \right\},$$

where $\Gamma_{[a]}$ is the stabilizer of $[a]$ in $\Gamma_0(q)$.

We then consider, for integers $m \geq 0$, the Poincaré series

$$U_m(z, [a]; s) = \sum_{\gamma \in \Gamma_{[a]} \backslash \Gamma_0(q)} \left(\operatorname{Im}(V_a^*)^{-1} \gamma(z) \right)^s e\left(m(V_a^*)^{-1} \gamma(z) \right),$$

$$(z = x + iy, \; y > 0.)$$

The Fourier expansion of $U_m(z, [a_1]; s)$ around the cusp $[a_2]$ is as follows:

$$U_m(V_{a_2}^*(z), [a_1]; s)$$

$$= \delta_{a_1, a_2} e(m(z + c_{a_1, a_2})) y^s$$

$$+ y^{1-s} \sum_{n=-\infty}^{\infty} e(nx) \sum_{\substack{r=1 \\ ((a_1, b_2)(a_2, b_1), r)=1}}^{\infty} \frac{S(\overline{(a_2, b_1)}m, \overline{(a_1, b_2)}n ; (a_1, a_2)(b_1, b_2)r)}{((a_1, a_2)r\sqrt{b_1 b_2})^{2s}}$$

$$\times \int_{-\infty}^{\infty} \exp\left(-2\pi i n y \xi - \frac{2\pi m}{((a_1, a_2)r\sqrt{b_1 b_2})^2 y(1 - i\xi)}\right) \frac{d\xi}{(1 + \xi^2)^s}. \quad (2.1)$$

Here δ_{a_1, a_2} is the Kronecker delta, c_{a_1, a_2} a certain real number, S the Kloosterman sum; and the bars denote congruence inverses mod $(a_1, a_2)(b_1, b_2)r$. In particular, we have the following Fourier expansion of the Eisenstein series $E(z, [a]; s) = U_0(z, [a]; s)$: for any combination of cusps $[a_1]$ and $[a_2]$

$$E(V_{a_2}^*(z), [a_1]; s)$$

$$= \delta_{a_1, a_2} y^s + \sqrt{\pi} y^{1-s} \frac{\Gamma(s - \frac{1}{2})}{\Gamma(s)} \phi_0(s; a_1, a_2)$$

$$+ 2y^{\frac{1}{2}} \frac{\pi^s}{\Gamma(s)} \sum_{n \neq 0} |n|^{s - \frac{1}{2}} \phi_n(s; a_1, a_2) K_{s - \frac{1}{2}}(2\pi |n| y) e(nx), \quad (2.2)$$

where K_ν is the K-Bessel function of order ν and

$$\phi_n(s; a_1, a_2)$$

$$= \frac{\sigma_{1-2s}(n, \chi_0)}{L(2s, \chi_0)((a_1, b_2)(a_2, b_1))^s} \prod_{p | (a_1, a_2)(b_1, b_2)} \left(\sigma_{1-2s}(n_p)(1 - p^{-2s}) - 1\right)$$

with $n_p = (n, p^\infty)$.

Next, let $\{\lambda_j = \frac{1}{4} + \kappa_j^2 > 0 : j \geq 1\} \cup \{0\}$ be the discrete spectrum of the hyperbolic Laplacian over $\Gamma_0(q)$. Let φ_j be the Maass wave corresponding to λ_j, so that the set $\{\varphi_j\}$ forms an orthonormal system. The Fourier expansion of $\varphi_j(z)$ around the cusp $[a]$ is denoted by

$$\varphi_j(V_a^*(z)) = \sqrt{y} \sum_{n \neq 0} \rho_j(n, [a]) K_{i\kappa_j}(2\pi |n| y) e(nx).$$

Also let $\{\varphi_{j,k} : 1 \leq j \leq \vartheta_q(k)\}$ be an orthonormal base of the space of holomorphic cusp forms of the even weight $2k$ with respect to $\Gamma_0(q)$. The Fourier expansion of $\varphi_{j,k}(z)$ around the cusp $[a]$ is denoted by

$$\varphi_{j,k}(V_a^*(z)) = \sum_{n > 0} \rho_{j,k}(n, [a]) n^{k - \frac{1}{2}} e(nz).$$

We may now state a version of Kuznetsov's sum formulas (cf. Theorem 1 in [8]):

Lemma. *Let $m > 0$, $n > 0$. Let $\phi(x)$ be sufficiently smooth for $x \geq 0$ and of rapid decay as x tends either to $+0$ or to $+\infty$. Then we have*

$$\sum_{\substack{r=1 \\ ((a_1,b_2)(a_2,b_1),r)=1}}^{\infty} \frac{S\big(\overline{(a_2,b_1)}m,\, \pm\overline{(a_1,b_2)}n\,;\, (a_1,a_2)(b_1,b_2)r\big)}{(a_1,a_2)r\sqrt{b_1 b_2}}\, \phi\!\left(\frac{4\pi\sqrt{mn}}{(a_1,a_2)r\sqrt{b_1 b_2}}\right)$$

$$= \sum_{j=1}^{\infty} \frac{1}{\cosh \pi \kappa_j}\, \overline{\rho_j\big(m,[a_1]\big)}\, \rho_j\big(\pm n, [a_2]\big)(\phi)^{\pm}(\kappa_j)$$

$$+ \tfrac{1}{2}(1 \pm 1) \sum_{k=1}^{\infty} \frac{(2k-1)!}{\pi^{2k+1}2^{4k-2}} \sum_{j=1}^{\vartheta_q(k)} \overline{\rho_{j,k}\big(m,[a_1]\big)}\rho_{j,k}\big(n,[a_2]\big)(\phi)^{+}\big((\tfrac{1}{2}-k)i\big)$$

$$+ \frac{1}{\pi} \sum_{q=ab} \int_{-\infty}^{\infty} \Bigg\{ \frac{\sigma_{2ir}(m,\chi_0)\sigma_{-2ir}(n,\chi_0)(n/m)^{ir}}{\big|L(1+2ir,\chi_0)\big|^2 \big((a,b_1)(a_1,b)\big)^{\frac{1}{2}-ir}\big((a,b_2)(a_2,b)\big)^{\frac{1}{2}+ir}}$$

$$\times \prod_{p|(a,a_1)(b,b_1)} \big(\sigma_{2ir}(m_p)(1-p^{-1+2ir})-1\big)$$

$$\times \prod_{p|(a,a_2)(b,b_2)} \big(\sigma_{-2ir}(n_p)(1-p^{-1-2ir})-1\big)(\phi)^{\pm}(r)\Bigg\}dr,$$

$$(2.3)$$

where

$$(\phi)^{+}(r) = \frac{\pi i}{2 \sinh \pi r} \int_0^{\infty} \big(J_{2ir}(x) - J_{-2ir}(x)\big)\frac{\phi(x)}{x}\, dx,$$

$$(\phi)^{-}(r) = 2 \cosh \pi r \int_0^{\infty} K_{2ir}(x)\frac{\phi(x)}{x}\, dx$$

with J_ν being the J-Bessel function of order ν.

This is a consequence of (2.1) and (2.2). For a simple proof see our recent work [6]; there the case of the full modular group is treated, but the extension to the above situation is immediate. Also, the condition on ϕ can be relaxed considerably, as can be found in [6].

Next, as usual, we take into account the action of Hecke operators $T(n)$ over φ_j. We may assume that for all j and $n \geq 1$, $(n,q) = 1$, there exists a real number $t_j(n)$ such that

$$(T(n)\varphi_j)(z) = \frac{1}{\sqrt{n}} \sum_{ad=n} \sum_{b=1}^{d} \varphi_j\left(\frac{az+b}{d}\right) = t_j(n)\varphi_j(z).$$

Also, each φ_j can be taken for an eigenfunction of both the reflection operator $z \mapsto -\bar{z}$ and the Fricke involution $z \mapsto -1/(qz)$; so

$$\varphi_j(-\bar{z}) = \varepsilon_j \varphi_j(z), \qquad \varphi_j\big(-1/(qz)\big) = \eta_j \varphi_j(z)$$

with $\varepsilon_j = \pm 1$, $\eta_j = \pm 1$. We then consider the Fourier expansion of φ_j around the cusp at infinity:

$$\varphi_j(z) = \sqrt{y} \sum_{n \neq 0} \rho_j(n) K_{i\kappa_j}\big(2\pi|n|y\big) e(nx).$$

In this we have

$$\rho_j(n) = \rho_j\big(n, [q]\big)$$

and

$$\rho_j(n) = t_j(n_1)\rho_j(n_2), \quad \rho_j(-n) = \varepsilon_j\rho_j(n), \qquad (2.4)$$

where $n = n_1 n_2$ with $(n_1, q) = 1$, $n_2 \,|\, q^\infty$; the first assertion is equivalent to $V_q^* \in \Gamma_0(q)$.

We shall need the automorphic L-function attached to φ_j:

$$L_j(s) = \sum_{n=1}^{\infty} \frac{\rho_j(n)}{n^s},$$

which converges to an entire function, satisfying the functional equation

$$L_j(s) = \frac{\eta_j}{\pi} \left(\frac{\sqrt{q}}{2\pi}\right)^{1-2s} \Gamma(1-s+i\kappa_j)\Gamma(1-s-i\kappa_j)$$
$$\times \big(\varepsilon_j \cosh \pi\kappa_j - \cos \pi s\big) L_j(1-s). \quad (2.5)$$

We shall need also the χ-twist of the Hecke series attached to φ_j:

$$H_j(s, \chi) = \sum_{n=1}^{\infty} \frac{\chi(n)t_j(n)}{n^s}.$$

It converges to an entire function satisfying the functional equation

$$H_j(s, \chi) = \frac{1}{\pi} \left(\frac{q}{2\pi}\right)^{1-2s} \Gamma(1 - s + i\kappa_j)\Gamma(1 - s - i\kappa_j)$$
$$\times \big(\varepsilon_j \cosh \pi\kappa_j - \chi(-1)\cos \pi s\big) H_j(1 - s, \chi). \quad (2.6)$$

The first relation in (2.4) and the multiplicative property of Hecke eigenvalues yield the following identity, which is an essential tool in our argument: in the region of absolute convergence we have

$$\sum_{n=1}^{\infty} \frac{\sigma_a(n, \chi)\rho_j(n)}{n^s} = \frac{H_j(s - a, \chi)L_j(s)}{L(2s - a, \chi)}. \qquad (2.7)$$

Similarly we consider the action of Hecke operators $T_k(n)$ over holomorphic cusp forms: We may assume that each $\varphi_{j,k}$ is an eigenfunction of $T_k(n)$ when $(n, q) = 1$, so that

$$(T_k(n)\varphi_{j,k})(z) = \frac{1}{\sqrt{n}} \sum_{ad=n} \left(\frac{a}{b}\right)^k \sum_{b=1}^{d} \varphi_{j,k}\left(\frac{az+b}{d}\right) = t_{j,k}(n)\varphi_{j,k}(z)$$

with a certain real number $t_{j,k}(n)$. The Fourier coefficients $\rho_{j,k}(n)$ of $\varphi_{j,k}$ around the point at infinity are defined by the expansion

$$\varphi_{j,k}(z) = \sum_{n=1}^{\infty} \rho_{j,k}(n)n^{k-\frac{1}{2}}e(nz).$$

As before, we have

$$\rho_{j,k}(n) = \rho_{j,k}(n, [q]), \quad \rho_{j,k}(n) = t_{j,k}(n_1)\rho_{j,k}(n_2),$$

where n_1 and n_2 are as in (2.4). The L-function $L_{j,k}(s)$ attached to $\varphi_{j,k}$ is defined by

$$L_{j,k}(s) = \sum_{n=1}^{\infty} \frac{\rho_{j,k}(n)}{n^s},$$

which converges to an entire function and satisfies the functional equation

$$L_{j,k}(s) = (-1)^k \eta_{j,k} \left(\frac{\sqrt{q}}{2\pi}\right)^{1-2s} \frac{\Gamma\left(k + \frac{1}{2} - s\right)}{\Gamma\left(k - \frac{1}{2} + s\right)} L_{j,k}(1-s). \tag{2.8}$$

Here $\eta_{j,k} = \pm 1$ is such that $\varphi_{j,k}\left(-1/(qz)\right) = \eta_{j,k}\varphi_{j,k}(z)$, which we may assume. The χ-twist of the Hecke series attached to $\varphi_{j,k}$ is

$$H_{j,k}(s, \chi) = \sum_{n=1}^{\infty} \frac{\chi(n)t_{j,k}(n)}{n^s}.$$

It is an integral function satisfying the functional equation

$$H_{j,k}(s, \chi) = (-1)^k \chi(-1)\left(\frac{q}{2\pi}\right)^{1-2s} \frac{\Gamma\left(k + \frac{1}{2} - s\right)}{\Gamma\left(k - \frac{1}{2} + s\right)} H_{j,k}(1-s, \chi). \tag{2.9}$$

Finally we have the following analogue of the identity (2.7): in the region of absolute convergence,

$$\sum_{n=1}^{\infty} \frac{\sigma_a(n, \chi)\rho_{j,k}(n)}{n^s} = \frac{H_{j,k}(s - a, \chi)L_{j,k}(s)}{L(2s - a, \chi)}. \tag{2.10}$$

3. Explicit formula

Now we apply the above Lemma 1 to $Y_1^{\pm}(u,v,w,z;\chi)$, observing that the case $a_1 = a_2 = q$ is relevant. The condition $(u,v,w,z) \in \mathcal{E}_\beta$ ensures that all sums and integrals involved in this procedure are absolutely convergent; and we may appeal to (2.7) and (2.10), with which we exploit the multiplicative structure of our problem.

After manipulating some intricate Euler products, we reach the expression

$$\{Y_1^+ + Y_1^-\}(u,v,w,z;\chi)$$

$$= \frac{\chi(-1)}{\tau(\chi)} \left(\frac{q}{2\pi}\right)^{w-z+1} \sum_{j=1}^{\infty} \frac{1}{\cosh \pi \kappa_j} \{(\psi_+)^+ + \varepsilon_j (\psi_-)^-\}(\kappa_j; u,v,w,z;\chi)$$

$$\times \overline{L_j}\left(\frac{u+v+w+z-1}{2}\right) L_j\left(\frac{u+w-v-z+1}{2}\right) H_j\left(\frac{v+w-u-z+1}{2},\chi\right)$$

$$+ \frac{\chi(-1)}{\tau(\chi)} \left(\frac{q}{2\pi}\right)^{w-z+1} \sum_{k=1}^{\infty} \frac{(2k-1)!}{\pi^{2k+1}2^{4k-2}} \sum_{j=1}^{\vartheta_q(k)} (\psi_+)^+ \left((\tfrac{1}{2}-k)i; u,v,w,z;\chi\right)$$

$$\times \overline{L_{j,k}}\left(\frac{u+v+w+z-1}{2}\right) L_{j,k}\left(\frac{u+w-v-z+1}{2}\right) H_{j,k}\left(\frac{v+w-u-z+1}{2},\chi\right)$$

$$+ \frac{\chi(-1)}{\pi q \tau(\chi)} \left(\frac{q}{2\pi}\right)^{w-z+1} \int_{-\infty}^{\infty} \frac{\{(\psi_+)^+ + (\psi_-)^-\}(r; u,v,w,z;\chi)}{|L(1+2ir,\chi_0)|^2}$$

$$\times \zeta\left(\frac{u+v+w+z-1}{2}+ir\right) \zeta\left(\frac{u+v+w+z-1}{2}-ir\right)$$

$$\times \zeta\left(\frac{u+w-v-z+1}{2}+ir\right) \zeta\left(\frac{u+w-v-z+1}{2}-ir\right)$$

$$\times L\left(\frac{v+w-u-z+1}{2}+ir,\chi\right) L\left(\frac{v+w-u-z+1}{2}-ir,\chi\right)$$

$$\times \prod_{p|q} \begin{bmatrix} 1+p^{1-u-w} - p^{\frac{1}{2}(v+z-u-w-1)+ir} \\ -p^{\frac{1}{2}(v+z-u-w-1)-ir} - p^{\frac{1}{2}(1-u-v-w-z)+ir} \\ -p^{\frac{1}{2}(1-u-v-w-z)-ir} + p^{-u-w} + p^{-1} \end{bmatrix} dr, \qquad (3.1)$$

where we have used an obvious convention.

To give precise expressions to $(\psi_+)^+$ and $(\psi_-)^-$, we introduce

$$\Xi(\xi; u,v,w,z) = \frac{1}{2\pi i} \int_{-i\infty}^{i\infty} \frac{\Gamma\left(\frac{1}{2}(u+v+w+z-1)+\xi-s\right)}{\Gamma\left(\frac{1}{2}(3-u-v-w-z)+\xi+s\right)}$$

$$\times \Gamma(s-u-z+1)\Gamma(s-v-z+1) W^*(s,z)\, ds.$$

Here the path is curved to ensure that the poles of

$$\Gamma\big(\tfrac{1}{2}(u+v+w+z-1)+\xi-s\big), \quad \Gamma(s-u-z+1)\Gamma(s-v-z+1)W^*(s,z)$$

are separated by the path to the right and the left, respectively; it is assumed that the variables ξ, u, v, w, z are such that the path can be drawn. Then we have, in (3.1),

$$(\psi_+)^+(r;u,v,w,z;\chi) = \frac{\pi i}{4\sinh \pi r}\Big(e\big(\tfrac{1}{4}(u-v)\big) + \chi(-1)e\big(-\tfrac{1}{4}(u-v)\big)\Big)$$
$$\times \big\{\Xi(ir;u,v,w,z) - \Xi(-ir;u,v,w,z)\big\},$$

$$(\psi_-)^-(r;u,v,w,z;\chi)$$

$$= -\frac{\chi(-1)\pi}{4\sinh \pi r}\left[\begin{array}{l}\left\{\begin{array}{l} e\big(\tfrac{1}{4}(w-z+2ir)\big) \\ -\chi(-1)e\big(-\tfrac{1}{4}(w-z+2ir)\big)\end{array}\right\}\Xi(ir;u,v,w,z) \\[2mm] - \left\{\begin{array}{l} e\big(\tfrac{1}{4}(w-z-2ir)\big) \\ -\chi(-1)e\big(-\tfrac{1}{4}(w-z-2ir)\big)\end{array}\right\}\Xi(-ir;u,v,w,z)\end{array}\right]$$

as well as

$$(\psi_+)^+\Big(\big(\tfrac{1}{2}-k\big)i;u,v,w,z;\chi\Big)$$
$$= (-1)^k\tfrac{1}{2}\pi\Big(e\big(\tfrac{1}{4}(u-v)\big) + \chi(-1)e\big(-\tfrac{1}{4}(u-v)\big)\Big)\Xi\big(k-\tfrac{1}{2};u,v,w,z\big)$$

(cf. p. 201 of [4]).

According to Lemma 6 in [4], though the designation of variables is different, $\Xi(\xi;u,v,w,z)$ is meromorphic over the whole of $\mathbb{C}^5$. Moreover, if $|\xi|$ tends to infinity in any fixed vertical strip, we have, for any fixed $A > 0$,

$$\Xi(\xi;u,v,w,z) \ll |\xi|^{-A}$$

uniformly for bounded (u,v,w,z); and the same holds when $\mathrm{Re}(\xi)$ tends to $+\infty$ in any fixed horizontal strip. Hence we see that (3.1) gives a meromorphic continuation of $\{Y_1^+ + Y_1^-\}(u,v,w,z;\chi)$ to the whole of $\mathbb{C}^4$. This proves that the decomposition (1.2) holds throughout $\mathbb{C}^4$.

In (3.1) the contribution of the holomorphic and non-holomorphic cusp-forms are easily seen to be regular at the point P_T. Also, if q has more than two prime factors, then the contribution of the continuous spectrum is regular at P_T. Otherwise it is singular at P_T. However, the singularity can be separated by appropriately shifting the contour in the relevant integral in (3.1), and when brought into the relation (1.2) it has to be cancelled out because of an obvious reason. In this way we are led to our main result:

Theorem 1. *For any* $T > 0$, $G > 0$ *there exists an* $M(T, G; D)$ *which is* $O(\log^2 T)$ *as* T *tends to infinity, and*

$$I(T, G; D)$$

$$= M(T, G; D) + \sum_{j=1}^{\infty} \frac{1}{\cosh \pi \kappa_j} \left| L_j\left(\tfrac{1}{2}\right) \right|^2 H_j\left(\tfrac{1}{2}, \chi\right) \Theta_\chi(\kappa_j; T, G)$$

$$+ \operatorname{Re}\{\tau(\chi)\} \sum_{k=1}^{\infty} \frac{(4k-1)!}{\pi^{4k+1} 2^{8k-2}} \Lambda\left(2k - \tfrac{1}{2}; T, G\right) \sum_{j=1}^{\vartheta_q(2k)} \left| L_{j,2k}\left(\tfrac{1}{2}\right) \right|^2 H_{j,2k}\left(\tfrac{1}{2}, \chi\right)$$

$$+ \frac{d(|D|)}{\pi |D|} \int_{-\infty}^{\infty} \left| \frac{L\left(\tfrac{1}{2} + ir, \chi_0\right)}{L\left(1 + 2ir, \chi_0\right)} \right|^2 \left| \zeta_D\left(\tfrac{1}{2} + ir\right) \right|^2 \Theta_\chi(r; T, G) \, dr. \quad (3.2)$$

Here d is the divisor function, and

$$\Theta_\chi(r; T, G) = \frac{1}{2 \sinh \pi r} \operatorname{Re}\left[\tau(\chi) \left\{ \begin{matrix} (e^{\pi r} - \chi(-1)e^{-\pi r}) \\ + i(1 + \chi(-1)) \end{matrix} \right\} \Lambda(ir; T, G) \right]$$

with

$$\Lambda(\xi; T, G) = \frac{\Gamma\left(\tfrac{1}{2} + \xi\right)^2}{\Gamma(1 + 2\xi)} \int_0^{\infty} \frac{\cos\left(T \log(1 + x)\right)}{x^{\frac{1}{2} - \xi}(1 + x)^{\frac{1}{2}}}$$

$$\times F\left(\tfrac{1}{2} + \xi, \tfrac{1}{2} + \xi; 1 + 2\xi; -x\right) \exp\left(-\left(\tfrac{1}{2} G \log(1 + x)\right)^2\right) dx,$$

where F is the hypergeometric function.

The main term $M(T, G; D)$ could be computed explicitly as in the corresponding part of [4]. We note that the ε_j in (3.1) is removed in (3.2), since we have

$$H_j\left(\tfrac{1}{2}, \chi\right) = 0 \quad \text{if} \quad \varepsilon_j = -1, \quad (3.3)$$

which is a consequence of the functional equation (2.6). Also we have used the fact that

$$H_{j,k}\left(\tfrac{1}{2}, \chi\right) = 0 \quad \text{if} \quad (-1)^k \chi(-1) = -1, \quad (3.4)$$

which follows from (2.9). Further, it should be worth remarking that despite the presence of ζ_D the last term in (3.2) is, in fact, negligible with respect to both D and T, provided $T^{\frac{1}{2}} \leq G \leq T/\log T$, say.

The identity (3.2) exhibits a relation between ζ_D and the discrete spectrum of the hyperbolic Laplacian over $\Gamma_0(q)$. As we did in [7], we can enhance this relation by considering the meromorphic continuation of

$$Z(\xi, D) = \int_1^{\infty} \left| \zeta_D\left(\tfrac{1}{2} + it\right) \right|^2 \frac{dt}{t^\xi}.$$

The argument developed in [7] can be applied to $Z(\xi, D)$, too, without much alteration, though there are some non-trivial subtleties related to the nature of the underling group that have been pointed out in the above. Thus we may state here the following assertion without proof:

Theorem 2. *The function $Z(\xi, D)$ is meromorphic over the entire complex plane. In particular, in the half-plane $\mathrm{Re}(\xi) > 0$ it has a pole of order three at $\xi = 1$ and infinitely many simple poles of the form $\frac{1}{2} \pm \kappa i$; all other poles are of the form $\frac{1}{2}\rho$. Here $\kappa^2 + \frac{1}{4}$ is in the discrete spectrum of the hyperbolic Laplacian over $\Gamma_0(|D|)$, and ρ is a complex zero of the Riemann zeta-function. More precisely, the residue of the pole at $\frac{1}{2} + \kappa i$ is equal to*

$$\overline{\tau(\chi)} \cdot \frac{e^{\pi\kappa} - \chi(-1)e^{-\pi\kappa} - i\big(1 + \chi(-1)\big)}{2(\sinh \pi\kappa - i)}$$

$$\times \left(\frac{\pi}{2}\right)^{\frac{1}{2}} \left(2^{-i\kappa} \frac{\Gamma\big(\frac{1}{2}(\frac{1}{2} - i\kappa)\big)}{\Gamma\big(\frac{1}{2}(\frac{1}{2} + i\kappa)\big)}\right)^3 \Gamma(2i\kappa) M_D(\kappa),$$

where

$$M_D(\kappa) = \sum_{\kappa_j = \kappa} \left|L_j\big(\tfrac{1}{2}\big)\right|^2 H_j\big(\tfrac{1}{2}, \chi\big).$$

It should be remarked that there might be poles in the interval $\frac{1}{4} < \xi < \frac{3}{4}$ because of the yet unremoved possibility of the existence of exceptional eigenvalues. Naturally the above assertion depends on a non-vanishing type of theorem that

$$M_D(\kappa) \neq 0 \quad \text{for infinitely many } \kappa.$$

In fact we have more precisely

$$\sum_{\kappa \leq K} \frac{1}{\cosh \pi\kappa} M_D(\kappa) = \big(1 + o(1)\big) 4\pi^{-\frac{3}{2}} L^2(1, \chi) K^2 \log K, \qquad (3.5)$$

as K tends to infinity, which obviously yields the last assertion. This asymptotic formula can be established in much the same way as Theorem 3 in [3]; so we may restrict ourselves to the salient points of the proof.

We observe first that in dealing with the product $\left|L_j\big(\tfrac{1}{2}\big)\right|^2 H_j\big(\tfrac{1}{2}, \chi\big)$ we may assume that $\varepsilon_j = 1$ and $\eta_j = 1$. The former is due to (3.3), and the latter to the fact that $L_j\big(\tfrac{1}{2}\big) = 0$ if $\varepsilon_j \eta_j = -1$, which follows from (2.5). Then we replace the factor $\overline{L_j\big(\tfrac{1}{2}\big)}$ of the product by an approximation in Dirichlet polynomials of length $\ll \kappa_j \sqrt{q}$ as (4.9) in [3] indicates. Thus we are led to the sum

$$\sum_{j=1}^{\infty} \frac{\varepsilon_j}{\cosh \pi\kappa_j} \overline{\rho_j(f)} L_j\big(\tfrac{1}{2}\big) H_j\big(\tfrac{1}{2}, \chi\big) h(\kappa_j) \quad \text{for} \quad f \geq 1, \qquad (3.6)$$

where the function h is as in section 2 in [3]. Further, in this we replace $L_j(\frac{1}{2})H_j(\frac{1}{2},\chi)$ by $L_j(u)H_j(v,\chi)$ with complex variables u, v. The resulting infinite sum is an entire function over $\mathbb{C}^2$, and we transform it with the aid of (2.7) in an appropriate range of the variables. We then come to the stage of appealing to the Kuznetsov-Bruggemann trace formula for $\Gamma_0(q)$; we need the version for the opposite sign case, a simple proof of which can be obtained by modifying slightly the argument of section 5 in [6]. We thus get a sum of Kloosterman sums, whose moduli are all divisible by q. Finally we transform the sum by using (1.4), and take (u,v) to be $(\frac{1}{2},\frac{1}{2})$. The end result of this procedure is an analogue of the Lemma in [3] for our sum (3.6). Thus (3.6) is related to the additive divisor problems

$$\sum_{n=1}^{N} a(n)a(n+f), \qquad \sum_{n=1}^{f-1} a(n)a(f-n),$$

where $a(n)$ is the number of integral ideals in $\mathbb{Q}(\sqrt{D})$ with the norm n. The rest of the proof of (3.5) is quite similar to the corresponding part of [3].

Theorems 1 and 2 are both related to the nature of the error term in the asymptotic formula for the unweighted mean

$$\int_0^T \left| \zeta_D(\tfrac{1}{2}+it) \right|^2 dt,$$

the details of which are to be discussed in our future work.

Concluding remarks

1. We may consider more generally the mean square

$$\frac{1}{G\sqrt{\pi}} \int_{-\infty}^{\infty} \left| \zeta(\tfrac{1}{2}+i(T+t))L(\tfrac{1}{2}+i(T+t),\chi) \right|^2 \exp\left(-(t/G)^2\right) dt$$

for *any* primitive Dirichlet character χ, as we have already shown in a special instance in our note [5]. There is a difficulty in extending the Theorem in [5] to the cases of characters with arbitrary composite conductors. This is mainly due to the lack of precise description of the Fourier expansion of Eisenstein series for general $\Gamma_0(q)$; if q is square-free we have such a result, as (2.2) shows.

2. A more interesting problem is to find an explicit formula for

$$\frac{1}{G\sqrt{\pi}} \int_{-\infty}^{\infty} \left| L\left(\tfrac{1}{2} + i(T+t), \chi\right) \right|^4 \exp\left(-(t/G)^2\right) dt$$

with an arbitrary primitive χ. It should be noted that this has already been reduced to a problem about sums of Kloosterman sums. In fact, an argument similar to the one developed in our recent work [8] yields that the relevant sum is

$$\sum_{a=1}^{q} \overline{\chi}(a)\chi(a+b) \sum_{\substack{k=1 \\ (k,q)=1}}^{\infty} \frac{1}{k} S\left(\overline{q}m, \pm\overline{q}bn; k\right) S\left(\overline{k}m, \pm\overline{k}an; q\right) \psi\left(\frac{4\pi\sqrt{bmn}}{qk}\right),$$

where $b, m, n \geq 1$, $\chi \bmod q$, $\overline{q}q \equiv 1 \bmod k$, $\overline{k}k \equiv 1 \bmod q$; and ψ is similar to $\psi_\pm$ above. The remaining problem is to produce these Kloosterman sums by a suitable choice of Poincaré series over a certain congruence group, to which we shall return elsewhere.

References

1 J.-M. Deshouillers and H. Iwaniec: Kloosterman sums and Fourier coefficients of cusp forms. *Invent. Math.* **70**, 219–288 (1982).
2 D.A. Hejhal: The Selberg Trace Formula for $PSL(2,\mathbb{R})$ II (*Lect. Notes in Math.* **1001**). Springer (1983).
3 Y. Motohashi: Spectral mean values of Maass wave form L-functions. *J. Number Theory* **42**, 258–284 (1992).
4 Y. Motohashi: An explicit formula for the fourth power mean of the Riemann zeta-function. *Acta Math.* **170**, 181–220(1993).
5 Y. Motohashi: On the mean square of the product of the zeta- and L-functions. *Kokyuroku RIMS Kyoto Univ.* **837**, 57–62 (1993).
6 Y. Motohashi: On Kuznetsov's trace formulae. In *Analytic Number Theory. Proceedings of a Conference in Honor of Heini Halberstam* Vol. II, 641–667 (B.C. Berndt, H.G. Diamond and A.J. Hildebrand, eds.). Birkhäuser (1996).
7 Y. Motohashi: A relation between the Riemann zeta-function and the hyperbolic Laplacian. *Ann. Scuola Norm. Sup. Pisa, Cl. Sci.* (Ser. 4) **22**, 299–313 (1995).
8 Y. Motohashi: The Riemann zeta-function and the Hecke congruence subgroups. *Kokyuroku RIMS Kyoto Univ.*, to appear.

19. Artin's Conjecture and Elliptic Analogues

M. Ram Murty

1. Introduction

A well-known conjecture of Emil Artin predicts that every natural number a unequal to unity or a perfect square is a primitive root $(\bmod\, p)$ for infinitely many primes p. In 1967, Hooley [8] proved this conjecture assuming the generalized Riemann hypothesis (GRH) for the Dedekind zeta functions of certain Kummer extensions. In fact, he establishes an asymptotic formula for the number of such primes up to x. He also remarks in [9] that to deduce a positive density of primes for which a is a primitive root $(\bmod\, p)$ it suffices to assume that the corresponding Dedekind zeta functions of the number fields $\mathbb{Q}(a^{1/q})$ do not vanish for $\mathrm{Re}(s) > 1 - 1/2e$.

In 1984, Rajiv Gupta and the author [2] proved that given three prime numbers a, b, c, one of

$$\left\{ ac^2,\, a^3b^2,\, a^2b,\, b^3c^2,\, b^2c,\, a^2c^3,\, ab^3,\, a^3bc^2,\, bc^3,\, a^2b^3c,\, a^3c,\, ab^2c^3,\, abc \right\}$$

is a primitive root $(\bmod\, p)$ for infinitely many primes p. Their method used a lower bound sieve inequality implied by a theorem of Fouvry and Iwaniec. In [5], the authors improved upon this to show that one of the seven numbers

$$\left\{ a,\, b,\, c,\, a^2b,\, ab^2,\, a^2c,\, ac^2 \right\}$$

is a primitive root for infinitely many primes p. By using the Chen-Iwaniec switching and the celebrated theorem of Bombieri, Friedlander and Iwaniec, Heath-Brown [6] refined the above result to show that in fact one of a, b, c is a primitive root $(\bmod\, p)$ for infinitely many primes p. The number of such primes obtained by the method is

$$\gg \frac{x}{\log^2 x}\, .$$

In [6], the result is slightly more general: given three non-zero integers a, b, c which are multiplicatively independent such that none of a, b, c, $-3ab$, $-3ac$, $-3bc$, abc is a perfect square, then one of a, b, c is a primitive root $(\bmod\, p)$ for infinitely many primes p. In this paper, we will show

Research supported by NSERC, FCAR and NSF Grant DMS 9304580.

Sieve Methods, Exponential Sums, and their Applications in Number Theory
Greaves, G.R.H., Harman, G., Huxley, M.N., Eds. ©Cambridge University Press, 1996

Theorem 1. *Let a, b, c be three non-zero integers which are multiplicatively independent such that none of a, b, c, −3ab, −3ac, −3bc, abc is a perfect square. Suppose further that for some $\epsilon > 0$, and each prime q the Dedekind zeta function of $\mathbb{Q}\left(a^{1/q}, b^{1/q}, c^{1/q}\right)$ has no zeroes in $\mathrm{Re}(s) > 1 - \epsilon$. Then one of a, b, c is a primitive root $(\bmod\, p)$ for a positive density of primes p.*

In 1976, Lang and Trotter [13] formulated elliptic analogues of the Artin primitive root conjecture. Suppose E is an elliptic curve over $\mathbb{Q}$ with a rational point of infinite order. A natural question is: how often does the prescribed point generate $E(\mathbb{F}_p)$, the group of points $(\bmod\, p)$?

More precisely, let a be a rational point of infinite order. The problem is to determine the density of primes p for which $E(\mathbb{F}_p)$ is generated by $\bar{a}$, the reduction of a $(\bmod\, p)$. (Here, in addition to primes of bad reduction, we may need to exclude primes dividing the denominators of the co-ordinates of a.) Such a point will be called a *primitive point* for these primes. Lang and Trotter conjectured that the density of primes for which a is a primitive point always exists. This is the elliptic analogue of Artin's primitive root conjecture.

Of course, situations may (and do) arise when the density is zero. In such a case the set of primes for which a is a primitive point is finite (and often empty). If this is the case, then it is so for obvious reasons. However, if the density is positive, then there are infinitely many such primes.

In considering the elliptic analogue, we see that two assertions are being made about a prime p for which a is a primitive point: first, that $E(\mathbb{F}_p)$ is cyclic and second, that it is generated by the image of a $(\bmod\, p)$. Is it even true that $E(\mathbb{F}_p)$ is cyclic infinitely often?

It was Serre [21] who pointed out the relevance of this question. In a course at Harvard in the fall of 1976, he proved that Hooley's method of proving Artin's primitive root conjecture can be adapted to show that the set of primes p for which $E(\mathbb{F}_p)$ is cyclic has a density, assuming the GRH for the Dedekind zeta functions of fields obtained by adjoining the l-division points of E to $\mathbb{Q}$. (These fields should be viewed as the elliptic analogues of the classical cyclotomic extensions.)

In 1980, the author [19] eliminated the use of GRH in Serre's argument for elliptic curves with complex multiplication (CM). More precisely, he proved that the number of primes p for which $E(\mathbb{F}_p)$ is cyclic is

$$\delta_E \frac{x}{\log x} + o\left(\frac{x}{\log x}\right)$$

where

$$\delta_E = \sum_{k=1}^{\infty} \frac{\mu(k)}{n(k)},$$

and $n(k) = [\,\mathbb{Q}(E[k]) : \mathbb{Q}\,]$ is the degree of the extension obtained by adjoining the k-division points of E to $\mathbb{Q}$. As noted by Serre, $\delta_E > 0$ if and only if E has an irrational 2-division point. That is, if E has the Weierstrass model

$$y^2 = x^3 + cx + d$$

we require that $x^3 + cx + d$ has an irrational root. It is not difficult to see that the density is positive in this case. Indeed, for l prime and greater than some constant C (say),

$$n(l) \asymp l^2$$

in the CM case and

$$n(l) \asymp l^4$$

in the non-CM case, by a celebrated theorem of Serre. Moreover, since $\mathbb{Q}\big(E[l]\big)$ contains the lth cyclotomic field, we easily see that

$$\delta_E \geq \tfrac{1}{2} \prod_{2 < l \leq C} \left(1 - \frac{1}{l-1}\right) \prod_{l > C} \left(1 - \frac{1}{n(l)}\right) > 0.$$

In the non-CM case, it is still unknown if $E(\mathbb{F}_p)$ is cyclic for a positive density of primes whenever E has an irrational 2-division point. However, Gupta and the author [3] proved that if E has an irrational 2-division point then the number of primes $p \leq x$ for which $E(\mathbb{F}_p)$ is cyclic is

$$\gg \frac{x}{\log^2 x}.$$

The method used to prove Theorem 1 can be utilised to prove

Theorem 2. *Let E be an elliptic curve over $\mathbb{Q}$ without complex multiplication. Suppose E has an irrational 2-division point. Suppose for some $\epsilon > 0$ and for each prime q that the Dedekind zeta function of the fields $\mathbb{Q}\big(E[q]\big)$ does not have any zeroes in the region $\mathrm{Re}(s) > 1 - \epsilon$ Then $E(\mathbb{F}_p)$ is cyclic for a positive density of prime numbers. That is, the number of such primes $p \leq x$ is $\geq cx/\log x$ for some positive constant c.*

Returning to the original conjecture of Lang and Trotter, it was recognized by Serre, Lang and Trotter that the method of Hooley cannot be adapted to the elliptic curve case owing to the large error terms introduced by the Chebotarev density theorem. This problem was, however, circumvented by Gupta and the author [4] in the CM case. They proved: if E is an elliptic curve over $\mathbb{Q}$, with CM, and a is a rational point of infinite order, then the number of primes p for which $E(\mathbb{F}_p)$ is generated by a has a density, assuming the GRH for the Dedekind zeta functions associated to extensions of the form $\mathbb{Q}\big(E[l], l^{-1}a\big)$. (These extensions are the elliptic analogues of the classical Kummer extensions.) If E is an elliptic curve over $\mathbb{Q}$ with CM by k, then k can be one of nine fields. If k is one of

$$\mathbb{Q}(\sqrt{-11}\,), \quad \mathbb{Q}(\sqrt{-19}\,), \quad \mathbb{Q}(\sqrt{-43}\,), \quad \mathbb{Q}(\sqrt{-67}\,), \quad \mathbb{Q}(\sqrt{-163}\,),$$

then it is shown on p. 30 of [4] that the density is positive. It is also shown to be positive in certain instances when k is one of the four remaining fields. At the end of their paper, they proved: there is a finite set S (which can be given explicitly) such that, for some $a \in S$, $E(\mathbb{F}_p) = \langle \bar{a} \rangle \pmod p$ for infinitely many primes p, provided the Mordell-Weil rank of $E(\mathbb{Q})$ is at least 6. They did not give the size of S, but an examination of their paper shows that $|S| = 2^{18}$.

We will show

Theorem 3. *Let E be an elliptic curve with CM. Suppose*

$$\{P_1, \ldots, P_6\}$$

are independent points of infinite order in $E(\mathbb{Q})$. Then, for infinitely many primes p, one of $P_1, \ldots, P_6$ generates a subgroup of $E(\mathbb{F}_p)$ of index bounded by 4.

Remark. This is a substantial improvement of the result of Gupta and Murty. Note that we assume that the rank of $E(\mathbb{Q})$ is at least 6. Mestre [14] has shown that there are infinitely many elliptic curves E defined over $\mathbb{Q}$ with j-invariant equal to zero (and hence with CM) such that this holds. Such curves are twists of the curve $y^2 = x^3 + 1$. By a more careful and detailed analysis, it should be possible to obtain index 1 in the case where E has an irrational 2-division point.

We will now give a brief description of the basic strategy involved in proving these theorems. Let us first consider Theorem 1. In [6] the lower bound sieve method combined with the Chen-Iwaniec switching method gave rise to $\gg x/\log^2 x$ primes $p \leq x$ such that either $p - 1 = 2^e q$ for some odd prime q or $p - 1 = 2^e q_1 q_2$ with $q_1 < q_2$ and q_1 satisfying $p^\alpha < q < p^\delta$ for some $\alpha > \frac{1}{4}$ and $\delta < \frac{1}{2}$. By imposing a suitable initial congruence condition, one can also ensure that none of a, b, c is a quadratic residue $\pmod p$ for these primes and that e is bounded. It is now clear that if

$p - 1 = 2^e q$ with q prime then each of a, b, c is a primitive root $(\bmod\, p)$ for all but finitely many such primes. So we only need to deal with the second case. By essentially the pigeonhole principle one can show that for almost all of these primes the subgroup generated by a, b, c $(\bmod\, p)$ cannot have order $< x^{1-\alpha}$, which would be the case if either q_1 or q_2 divides the index of $[\,\mathbb{F}_p^* : \langle a, b, c\rangle\,]$. But now there are only two possible orders for each of a, b, c. Two must have the same order by the pigeonhole principle again! Once more by this principle, this can happen for at most $O(x^{1-2\delta})$ such primes $p \le x$.

The essential ingredients in the proof are the sieve exponents $\alpha > \frac{1}{4}$ and $\delta < \frac{1}{2}$. As indicated on p. 34 of [6], the theorem of Bombieri, Friedlander and Iwaniec alluded to above gives $\alpha = .276$. Since all that is needed in the proof is an exponent greater than $\frac{1}{4}$, there is some room for introducing certain degrees of freedom. We relax the conditions on $p - 1$ and consider primes of the form $p - 1 = 2^e m q$ or $p - 1 = 2^e m q_1 q_2$ where q, q_1, q_2 are primes satisfying the same constraints above and $m < x^\epsilon$ for some small $\epsilon > 0$. This relaxation produces a positive proportion of primes satisfying these conditions. The argument briefly indicated above shows that for almost all these primes one of a, b, c has order divisible by $2^e q$ (in the first case) and $2^e q_1 q_2$ (in the second case). We still need to rule out the possibility that a prime divisor q of m can divide the index of one of $[\,\mathbb{F}_p^* : \langle a\rangle\,]$, $[\,\mathbb{F}_p^* : \langle b\rangle\,]$, $[\,\mathbb{F}_p^* : \langle c\rangle\,]$. This is done by the effective Chebotarev density theorem. This theorem implies that the number of primes $p \le x$ for which a prime q divides the index $[\,\mathbb{F}_p^* : \langle a\rangle\,]$ is

$$\frac{\operatorname{li} x}{q(q-1)} + O\left(x^{1-\epsilon}\right)$$

uniformly for $q < x^\epsilon$, if we are assuming a quasi-Riemann hypothesis of the type stated in Theorem 1. (The best unconditional result is due to Pappalardi ([17], p. 40) who has proved an asymptotic formula of the type

$$\frac{\operatorname{li} x}{q(q-1)} + O\left(\frac{x}{\log^A x}\right)$$

uniformly for $q < \log^{1-\eta} x$.) Invoking this result in the appropriate range leads to Theorem 1.

Theorems 2 and 3 are proved analogously using the elliptic analogues of the appropriate pigeonhole arguments. (See §6 below.)

It is a pleasure to thank C.S. Rajan, V. Kumar Murty, John Friedlander and Henryk Iwaniec for useful discussions, and the Institute for Advanced Study for its hospitality and financial support.

2. Sieve preliminaries

For a positive integer k, we define $\tau_k(n)$ as the the number of ways of writing n as a product of k positive integers. An arithmetical function $\lambda(q)$ is said to be of level Q and of order k if

$$\lambda(q) = 0 \quad \text{for} \quad q > Q, \qquad |\lambda(q)| \leq \tau_k(q).$$

The function λ is *well-factorable* if for any $Q_1, Q_2 \geq 1$ with $Q_1 Q_2 = Q$ there exist two arithmetical functions λ_1, λ_2 of levels Q_1, Q_2 and of order k such that $\lambda = \lambda_1 * \lambda_2$. (Here $*$ indicates the Dirichlet convolution of two arithmetical functions.) Well-factorable coefficients appear in the error term of the Rosser-Iwaniec linear sieve (see the lemma below). We make the convention that $\lambda(q)$ will always denote a well-factorable function of level Q and of order k. It is clear that if λ' is another arithmetical function of level Q' (with $Q' \leq Q$) and of order k' then the arithmetical function $\lambda * \lambda'$ is well-factorable of level QQ' and of order $k + k'$.

We now state the formula of the Rosser-Iwaniec sieve. For $\mathcal{A}$ a finite sequence of integers and $\mathcal{P}$ a set of prime numbers we are interested in evaluating, for $z \geq 2$,

$$S(\mathcal{A}, \mathcal{P}, z) = \#\big\{ a \in \mathcal{A} : \big(a, P(z)\big) = 1 \big\},$$

where

$$P(z) = \prod_{p \leq z; \, p \in \mathcal{P}} p.$$

If d is a squarefree integer with all its prime factors belonging to $\mathcal{P}$, we denote

$$\mathcal{A}_d = \big\{ a \in \mathcal{A} : d \,|\, a \big\}$$

and write its cardinality (as a multiset) as

$$|\mathcal{A}_d| = \frac{\omega(d)}{d} X + r_d(\mathcal{A}), \tag{1}$$

where $X > 1$ is independent of d and $\omega(d)$ is a multiplicative function satisfying

$$0 \leq \omega(p) < p \quad \text{for} \quad p \in \mathcal{P}. \tag{2}$$

One thinks of X as an approximation to the size of $\mathcal{A}$ and $r_1(\mathcal{A})$ as the error in the approximation. We define

$$V(z) = \prod_{p < z; \, p \in \mathcal{P}} \left(1 - \frac{\omega(p)}{p} \right) \tag{3}$$

and suppose that

$$\frac{V(z_1)}{V(z_2)} \leq \frac{\log z_2}{\log z_1}\left(1 + \frac{K}{\log z_1}\right) \quad \text{for} \quad z_2 \geq z_1 \geq 2$$

where K is some constant > 1. Then

Lemma 1. (Rosser-Iwaniec sieve) *Let* $0 < \epsilon < \frac{1}{8}$, $2 \leq z \leq Q^{1/2}$. *Under* (1), (2), (3) *above, we have*

$$S(\mathcal{A}, \mathcal{P}, z) \leq XV(z)\big(F(\log Q / \log z) + E\big) + \sum_{l < L}\sum_{q | P(z)} \lambda_l^+(q) r_q(\mathcal{A}),$$

$$S(\mathcal{A}, \mathcal{P}, z) \geq XV(z)\big(f(\log Q / \log z) - E\big) - \sum_{l < L}\sum_{q | P(z)} \lambda_l^-(q) r_q(\mathcal{A}).$$

Here, L depends only on ϵ, and λ_l^+, λ_l^- are well-factorable functions of order 1 and of level Q. The constant E satisfies

$$E = O\left(\epsilon + \epsilon^{-8} e^K / \log^{\frac{1}{3}} Q\right).$$

The continuous functions $F(s)$ and $f(s)$ are defined recursively by

$$F(s) = 2e^\gamma / s, \quad f(s) = 0 \qquad \text{for} \quad s \leq 2,$$
$$(sF(s))' = f(s - 1), \quad (sf(s))' = F(s - 1) \quad \text{for} \quad s > 2,$$

and γ denotes Euler's constant.

Proof. See [10] and [11].

Lemma 2. *Let* $(u, v) = 1$. *For any q such that* $(q, v) = 1$, *define u^* to be the solution of the congruences $u^* \equiv u \pmod{v}$, $u^* \equiv 1 \pmod{q}$. Suppose that $\epsilon > 0$ and $A > 0$. Then, for any well-factorable function λ of level $x^{4/7-\epsilon}$, one has*

$$\sum_{(q,v)=1} \lambda(q)\left(\pi(x, qv, u^*) - \frac{\text{li}\,x}{\phi(qv)}\right) \ll \frac{x}{\log^A x},$$

where the implied constant depends only on u, v, ϵ and A.

Proof. This is Heath-Brown's ([6], p. 29) slight adjustment of Theorem 10 of [1].

Lemma 3. *Let*
$$\pi(X; a, d, l) = \sum_{\substack{ap \leq X \\ ap \equiv l \,(\mathrm{mod}\, d)}} 1$$

and let $f(a)$ be a real-valued function satisfying the conditions

$$\sum_{n \leq x} |f(n)| \ll x \log^{c_1} x, \qquad \sum_{n \leq x} \sum_{d|n} |f(d)| \ll x \log^{c_2} x,$$

where c_1, c_2 are positive constants. Given $A > 0$, there is a $B = B(A, c_1, c_2)$ such that

$$\sum_{d \leq \frac{\sqrt{X}}{\log^B X}} \max_{y \leq X} \max_{(l,d)=1} \left| \sum_{\substack{a \leq X^{1-\epsilon} \\ (a,d)=1}} f(a) \left(\pi(y; a, d, l) - \frac{\pi(y; a, 1, 1)}{\phi(d)} \right) \right| \ll \frac{X}{\log^A X}$$

and $0 < \epsilon < 1$.

Proof. This is Theorem 3, combined with the remark on p. 281 of [16].

Lemma 4. *Fix a prime $p_1 < x^{1/2}$. Then the number of primes $p \leq x$ such that $p - 1 = 4 p_1 p_2$ with p_2 a prime is*

$$\ll \frac{x}{p_1 \log^2 x}.$$

Proof. This is a standard application of Brun's sieve (see e.g. Theorem 3.12 of [7]).

3. An application of the lower bound sieve

Our goal in this section and the next is to set up the apparatus to prove Theorem 4 stated below. We begin as in [6].

Let $K = 2^k$ with $k = 1$, 2 or 3 and let u and v be coprime integers such that $K|(u-1)$, $16|v$ and $((u-1)/K, v) = 1$. Fix an integer m coprime to v. Define

$$\mathcal{A}_m = \left\{ \frac{p-1}{Km} : p \leq x, \; p \equiv u \;(\mathrm{mod}\, v), \; p \equiv 1 \;(\mathrm{mod}\, m) \right\};$$

$\mathcal{P}_m$ will denote the set of odd primes coprime to vm. We will use Lemma 1 to derive a lower bound for $S(\mathcal{A}_m, \mathcal{P}_m, z)$. Indeed, if q is coprime to vm, we may write

$$\#\{a \in \mathcal{A}_m : q|a\} = \pi(x, vqm, u^*) = \frac{\mathrm{li}\, x}{\phi(qvm)} + r_{vqm} \quad \text{(say)}.$$

The conditions $u \equiv 1 \pmod{K}$, $v \equiv 0 \pmod{K}$ and $(q, v) = 1$ imply that if $p \equiv u \pmod{v}$ then the conditions $p \equiv 1 \pmod{Kq}$ and $p \equiv 1 \pmod{q}$ are equivalent. Define

$$X = \frac{\operatorname{li} x}{\phi(v)\phi(m)}, \qquad \omega(d) = \frac{d}{\phi(d)}.$$

Applying Lemma 1 yields, for $z = x^{\alpha}$, $Q = x^{\mu}$,

$$S\left(\mathcal{A}_m, \mathcal{P}_m, x^{\alpha}\right) \geq X \prod_{\substack{p \leq x^{\alpha} \\ p \nmid vm}} \left(1 - \frac{1}{p-1}\right) \left\{f(\mu/\alpha) - \epsilon\right\} - \sum_{l < L} \sum_{q \mid P_m(x^{\alpha})} \lambda_l^{-}(q) r_{qm},$$

for some well-factorable functions λ_l^{-} of level x^{μ}, where L depends only on ϵ.

Fix an integer N and sum the above inequality over $m \leq z$ with $(m, N) = 1$ and $z \leq x^{\epsilon_1}$ (with ϵ_1 sufficiently small and to be chosen later) to obtain

$$\sum_{\substack{m \leq z \\ (m,N)=1}} S\left(\mathcal{A}_m, \mathcal{P}_m, x^{\alpha}\right) \geq \sum_{m \leq z} \frac{\operatorname{li} x}{\phi(vm)} \prod_{\substack{p \leq x^{\alpha} \\ p \nmid vm}} \left(1 - \frac{1}{p-1}\right) \left\{f\left(\tfrac{4}{7}\alpha\right) - \epsilon\right\}$$
$$+ O\left(\frac{x}{\log^A x}\right),$$

by Lemma 1 and Lemma 2 and noting that summing over m introduces only new well-factorable functions of essentially the same level. The sum on the right is an enumeration of disjoint sets and is easily evaluated:

$$\frac{1}{\phi(vm)} \prod_{\substack{p \leq x^{\alpha} \\ p \nmid vm}} \left(1 - \frac{1}{p-1}\right) = \frac{1}{\phi(vm)} \prod_{2 < p \mid vm} \frac{p-1}{p-2} \prod_{2 < p \leq x^{\alpha}} \left(1 - \frac{1}{p-1}\right),$$

and it is not difficult to see by standard methods of analytic number theory that the sum is, for some constant c_1 (which may depend on v),

$$\geq c_1 \frac{\phi(N)}{N} \frac{x \log z}{\log^2 x} \left\{\frac{2e^{-\gamma}}{\alpha} f\left(\tfrac{4}{7}\alpha\right) - \epsilon\right\} + O\left(\frac{x}{\log^A x}\right).$$

4. The upper bound sieve

As in [6], we employ the Chen-Iwaniec switching method. We now consider

$$\mathcal{B}_m = \left\{1 + Kmp_1p_2p_3 \leq x \ : \ p_i \geq x^\alpha, \ 1 + Kmp_1p_2p_3 \equiv u \pmod{v}\right\},$$

where the different orderings of p_1, p_2, p_3 are to be counted as distinct so that $\mathcal{B}_m$ is a multiset. If $(q, mv) = 1$ then

$$\#\left\{b \in \mathcal{B}_m : q \mid m\right\} = \#\left\{p_1p_2p_3 \leq y \ : \ p_i \geq x^\alpha, \ p_1p_2p_3 \equiv l \pmod{mqv/K}\right\},$$

where $y = (x-1)/Km$ and l is the common solution of

$$Kml + 1 \equiv u \pmod{v} \qquad Kl + 1 \equiv 0 \pmod{q}.$$

Let

$$g_m(a) = \#\left\{mp_2p_3 = a : p_2, p_3 \geq x^\alpha\right\}$$

and note that $g_m(a) \leq \tau_3(a)$. By the upper bound sieve (Lemma 1) we obtain

$$S\left(\mathcal{B}_m, \mathcal{P}_m, x^{\frac{1}{2}-\epsilon'}\right) \leq Y \prod_{\substack{p \leq x^{1/2-\epsilon'} \\ p \nmid vm}} \left(1 - \frac{\omega(p)}{p}\right)(F(1) + \epsilon) + R_m,$$

where

$$Y = \frac{1}{\phi(mv/K)} \sum_{a \leq yx^{-\alpha}} g_m(a)\left(\pi(y/a) - \pi(x^\alpha)\right),$$

$$F(1) = 2e^\gamma, \qquad R_m = \sum_{\substack{q \leq x^{1/2}\log^{-A} x \\ (q, mv)=1}} \lambda_l^+(q) r_q(\mathcal{B}_m),$$

$$r_q(\mathcal{B}_m) = \sum_{a \leq yx^{-\alpha}} g_m(a)\left(E(y; a, mqv/K, l) - E(ax^\alpha; a, mqv/K, l)\right)$$

$$E(y; a, mqv/K, l) = \pi(y; a, mqv/K, l) - \frac{\pi(y/a)}{\phi(mqv/K)}.$$

Again, we sum over $m \leq z$, $(m, N) = 1$:

$$\sum_{m \leq z} S\left(\mathcal{B}_m, \mathcal{P}_m, x^{\frac{1}{2}-\epsilon'}\right),$$

and find that the main term of the upper bound is

$$\leq c_1 \frac{\phi(N)x \log z}{N \log^2 x} \left(8e^{-\gamma} IF(1) + \epsilon'\right),$$

where

$$I = \int_\alpha^{1-2\alpha} \log\left(\frac{1-\alpha-\theta}{\alpha}\right) \frac{d\theta}{\theta(1-\theta)}.$$

The error term is

$$\sum_{m\leq z} \sum_{l<L} \sum_{q<x^{1/2-\epsilon'}} \lambda_l^+(q) r_q(\mathcal{B}_m),$$

and we can apply Pan's theorem (Lemma 3) with

$$f(a) = \sum_{m\leq z} g_m(a)$$

which is bounded by a divisor function. Hence for some $\epsilon_1 > 0$ we have established the following:

Theorem 4. *Let $(u,v) = 1$ and N be fixed integers and $z \leq x^{\epsilon_1}$. The number of primes $p \leq x$ such that (i) $p \equiv u \pmod v$ (ii) every odd prime divisor q of $p-1$ satisfies one of the conditions*

$$q \geq x^{\frac{1}{4}+\epsilon} \quad or \quad q \mid m \text{ with } m \leq z, \ (m,N) = 1$$

is at least

$$\gg \frac{\phi(N)x \log z}{N \log^2 x}.$$

Proof. As on p. 31 of [6], the quantity we want to enumerate is at least

$$\sum_{\substack{m\leq z \\ (m,\bar{N})=1}} S(\mathcal{A}_m, \mathcal{P}_m, x^\alpha) - \tfrac{1}{6} S(\mathcal{B}_m, \mathcal{P}_m, x^{\frac{1}{2}}) + O(x^{1-\alpha}).$$

By the lower bound obtained in section 3 and the upper bound obtained above, we are done after analyzing the constants. But this is the same analysis as on p. 34 of [6].

A similar result can be stated with $p-1$ replaced by $p+1$ in Theorem 4. This will be useful in considering elliptic analogues of the Artin primitive root conjecture in §9.

We can refine Theorem 4 to ensure that if $p-1$ has at least two prime divisors $q_2 > q_1 > x^{1/4+\epsilon}$ then $q_1 < x^{1/2-\delta}$ for some $\delta > 0$. To see this, for fixed q_1 let us enumerate the number of primes $p \leq x$ such that $p - 1 = q_1 n$ and n is free of prime factors in the interval $(z, x^{1/2})$. By Brun's sieve, the number of such primes is

$$\ll \frac{x \log z}{q_1 \log^2 x} .$$

Summing over $x^{1/2-\delta} < q_1 < x^{1/2}$, we get an estimate of

$$\frac{\delta x \log z}{\log^2 x}$$

for the number of such primes. Choosing δ sufficiently small yields

Theorem 4*. *Let* $(u, v) = 1$ *and* N *be fixed integers. Let* ϵ_1 *be as in Theorem 4 and let* $z \leq x^{\epsilon_1}$. *There is a* $\delta = \delta(\epsilon_1) > 0$ *such that there are at least*

$$\gg \frac{\phi(N) x \log z}{N \log^2 x}$$

primes $p \leq x$ *such that* (i) $p - 1$ *has no prime factor in the interval* $(x^{1/2-\delta}, x^{1/2})$ (ii) $p \equiv u \pmod{v}$ (iii) *every odd prime divisor* q *of* $p - 1$ *satisfies one of the conditions*

$$q > x^{\frac{1}{4}+\epsilon}, \quad \text{or} \quad q \mid m \text{ with } m \leq z, \ (m, N) = 1.$$

If E is an elliptic curve defined over $\mathbb{Q}$ with complex multiplication by an imaginary quadratic field F, then F is one of nine fields of class number one. If p is a prime of good reduction for E and inert in F, then $|E(\mathbb{F}_p)| = p + 1$. Such primes are determined by congruence conditions modulo the discriminant of F and form a set of Dirichlet density $\frac{1}{2}$. We will sieve the sequence

$$\mathcal{S} = \big\{ p + 1 : p \leq x, \ p \text{ inert in } F \big\}$$

and apply the same reasoning to deduce

Theorem 5. *Let* F *be one of the nine imaginary quadratic fields with class number one. Let* N *be a fixed integer. There is an* $\epsilon_1 > 0$ *and a* $\delta = \delta(\epsilon_1) > 0$ *such that at least*

$$\gg \frac{\phi(N) x \log z}{N \log^2 x}$$

primes $p \leq x$ *have the following properties:* (i) $p + 1$ *has no prime factor in the interval* $(x^{1/2-\delta}, x^{1/2})$ (ii) p *is inert in* F (iii) *every odd prime divisor* q *of* $p + 1$ *satisfies one of the conditions*

$$q > x^{\frac{1}{4}+\epsilon} \quad \text{or} \quad q \mid m \text{ with } m \leq z, \ (m, N) = 1 \quad \text{where} \quad z \leq x^{\epsilon_1}.$$

In addition, the power of 2 dividing $p + 1$ *is bounded.*

5. Results on the Chebotarev density theorem

We record in this section two lemmas derived from the effective Chebotarev density theorem (see [15], [22], and [12]).

Lemma 5. *Let a be squarefree and q an odd prime. The number of primes $p \leq x$ such that $q \mid [\mathbb{F}_p^* : \langle a \rangle]$ is*

$$\frac{\operatorname{li} x}{q(q-1)} + O\left(\frac{x}{\log^A x}\right),$$

for any $A > 0$ and uniformly for $q \leq \log^{1/4} x$. Suppose that for some $\epsilon > 0$ and each prime q the Dedekind zeta function of $\mathbb{Q}(a^{1/q})$ has no zeroes for $\operatorname{Re}(s) > 1 - \epsilon$. Then the number of primes $p \leq x$ such that $q \mid [\mathbb{F}_p^* : \langle a \rangle]$ is

$$\frac{\operatorname{li} x}{q(q-1)} + O\left(x^{1-\epsilon/2}\right)$$

uniformly for $q \leq x^\epsilon$.

Proof. See p. 243 of [20] for the first part of the assertion in the lemma. For the second part, one derives it easily by the standard method (for instance as in [9]). As remarked earlier, Pappalardi [17] has extended the range of validity of the first part of the lemma.

Lemma 6. *Let E be an elliptic curve defined over $\mathbb{Q}$ without CM. Let q be a prime such that $q > c(E)$, where $c(E)$ depends only on E. Then the number of primes $p \leq x$ such that $E(\mathbb{F}_p)$ contains a subgroup of type (q,q) is*

$$\frac{\operatorname{li} x}{(q^2-1)(q^2-q)} + O\left(\frac{x}{\log^A x}\right),$$

for any $A > 0$, uniformly for $q \leq (\log x)^{1/6}$. If we assume that for each prime q the Dedekind zeta function of $\mathbb{Q}(E[q])$ has no zeroes in $\operatorname{Re}(s) > 1-\epsilon$, then the number of primes $p \leq x$ such that $E(\mathbb{F}_p)$ contains a subgroup of type (q,q) is

$$\frac{\operatorname{li} x}{(q^2-1)(q^2-q)} + O\left(x^{1-\epsilon/2}\right)$$

uniformly for $q \leq x^\epsilon$.

Proof. This is again a direct application of the effective Chebotarev density theorem. The proofs are analogous to those in the previous lemma.

Lemma 7. *Let E be an elliptic curve defined over $\mathbb{Q}$ and with CM. Let a be a point of infinite order in $E(\mathbb{Q})$. There is a constant $c_2(E)$ such that for a prime $q > c_2(E)$ the number of primes $p \leq x$ such that $q \mid [E(\mathbb{F}_p) : \langle a \rangle]$ is*

$$\frac{\mathrm{li}\, x}{n(q)} + O\left(\frac{x}{\log^A x}\right),$$

for any $A > 0$, uniformly for $q \leq \log^{1/5} x$ and $n(q) \asymp q^4$. If, for some $\epsilon > 0$ and each prime q, the Dedekind zeta function of $\mathbb{Q}(E[q], q^{-1}a)$ has no zero in $\mathrm{Re}(s) > 1 - \epsilon$ then the number of primes $p \leq x$ such that $q \mid [E(\mathbb{F}_p) : \langle a \rangle]$ is

$$\frac{\mathrm{li}\, x}{n(q)} + O\left(x^{1-\epsilon/2}\right)$$

uniformly for $q \leq x^\epsilon$.

Proof. Again, we apply the effective Chebotarev density theorem as in [2], [3] and [4].

6. Further lemmata

In this section, we formalize the pigeonhole principle in a quantitative manner as in [2] and [4].

Lemma 8. *Let $a_1, \ldots, a_r$ be r multiplicatively independent integers. The number of primes p such that the image of $\langle a_1, \ldots, a_r \rangle$ $(\mathrm{mod}\, p)$ has order not exceeding y is $O\left(y^{1+1/r}\right)$.*

Below we discuss the elliptic analogues of Lemma 8.

Suppose we have a free subgroup Γ of rational points of rank (over $\mathbb{Z}$) equal to r. Let $P_1, \ldots, P_r$ be r independent generators of Γ. We will make use of the canonical height pairing of Néron and Tate to estimate the number of primes p for which the image Γ_p of Γ $(\mathrm{mod}\, p)$ is small. Such a situation arises naturally as follows. Suppose that $q \mid [E(\mathbb{F}_p) : \Gamma_p]$ and that $q > z$. For primes $p \leq x$ this means that

$$|\Gamma_p| \leq \frac{x}{z}.$$

Thus, if z is large, the image of Γ $(\mathrm{mod}\, p)$ is small. If we can show that the number of primes satisfying the above inequality is small, we can conclude that for most primes $q \nmid [E(\mathbb{F}_p) : \Gamma_p]$ with $q > z$. This is our basic strategy.

Recall that the canonical height pairing of Néron and Tate is a positive semidefinite bilinear pairing on $E(\overline{\mathbb{Q}})$ with the property that $\langle P, P \rangle = 0$ if and only if P is a torsion point. This height pairing is related to the naïve height of Weil in the following way. If $P = (x, y) \in E(\mathbb{Q})$ then, writing $x = r/s$ with r, s coprime integers, we define the x-height of P as

$$h_x(P) = \log \max(|r|, |s|).$$

Observe that the image of P (mod p) is the identity element if and only if $p \mid s$. Since the number of prime divisors of s is bounded by $2 \log |s|$ we note that the number of primes for which P reduces to the identity element on $E(\mathbb{F}_p)$ is bounded by $2h_x(P)$. (Recall that the identity element on E is the point at infinity.) If we let $H(P) = \langle P, P \rangle$ then, for $P \in E(\mathbb{Q})$,

$$H(P) = h_x(P) + O(1),$$

where the implied constant depends only on E. So we can use $H(P)$ as an upper bound for the number of primes p for which P reduces to the identity on $E(\mathbb{F}_p)$.

Lemmas 9 and 10 appear in [4] and are reproduced here for the sake of completeness.

Lemma 9. *The number of r-tuples of integers $(n_1, \ldots, n_r)$ satisfying*

$$H(n_1 P_1 + \cdots + n_r P_r) \leq x$$

is

$$\frac{(\pi x)^{r/2}}{\sqrt{R}\Gamma(\frac{1}{2}r + 1)} + O\left(x^{\frac{1}{2}(r-1)+\epsilon}\right)$$

where $R = \det(\langle P_i, P_j \rangle)$.

Proof. We want to determine the integer solutions of

$$\left\langle \sum_{i=1}^r n_i P_i, \sum_{i=1}^r n_i P_i \right\rangle \leq x,$$

which is the same as

$$\sum_{i,j} n_i n_j \langle P_i, P_j \rangle \leq x.$$

This is tantamount to counting lattice points in the r-dimensional ellipsoid determined by the above quadratic form. By a result of Walfisz [23] the number of such lattice points is given by the expression stated in the lemma. (Note that in [4], the reference to Walfisz [23] appears with an incorrect year.)

Lemma 10. *The number of primes p such that $|\Gamma_p| \leq y$ is $O(y^{1+2/r})$.*

Proof. Consider the set S of all r-tuples $(n_1, \ldots, n_r)$ satisfying

$$H(n_1 P_1 + \cdots + n_r P_r) \leq C y^{2/r},$$

where C is any constant chosen so that

$$\frac{(C\pi)^{r/2}}{\sqrt{R}\Gamma(\frac{1}{2}r + 1)} > 1$$

with C greater than the O- constant implied by Lemma 3. Then by Lemma 3 the number of elements of S is $> y$. If p is a prime such that $|\Gamma_p| < y$

then we must have for two distinct r-tuples $(n_1, \ldots, n_r)$ and $(m_1, \ldots, m_r)$ the congruence

$$n_1 P_1 + \cdots + n_r P_r \equiv m_1 P_1 + \cdots + m_r P_r \pmod{p}.$$

Since $P_1, \ldots, P_r$ are $\mathbb{Z}$-independent, the point

$$Q = \sum_{i=1}^{r} (n_i - m_i) P_i$$

is non-zero and the above congruence $\pmod{p}$ implies that p divides the denominator of Q. As remarked above, the number of such primes is bounded by $2h_x(Q)$ because a natural number n has at most $2 \log n$ prime factors. Moreover Q is not a torsion point since $P_1, \ldots, P_r$ are independent over $\mathbb{Z}$. Therefore, $H(Q) \neq 0$ and

$$2h_x(Q) \ll H(Q).$$

Since $H(Q) \leq 2Cy^{2/r}$ the number of such Q is $O(y)$ by Lemma 3. Since each Q gives rise to only $O(y^{2/r})$ primes dividing the denominator of Q, the total number of prime factors satisfying $|\Gamma_p| < y$ is $O(y^{1+2/r})$, as desired.

7. Primitive roots

We are now ready to prove Theorem 1. Consider the primes enumerated by Theorem 4* with u and v chosen so that $p \equiv u \pmod{v}$ implies each of a, b, c is a quadratic non-residue $\pmod{p}$. This can be done because

$$\sum_{p \leq x} \left\{ 1 - \left(\frac{-3}{p} \right) \right\} \left\{ 1 - \left(\frac{a}{p} \right) \right\} \left\{ 1 - \left(\frac{b}{p} \right) \right\} \left\{ 1 - \left(\frac{c}{p} \right) \right\}$$

is asymptotic to $\pi(x)$ as $x \to \infty$, in view of the conditions imposed on a, b and c. Let us fix y and let

$$N = \prod_{p \leq y} p.$$

With $u \pmod{v}$ and N as above and $z = x^{\epsilon/3}$ we consider the primes enumerated by Theorem 4*. If p is one of these primes, then two cases arise: $p - 1 = 2^e m q$ or $p - 1 = 2^e m q_1 q_2$, with q, q_1, q_2 primes satisfying the various conditions of Theorem 4*. If q divides the index of $\langle a \rangle, \langle b \rangle, \langle c \rangle$ $\pmod{p}$ then by Lemma 8 the number of such primes is $O(x^{2\epsilon})$. If q_1 or q_2 divides the index of $\langle a, b, c \rangle$ $\pmod{p}$ again by Lemma 8 (and $r = 3$) the number of such primes is $O(x^{1-\epsilon'})$. Hence the order of $\langle a, b, c \rangle$ $\pmod{p}$ is divisible by q in the first case and $q_1 q_2$ in the second case. Also 2^e divides the order of each of a, b, c $\pmod{p}$ because none of a, b, c is a quadratic residue $\pmod{p}$. Suppose in the second case each of the orders of a, b, c $\pmod{p}$ is not divisible by q_1. Then the subgroup generated by a, b, c

$(\bmod p)$ has order $< x^{3/4-\epsilon}$, and by Lemma 8 (and $r = 3$) the number of such primes is $O(x^{1-\epsilon})$. So we may suppose that one of the orders of a, b, c is divisible by q_1. Suppose without loss of generality that the order of a $(\bmod p)$ is divisible by q_1. If $q_2 \mid [\mathbb{F}_p^* : \langle a \rangle]$ then, noting that $q_1 < x^{1/2-\delta}$ and again applying Lemma 8 with $r = 1$, we deduce that the number of such primes is $O(x^{1-2\delta})$. Thus we have

$$\geq \frac{c_1 \phi(N) x \log z}{N \log^2 x}$$

primes $p \leq x$ such that if $l \mid [\mathbb{F}_p^* : \langle a \rangle]$ then $l \mid m$, where $(m, N) = 1$ as in Theorem 4*. Now we invoke Lemma 5. The number of primes $p \leq x$ such that $l \mid [\mathbb{F}_p^* : \langle a \rangle]$ is

$$\frac{\operatorname{li} x}{l(l-1)} + O(x^{1-\epsilon}).$$

We sum this in the range $y < l < x^{\epsilon/3}$. This gives an estimate of

$$\ll \frac{\operatorname{li} x}{y} + O(x^{1-\epsilon}).$$

Since

$$\frac{\phi(N)}{N} = \prod_{p \leq y} \left(1 - \frac{1}{p}\right) \sim \frac{e^{-\gamma}}{\log y}$$

by Mertens' theorem, we can choose y sufficiently large so that

$$\frac{c_1 \epsilon e^{-\gamma}}{3 \log y} \gg \frac{1}{y}.$$

This completes the proof of Theorem 1.

8. Cyclicity of $E(\mathbb{F}_p)$

We now prove Theorem 2 and proceed as in [3] with minor variations. Let us observe that if $E(\mathbb{F}_p)$ contains a subgroup of type (q, q) then $q \mid p - 1$. This is because the field obtained by adjoining the q-division points of E to $\mathbb{Q}$ contains the cyclotomic field of the q-th roots of unity, by the theory of the Weil pairing. Moreover, the condition that $E(\mathbb{F}_p)$ contains a subgroup of type (q, q) is equivalent to the condition that p splits completely in $\mathbb{Q}(E[q])$. Now fix y, and set

$$N = \prod_{p \leq y} p$$

and $z = x^{\epsilon/3}$. Apply Theorem 4* and partition each of the primes enumerated into disjoint sets S_a according to the value of

$$a_p(E) = p + 1 - \#E(\mathbb{F}_p).$$

In each S_a we count the number of primes p such that $E(\mathbb{F}_p)$ is not cyclic. If, for a prime $q > x^{1/4+\epsilon}$, $E(\mathbb{F}_p)$ contains a (q, q) group then q is uniquely determined, for otherwise the size of $E(\mathbb{F}_p)$ would be greater than $x^{1+2\epsilon}$. In addition $p \equiv a - 1 \pmod{q^2}$. By Hasse's inequality $|a_p| \leq 2\sqrt{p}$ and so we can never have $p = a - 1$. Thus, the number of such primes is

$$\ll \frac{x}{q^2} \ll x^{\frac{1}{2}-2\epsilon}.$$

Summing this over $|a| \leq 2\sqrt{x}$ gives a total estimate of $O(x^{1-2\epsilon})$ for the number of such primes enumerated by Theorem 4*. After eliminating these primes from our enumeration we infer that if $E(\mathbb{F}_p)$ is not cyclic then p splits completely in some $\mathbb{Q}(E[l])$ with $l \leq x^{\epsilon/3}$ and $(l, N) = 1$. By Lemma 6, the number of such primes on the quasi-Riemann hypothesis is

$$\ll \frac{\mathrm{li}\,x}{l^4} + O(x^{1-\epsilon})$$

and we sum this over $y < l < x^{\epsilon/3}$ to deduce an estimate of

$$\ll \frac{\mathrm{li}\,x}{y}$$

such primes. Again, choosing y sufficiently large ensures that we have a positive density of primes for which $E(\mathbb{F}_p)$ is cyclic.

9. Primitive points on elliptic curves

We use Theorem 5 with $z = 1$ and $N = 1$. Let p be a prime enumerated by Theorem 5. Suppose first that $p + 1 = 2^e p_1$. Then, $(\bmod\, p)$, each one of $P_1, \ldots, P_6$ has order dividing $2^e p_1$. If the order divides 2^e, then p divides the 2^e-division polynomial evaluated at each of the P_i. Since e is bounded, there are only finitely many such p. So for p sufficiently large, the order divides p_1 and so the index is bounded by 2^e in this case.

Now suppose $p + 1 = 2^e p_1 p_2$ with $p^\alpha < p_1 < p_2$. If the theorem is false then for all p sufficiently large enumerated by Lemma 1 we must have that the order of each P_i divisible by p_1 or p_2 but not both. If the orders are all divisible by p_2, then the subgroup generated by $P_1, \ldots, P_6 \pmod{p}$ has size $\leq 4p_2 \leq 4p^{1-\alpha}$. By Lemma 4, the number of such primes $p \leq x$ is $O(x^{4(1-\alpha)/3})$. Since $\alpha > 1/4$, the number of such primes is $o(x/\log^2 x)$. By the same reasoning, we conclude that the number of primes for which all the orders are divisible by p_1 is also negligible. Hence, for at least $\gg x/\log^2 x$ primes enumerated by $\mathcal{A}$, the order of each P_i is divisible by p_1 or p_2 and both possibilities occur. Take an element P_1 (say) whose order $(\bmod\, p)$

is divisible by p_1. If $p_1 < x^{1/3-\epsilon}$ then an application of Lemma 4 with $r = 1$ shows the number of such primes is $O(x^{1-\epsilon})$. So we may suppose $p_1 > x^{1/3-\epsilon}$. We can in fact suppose $p_1 > x^{1/3+\epsilon}$ because by Lemma 2 the number of primes $p \leq x$ such that $p + 1 = 2^e p_1 p_2$ with

$$x^{\frac{1}{3}-\epsilon} < p_1 < x^{\frac{1}{3}+\epsilon} \quad \text{and} \quad p_2 > x^{\frac{1}{2}-\delta}$$

for some $\delta > 0$ is, by Lemma 2,

$$\ll \frac{x}{p_1 \log^2 x}.$$

Summing this over the range $x^{1/3-\epsilon} < p_1 < x^{1/3+\epsilon}$ gives a contribution of $o(\epsilon x / \log^2 x)$, which is negligible for sufficiently small ϵ. So we may suppose that

$$x^{\frac{1}{3}+\epsilon} < p_1 < p_2.$$

Let us say that p in $\mathcal{A}$ has type (s_1, s_2) if s_1 of $P_1, \ldots, P_6$ have order divisible by p_1 and s_2 have order divisible by p_2. Then we can partition $\mathcal{A}$ according to five types: $(1,5)$, $(2,4)$, $(3,3)$, $(4,2)$, and $(5,1)$.

We now consider each of the five types. If the type of p is $(1,5)$ then five independent points generate a group $(\bmod p)$ of order $O(p_2) = O(x^{2/3-\epsilon})$, and by Lemma 4 the number of such primes is $O(x^{14/15-\epsilon})$. If the type is $(2,4)$ then four independent points generate a group of order $O(p_2) = O(x^{2/3-\epsilon})$ and again by Lemma 4 the number of such primes is $O(x^{1-\epsilon})$. If the type is $(3,3)$, we have three independent points generating a group of order $p_1 < x^{1/2}$ and by Lemma 4 the number of such primes is $O(x^{5/6})$. The remaining two cases are similarly handled. This completes the proof.

Remark. It is clear that if we use Theorem 5 with $z = x^\epsilon$ and invoke Lemma 7 with a quasi-Riemann hypothesis, the assertion made in Theorem 3 holds for a positive density of primes.

10. Numerical examples

In [18], Quer produces three elliptic curves with complex multiplication such that the $\mathbb{Z}$ rank of the group of rational points is 12. More precisely, let

$$D_1 = -408\,368\,221\,541\,174\,183$$
$$D_2 = -3\,082\,320\,147\,153\,282\,331$$
$$D_3 = -3\,161\,659\,186\,633\,662\,283$$

and put

$$E_i : \qquad Y^2 = X^3 + 16D_i$$

for $i = 1, 2, 3$. Then, $\text{rank}_{\mathbb{Z}} E_i(\mathbb{Q}) = 12$. Quer also gives explicit generators. It is remarkable that these generators all have integral co-ordinates.

M. R. Murty

References

1 E. Bombieri, J.B. Friedlander, and H. Iwaniec: Primes in arithmetic progressions to large moduli. *Acta Math.* **370** (1986), 203–251.

2 R. Gupta and M. Ram Murty: A remark on Artin's conjecture. *Invent. Math.* **78** (1984), 127–130.

3 R. Gupta and M. Ram Murty: Cyclicity and generation of points (mod *p*) on elliptic curves. *Invent. Math.* **101** (1990), 225–235.

4 R. Gupta and M. Ram Murty: Primitive points on elliptic curves. *Compositio Math.* **58** (1986), 13–44.

5 R. Gupta, M. Ram Murty and V. Kumar Murty: The Euclidean algorithm for *S*-integers. In *Number Theory*, 189–201 (H. Kisilevsky and J. Labute, eds.). (Canadian Mathematical Society Conference Proceedings **7**) (1987).

6 D.R. Heath-Brown: Artin's conjecture for primitive roots. *Quart. J. Math. Oxford* (2) **37** (1986), 27–38.

7 H. Halberstam and H.-E. Richert: *Sieve Methods.* Academic Press (1974).

8 C. Hooley: On Artin's conjecture. *J. Reine Angew. Math.* **225** (1967), 209–220.

9 C. Hooley: *Applications of Sieve Methods.* Cambridge University Press (1976).

10 H. Iwaniec: Rosser's sieve. *Acta Arith.* **36** (1980), 171–202.

11 H. Iwaniec: A new form of the error term in the linear sieve. *Acta Arith.* **37** (1980), 307–320.

12 J. Lagarias and A. Odlyzko: Effective versions of the Chebotarev density theorem. In *Algebraic Number Fields*, 409–464 (A. Fröhlich, ed.). Academic Press (1977).

13 S. Lang and H. Trotter: Primitive points on elliptic curves. *Bull. Amer. Math. Soc.* **83** (1977), 289–292.

14 J.-F. Mestre: Rang de courbes elliptiques d'invariant donné. *C.R. Acad. Sci. Paris* **314** (1992), 919–922.

15 M. Ram Murty, V. Kumar Murty and N. Saradha: Modular forms and the Chebotarev density theorem. *Amer. J. Math.* **110** (1988), 253–281.

16 C.-D. Pan: A new mean value theorem and its applications. In *Recent Progress in Analytic Number Theory* Vol. 1, 275–287. (H. Halberstam and C. Hooley, eds.). Academic Press (1981).

17 F. Pappalardi: On Artin's conjecture for primitive roots (Ph.D. thesis). McGill University (1993).

18 J. Quer: Corps quadratiques de 3-rang 6 et courbes elliptiques de rang 12. *C.R. Acad. Sci. Paris* **305** (1987), 215–218.

19 M. Ram Murty: On Artin's conjecture. *J. Number Theory* **16** (1983), 147–168.

20 M. Ram Murty: An analogue of Artin's conjecture for abelian extensions. *J. Number Theory* **18** (1984), 241–248.

21 J.-P. Serre: Résumé des cours de 1977-78. *Annuaire du Collège de France* (1978), 67–70 (in Collected Papers, Vol. 3, 465–468 and 713).

22 J.-P. Serre: Quelques applications du théorème de densité de Chebotarev. *Publ. I.H.E.S.* no. 54 (1981), 123–201.

23 A. Walfisz: Über Gitterpunkte in mehrdimensionalen Ellipsoiden III. *Math. Zeit.* **27** (1927), 245–268.

Printed in the United States
By Bookmasters